ADVANCED TOPICS
IN SCIENCE AND TECHNOLOGY IN CHINA

ADVANCED TOPICS IN SCIENCE AND TECHNOLOGY IN CHINA

Zhejiang University is one of the leading universities in China. In Advanced Topics in Science and Technology in China, Zhejiang University Press and Springer jointly publish monographs by Chinese scholars and professors, as well as invited authors and editors from abroad who are outstanding experts and scholars in their fields. This series will be of interest to researchers, lecturers, and graduate students alike.

Advanced Topics in Science and Technology in China aims to present the latest and most cutting-edge theories, techniques, and methodologies in various research areas in China. It covers all disciplines in the fields of natural science and technology, including but not limited to, computer science, materials science, life sciences, engineering, environmental sciences, mathematics, and physics.

Liang Chen
Zeno Apostolides
Zong-Mao Chen

Global Tea Breeding

Achievements, Challenges and Perspectives

With 87 figures

ZHEJIANG UNIVERSITY PRESS
浙江大学出版社

Springer

Editors
Prof. Liang Chen
Tea Research Institute of the Chinese Academy of Agricultural Sciences
National Center for Tea Improvement
Hangzhou 310008, China
E-mail: liangchen@mail.tricaas.com

Prof. Zeno Apostolides
Department of Biochemistry
University of Pretoria
Pretoria 0002, South Africa
E-mail: zeno.apostolides@up.ac.za

Prof. Zong-Mao Chen
Tea Research Institute of the Chinese Academy of Agricultural Sciences
National Center for Tea Improvement
Hangzhou 310008, China
E-mail: chenzm@mail.tricaas.com

ISSN 1995-6819 e-ISSN 1995-6827
Advanced Topics in Science and Technology in China

ISBN 978-7-308-08274-7
Zhejiang University Press, Hangzhou

ISBN 978-3-642-31877-1 ISBN 978-3-642-31878-8 (eBook)
Springer Heidelberg New York Dordrecht London

Library of Congress Control Number: 2012942071

Printed on acid-free paper

Springer is part of Springer Science+Business Media (www.springer.com)

Preface

The tea plant, *Camellia sinensis* (L.) O. Kuntze, originated in the southwestern part of China. Tea is one of the most popular non-alcoholic and healthy beverage in the world. It contributes to massive wealth and job opportunities in many countries, including China, India, Kenya, Sri Lanka, etc. Up to date, tea plants have been cultivated in more than fifty countries in Asia, Africa, South America, Europe and Oceania. According to FAOSTAT (http://faostat.fao.org), the tea harvest acreage was 2,996 kilohectares; the production was 3,885 kilotonnes in 2009. Of this, about 83.8% is produced in Asia, 13.7% in Africa. Tea acreage and production have increased continuously in recent years. The increase is partially a result of the release and wide extension of clonal tea cultivars in the main tea producing countries. So far, more than 500 tea cultivars have been bred and released to the public in China, India, Sri Lanka, Kenya, Japan, Bangladesh Indonesia, and some other countries. Approximately half of the world tea acreage consists of clonal tea gardens.

The tea plant, as a unique crop, from cultivation to harvesting, does not fit into any typical cropping pattern. The production of tea, from processing to marketing, is also specific. The world tea community has lacked a monograph on the subject of tea germplasm and breeding for about 20 years. In recent years, the main tea producing countries have progressed significantly in the field of tea breeding. At the same time, they are facing big challenges owing to the progress of science and technology and the serious demands for tea quality and consistency. This book systematically and comprehensively expounds the achievements, challenges and perspectives of worldwide tea breeding. It consists of several specific topics, such as resistant breeding and molecular assistant selection in about ten different countries. The chapter for each country usually includes (1) a general introduction to the tea industry; (2) the collection, conservation, appraisal and utilization of tea germplasms; (3) conventional and molecular tea breeding and selection techniques; (4) the propagation and extension system for new cultivars; (5) future tendencies, strategies, opportunities and perspectives for world tea breeding which are well discussed. The authors are top tea breeders from China, India, Sri Lanka, Kenya, Japan, Turkey, Indonesia, Korea, Nigeria, etc., accounting for 90% of the world tea production.

We hope that this book will be useful not only to the world tea community to enhance the understanding, exchange and cooperation of tea genetics and breeding in the tea producing countries, to promote the progress of world tea breeding, to help us breed more desirable new tea cultivars to meet the demands of different markets and consumers in the world, but also as a reference for other woody perennial species.

The publication of the book is partially supported by the Hi-Tech Research and Development Program of China (863 Plan) (No. 2006AA10Z171), the Earmarked Fund for the China Agriculture Research System (CARS-023), the National Center for Tea Improvement in China, etc.

We would like to express sincere thanks to Dr. Satoshi Yamaguchi, Professor of Tamagawa University, Japan, Dr. Ya-Ya Shi, Fruit and Tea Research Institute of the Hubei Academy of Agricultural Sciences, China, Ms Zhi-lan Wang, Tea Research Institute of the Chinese Academy of Agricultural Sciences, China, for their constructive peer-reviews and revisions.

Liang Chen
Hangzhou, China
April, 2012

Contents

Contributors

Zeno Apostolides
Department of Biochemistry University of Pretoria, Pretoria 0002, South Africa

Liang Chen
Tea Research Institute of the Chinese Academy of Agricultural Sciences; National Center for Tea Improvement, 9 South Meiling Road, Hangzhou, Zhejiang 310008, China

Zong-Mao Chen
Tea Research Institute of the Chinese Academy of Agricultural Sciences, 9 South Meiling Road, Hangzhou, Zhejiang 310008, China

Suresh Chandra Das
Tocklai Experimental Station, Tea Research Association, Jorhat 785008, Assam, India

Sudripta Das
Tocklai Experimental Station, Tea Research Association, Jorhat 785008, Assam, India

Wen-Xia Dong
Tea Research Institute of the Chinese Academy of Agricultural Sciences, 9 South Meiling Road, Hangzhou, Zhejiang 310008, China

Sezai Ercisli
Department of Horticulture, Faculty of Agriculture, Ataturk University 25240 Erzurum, Turkey

M. Trixie K. Gunasekare
Plant Breeding Division, Tea Research Institute of Sri Lanka, Talawakelle, Sri Lanka

Mridul Hazarika
Tocklai Experimental Station, Tea Research Association, Jorhat 785008, Assam, India

Gerald O. Iremiren
Cocoa Research Institute of Nigeria, P.M.B. 5244, Ibadan, Nigeria

Byeong-Choon Jeong
Mokpo Experiment Station, National Institute of Crop Science, RDA, Muan 534-830, Korea

Samson M. Kamunya
Tea Research Foundation of Kenya, P.O. Box 820, Kericho 20200, Kenya

Abdul Qayyum Khan
Bangladesh Tea Research Institute, Srimangal-3210, Moulvibazar, Bangladesh

Heri Syahrian Khomaeni
Indonesia Research Institute for Tea and Cinchona (RITC), Gambung Village, P.O. Box 1013, Bandung City 40010, West Java, Indonesia

Richard C. Muoki
Tea Research Foundation of Kenya, P.O. Box 820, Kericho 20200, Kenya

Do-Van Ngoc
Northern Mountainous Agriculture and Forestry Science Institute—NOMAFSI, Phu Ho commune, Phu Tho town, Phu Tho province, Vietnam

Sui Ni
Faculty of Life Science and Biotechnology, Ningbo University, Ningbo 315211, China

Samuel S. Omolaja
Cocoa Research Institute of Nigeria, P.M.B. 5244, Ibadan, Nigeria

Young-Goo Park
Department of Forestry, Kyungpook National University, Daegu 702-701, Korea

Ram S. Pathak
Department of Crops, Horticulture and Soil Sciences, Egerton University, P.O. Box 536, Egerton, Njoro 20115, Kenya

Ram K. Sharma
Division of Biotechnology, Institute of Himalayan Bioresource Technology, Post Box 6, Palampur, Himachal Pradesh 17061, India

Bambang Sriyadi
Indonesia Research Institute for Tea and Cinchona (RITC) , Gambung Village, P.O. Box 1013, Bandung City 40010, West Java, Indonesia

Xiao-Ling Sun
Tea Research Institute of the Chinese Academy of Agricultural Sciences, 9 South Meiling Road, Hangzhou, Zhejiang 310008, China

Rohayati Suprihatini
Indonesia Research Institute for Tea and Cinchona (RITC) , Gambung Village, P.O. Box 1013, Bandung City 40010, West Java, Indonesia

Junichi Tanaka
NARO Institute of Crop Science, 2-1-18, Kannondai, Tsukuba, Ibaraki 305-8518, Japan

Francis N. Wachira
Tea Research Foundation of Kenya, P.O. Box 820, Kericho 20200, Kenya

Xin-Chao Wang
Tea Research Institute of the Chinese Academy of Agricultural Sciences; National Center for Tea Improvement, 9 South Meiling Road, Hangzhou, Zhejiang 310008, China

Ming-Zhe Yao
Tea Research Institute of the Chinese Academy of Agricultural Sciences; National Center for Tea Improvement, 9 South Meiling Road, Hangzhou, Zhejiang 310008, China

Li-Ping Zhao
Tea Research Institute of the Chinese Academy of Agricultural Sciences; National Center for Tea Improvement, 9 South Meiling Road, Hangzhou, Zhejiang 310008, China

Abbreviations

AFLP	Amplified fragment length polymorphism
AMOVA	Analysis of molecular variance
amsl	above mean sea level
bp	base pair
C	Catechin
CG	Catechin gallate
CNGTR	China National Germplasm Tea Repository
CTC	Crush, tear and curl
DUS	Distinctness, Uniformity and Stability
EC	Epicatechin
ECG	Epicatechin gallate
EGC	Epigallocatechin
EGCG	Epigallocatechin gallate
EST-SSR	Expressed sequence tag based simple sequence repeat
EST	Expressed sequence tag
FAO	Food and Agriculture Organization of the United Nations
GC	Gallocatechin
GCG	Gallocatechin gallate
GPS	Global Positioning System
gSSR	Genomic SSR
GUS	β-glucuronidase
ISSR	Inter simple sequence repeat
ITC	International Tea Committee
LD	Linkage disequilibrium
MAS	marker-assisted selection
NACTC	National Authentication Committee of Tea Cultivars
NCBI	National Center for Biotechnology Information
N_m	Gene flow
PCR	Polymerase chain reaction
PIC	Polymorphism information content
QTLs	Quantitative traits loci
RAPD	Random amplified polymorphic DNA

RFLP	Restriction fragment length polymorphism
RT	Reverse transcription
RTA	Ratio of tea polyphenols to amino acids
SNP	Single nucleotide polymorphism
SSR	Simple sequence repeat
STS	Sequence tagged site
TF	Theaflavin
TP	Tea polyphenols
TR	Thearubigin
UPASI	The United Planters' Association of Southern India
UPOV	International Union for the Protection of New Varieties of Plants

1

Delicious and Healthy Tea: An Overview

Zong-Mao Chen*, Liang Chen

Tea Research Institute of the Chinese Academy of Agricultural Sciences, 9 South Meiling Road, Hangzhou 310008, China
*E-mail: chenzm@mail.tricaas.com

Abstract: The tea plant, *Camellia sinensis* (L.) O. Kuntze, originated in the southwestern part of China, and has been cultivated there for approximately 5,000 years. Now tea plants are cultivated in 52 countries around the world. More than one-half of the population of the world consumes tea. The daily consumption of tea is approximately 3 billion cups all over the world. Tea, coffee and cocoa are the three most popular non-alcoholic beverages in the world.

1.1 Development of the Global Tea Industry

The discovery and drinking of tea originated during the “Shen-Nong” era in ancient China, around 5,000 years ago. Originally, tea was used as a medicine for various illnesses and it can be traced back as early as 2737 B.C. in ancient China (Yamanishi, 1995). Tea production has developed rapidly since the Tang Dynasty (618 – 907 A.D.) and has been accepted as a beverage. However, tea has achieved popularity in other parts of the world only since the middle of the 17th century. Commercial cultivation of tea gradually expanded to Indonesia, India and Sri Lanka until the middle of the 19th century (Chen & Yang, 2011). The history of tea cultivation in Africa is relatively short. The first record of cultivation in Africa was in 1850; however, the tea industry developed until the middle of the 20th century. Now, tea plants are distributed worldwide ranging in latitude from 43° N (Georgia) to 27° S (Argentina). It is now grown commercially in tropical, subtropical and temperate climatic regions of Asia, Africa and South America, and also in limited areas in North America, Australia and Europe. In 2010, the tea growing area in the world amounted to approximately 3,692 kilohectares and the total output amounted 4,162 kilotonnes (ITC, 2011). The average yield in the world now is around 1,110 kg/ha. The following 8 major tea producing countries, namely China, India, Kenya, Sri Lanka, Vietnam, Turkey, Indonesia and Japan, accounted for 88.9% of the world production (Table 1.1). China is the largest tea producing country once again in the world since 2005. The production in 2010 amounted to 1,475.1 kilotonnes and occupied 35.4% of the total world production. The total world export amounted to 1,728.8 kilotonnes in 2010 and was about 41.5% of total production. Kenya exports the most tea in the world, as 95% of production (ITC, 2011). Virtually all tea produced in Japan and about 74% of that produced in China is green tea. About 60% of consumers prefer black tea and the rest consume green tea and Oolong tea. Green tea is preferred in China, Japan and Middle East countries; Oolong tea is mainly consumed in the eastern part of China and in Japan. In terms of annual tea consumption per capita, Kuwait has the highest value at 2.86 kg (triennial average in the period of 2008 – 2010), followed by Ireland (2.31 kg), Qatar (2.04 kg), Turkey (2.02 kg), Afghanistan (2.01 kg) and United Kingdom (1.97 kg) (ITC, 2011).

Table 1.1 Tea production in the major tea producing countries

Country	1940	1950	1960	1970	1980	1990	2000	2010
China	100.0	62.5	136.0	136.0	303.7	540.1	683.3	1,475.1
India	210.4	278.0	321.0	418.5	569.5	720.3	980.8	966.4
Kenya	5.4	6.7	13.7	41.0	89.8	197.0	345.8	399.0
Sri Lanka	120.2	143.4	197.1	212.2	191.3	234.0	318.6	331.4
Vietnam	—	—	4.5	5.5	21.5	32.2	63.7	157.0
Turkey	—	0.2	5.9	33.4	95.8	126.7	155.0	148.0
Indonesia	—	35.3	46.0	44.0	98.6	145.1	137.5	129.2
Japan	58.2	41.7	78.9	91.1	102.3	89.9	93.0	93.0
Subtotal	494.2	567.8	803.1	981.7	1,472.5	2,085.3	2,777.7	3,699.1
World total	1,007.2	1,075.3	1,380.8	1,633.7	2,360.0	2,409.4	2,928.6	4,162.3

1.2 Botanical Characteristics

Although the tea plant is an ancient plant with a long history, the confusion of the nomenclature has continued for almost two centuries. As early as 1753, Linnaeus described the tea plant as *Thea sinensis*, and it was modified to *Camellia sinensis* in August of the same year. Since then, the genus name of *Thea* and *Camellia* has had a checkered history. In the second edition of *Species Plantarum*, Linnaeus abandoned the former name and described the two species separately: *Thea bohea* and *T. viridis*. Watt in India named it *Camellia thea* in 1907; Cohen-Stuart in Indonesia used a new name of *C. theifera*. Sealy in the UK (1958) also gave the same name, *Camellia sinensis* (L.) O. Kuntze and included two varieties: var. *sinensis* (small-leaf variety) and var. *assamica* (large-leaf variety). Since then, despite some papers contributing to the botanical name of the tea plant, *Camellia sinensis* (L.) O. Kuntze, uniformity has been achieved (Chen & Yu, 1994).

Botanically, tea plants belong to the order Theales, family Theaceae, genus *Camellia* L., section *Thea* (L.) Dyer. There are different numbers of species and varieties in the section *Thea* according to different taxonomic systems. Nevertheless, most of the cultivated varieties and cultivars of the tea plant belong to one species, *Camellia sinensis* (L.) O. Kuntze, including var. *sinensis*, var. *assamica* (Masters) Kitamura and var. *pubilimba* Chang. The tea plant is a perennial evergreen; the aerial portion of the tea plant is grown as a tree, semi-tree or shrub depending on the external environment. The var. *sinensis* tea plant usually grows into a shrub about 1 – 5 m high, characterized by more or less virgate stems. Leaves are small, hard and dark-green in color with a dull surface. The var. *assamica* tea plant is described as an erect tree with many branches, 8 – 12 m high. Leaves are 15 – 20 cm long, light-green in color with a glossy surface (Fig. 1.1) (Ming, 2000).

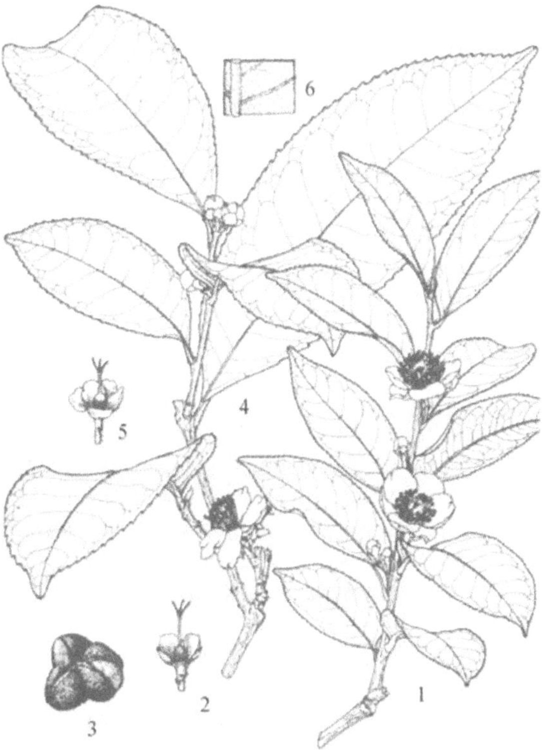

Fig. 1.1. *Camellia sinensis* (L.) O. Kuntze (1 – 3), *C. sinensis* var. *assamica* (Masters) Kitamura (4 – 6). 1: Flower and branch; 2: Gynoecium; 3: Fruit; 4: Flower and branch; 5: Gynoecium; 6: Under surface of leave

The fresh shoots are the economic harvest of the tea plant. The phyllotaxy of leaves on the shoot is alternate. The leaf pose on the stem can be erect, semi-erect, horizontal or drooping, according to the variety. Leaves are leathery in texture, with silvery or light-yellow colored pubescences on the under surface of tender leaves. There are 7 – 15 pairs of veins on the leaf. The lateral veins curve upward and connect with the upper veins, forming a close transporting network, which is characteristic of the leaves of the tea plant. Leaves are serrated at the margin. The first few new leaves at the flushing period of the tea shoot usually have a characteristic small size, being thick and brittle with a blunt apex; the petiole is wider and flat and called a fish-leaf, or in Indian terminology the *Janam*. Its position on the shoot is of the very greatest importance when considering the standards of plucking. The tea manufactured from the fish-leaf is of low quality. Sometimes the leaf primodium differentiates from the vegetative bud of the tea plant, ceasing growth prematurely instead of developing into a normal leaf. It is termed the dormant bud, or in Indian terminology the *Banjhi*. Normal tea shoots show the distinct periodicity of growth, i.e., after the development of several normal leaves, the *Banjhi* bud forms, thus completing a full periodic shoot growth rhythm.

Tea flowers are bisexual with a slight fragrance and are usually white in color. Their diameter is 2.0 – 5.0 cm. The morphology of the flower is one of the important indexes in the classification of the tea plant. The fruit of the tea plant is green in color, usually three-celled, thick-walled and shinny at first, but then duller and slightly rough later. Tea seed is brown in color, thin-shelled, about 1 – 2 cm in diameter, and round, semi-round in shape.

1.3 Genetics and Breeding

The tea plant is self-incompatible. Long-term allogamy makes it highly heterogeneous and consequently with broad genetic variation. The tea plant is usually diploid, with $2n = 30$ chromosomes. Natural triploid, tetraploid and aneuploid tea cultivars are also found. The genome size is estimated to be about 4.0 Giga bp. The transcriptome of the tea plant is partially sequenced using EST (expressed sequence tags) strategy and high-throughput DNA sequencing technology (Chen *et al*., 2009; Shi *et al*., 2011). The whole genome sequencing of the tea plant is still ongoing in China. It will benefit greatly the genetics and breeding of this crop.

Tea breeding was first started in ancient China. In the late 18th century, vegetative propagation methods using cutting and layering were developed in China. They accelerated the selection and breeding of new tea cultivars. Individual selection from landrace *jats* and natural populations and controlled hybridization are the main breeding methods for the tea plant to date. Recently, mutation breeding and molecular assisted selection (MAS) has been developed with the rapid development of new technologies. The main tea producing countries, such as

China, India, Sri Lanka, Kenya, Japan, etc., have bred and released hundreds of tea cultivars for the tea industry (Table 1.2). The made tea quality and yield increased significantly with the increase in the ratio of clonal tea acreage in the main tea producing countries.

Table 1.2 The national released tea cultivars and ratio of clonal tea gardens of some countries

Country	Year of the first release	National released cultivars	Ratio of clonal tea acreage (%)
China	1985	123 (17 *jats*)+158 local cultivars	46 (2010)
Japan	1953	54	92.1 (2004)
India	—	62+20 (Biclonal seed stocks)	60 (N.E. India, 2011)
Sri Lanka	1958	64	55 (2004)
Kenya	1964	50+11 (Biclonal seed stocks)	60 (2011)

1.4 Physiology and Biochemistry

The tea plant can be grown over a considerable range of conditions from temperate climates to hot, humid subtropics and tropics. However, the optimum mean daily ambient temperature for tea growth ranges between 20 °C and 30 °C. When the mean ambient temperature is higher than 30 °C, the growth of the tea plant is retarded. The tolerance of the tea plant to the minimum temperature varies with the varieties and cultivars, it generally ranges between –3 °C and –15 °C. Tea plants require not only certain amounts of rainfall, 1,000 – 1,700 mm annually, but also rainfall that is well-distributed during the whole year. Although the total rainfall in a year may be adequate for the production of green leaves in most of the tea producing areas in the world, the distribution of the rainfall in growing season may be inadequate in some areas. This can be regulated by irrigation. The optimum pH of soil for tea growing ranges between 4.5 and 6.5 (Chen & Yang, 2011; Zhen *et al.*, 2002).

The tea output per unit area is proportional to the coverage in the tea garden. For the purpose of obtaining the maximum productivity, a density of more than 12,000 – 20,000 bushes per ha in large-leaf cultivars and 45,000 – 60,000 bushes in small to median-leaf cultivars is recommended (Chen & Yang, 2011). The economic life of the tea plant is generally around 40 – 50 years. It is recommended to replant new clones when tea plants reach this age. However, such techniques as collar-pruning and heavy-pruning of old bushes are adopted.

The principle of fertilization is to compensate the nutrients removed by the crop and eluted by the rainfall. Generally, the level of nitrogen application is controlled at around 240 – 300 kg per ha and half amounts of potassium are added. The level of phosphorus is fixed at amounts of 60 – 90 kg P_2O_5 per ha and applied every 2 years. The need for microelements by the tea plant is few. It is not important in most of the tea producing countries; however, deficiencies were found in some particular instances. For example, a part of the soil in Sri Lanka and East African countries is zinc deficient; a part of the tea growing area in India, East Africa and Zaire is magnesium deficient. So, the application of microelement fertilizer showed significant effects in some instances (Chen & Yang, 2011; Zhen *et al.*, 2002).

Pruning is a "necessary evil" for the tea plant. The objects of pruning are to maintain the plant permanently in the younger phase, to stimulate the growth of shoots, and to build a rational height of the frame. In mature tea gardens, light-pruning and heavy-pruning should be done alternately. The best time for pruning is during a dormant period, because this is the time when the carbohydrates reserved within the tea plant are at a higher level.

The tea plant is a C_3 plant with high photorespiration. The utilization ratio of solar radiation energy is far lower than that of other C_4 plants. It was estimated that only 7% of photosynthetic products are used in the growth of the tea shoots, 9% in the formation of frame branches, and 84% is exhausted during respiration and other metabolic actions (Chen & Yang, 2011; Zhen *et al.*, 2002).

Fresh tea leaves and processed tea consist of a great number of substances which can be roughly divided into two categories: non-volatile compounds and volatile aroma compounds. The non-volatile that constitutes the major part of the tea solids include polyphenols, flavonols, amino acids, carbohydrates, organic acids, caffeine and purine derivatives, vitamins and many others. The physiological effects of caffeine are well known and documented. The most important and characteristic components in tea are the polyphenols and theanine. The total content of tea polyphenols expressed as a percentage of dry weight leaf is around 20% – 30%. They are the key compounds that determine the taste and color of infusion and have proved to have beneficial effects on human health. The most important compounds in tea polyphenols are the catechins. The content of catechins in tea is around 12% – 24% and represents more than 50% of the total amount of tea polyphenols. Six kinds of catechin compounds were isolated from tea. They are the various derivatives of catechins and gallic acids, including catechin (C), epicatechin (EC), gallocatechin (GC), epicatechin gallate (ECG), epigallocatechin (EGC), epigallocatechin gallate (EGCG). Generally, the content of the latter three catechins are relatively higher. The structures of main catechins and theaflavins are listed in Fig. 1.2. The catechin compounds condensed to theaflavin (TF) and thearubigin (TR) during the manufacturing process of black tea. The formation of these compounds makes the infusion orange-red in color. Theanine is an exclusive amino acid for the tea plant (Fig. 1.3); it accounts for about 60% – 70% of total amino acids in tea leaves. The site of theanine biosynthesis is the root and from there it is transferred to younger leaves. The pharmacological and physiological effects of theanine brought about a relaxing effect, the improvement of memory and also showed anticarcinogenic activity.

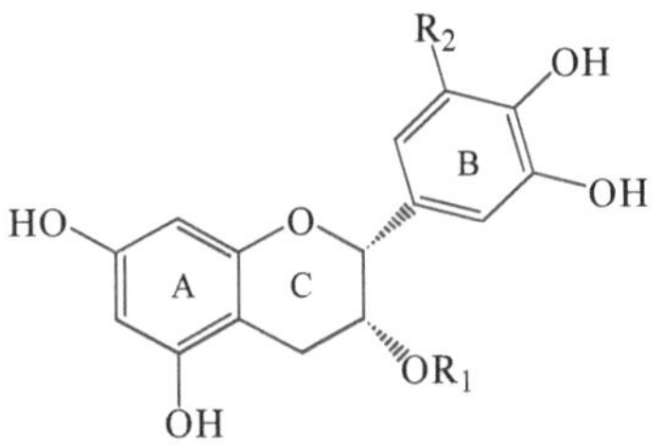

Epicatechin: $R_1 = R_2 = H$
Epigallocatechin: $R_1 = H$; $R_2 = OH$
Epicatechin-3-gallate: R_1 = Galloyl; $R_2 = H$
Epigallocatechin-3-gallate: R_1 = Galloyl; $R_2 = OH$

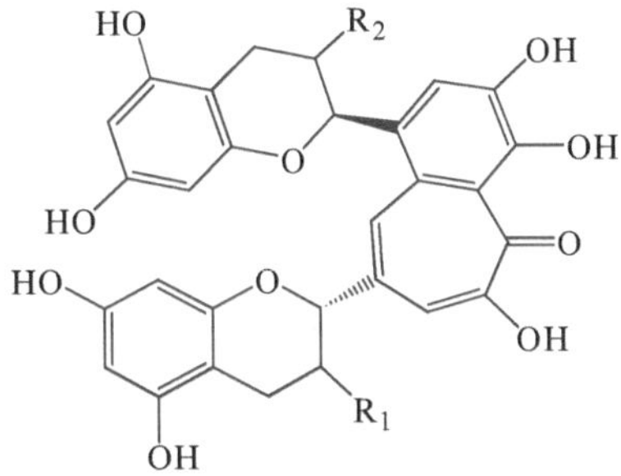

Theaflavin: $R_1 = R_2 = OH$
Theaflavin-3-gallate: R_1 = Galloyl; $R_2 = OH$
Theaflavin-3'-gallate: $R_1 = OH$; R_2 = Galloyl
Theaflavin-3,3'-digallate: $R_1 = R_2$ = Galloyl

Fig. 1.2. The structures of catechins and theaflavins

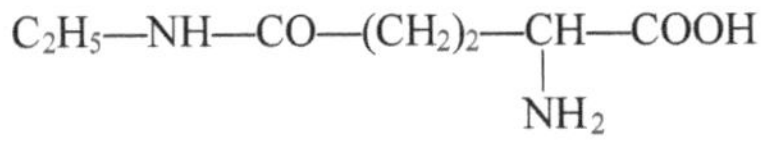

Fig. 1.3. The structure of theanine

The volatile aroma compounds play an important role in determining the quality of tea. Some of these compounds exist in the fresh leaves; however, most of them are formed during processing. Up to now, more than 600 flavor compounds have been identified in tea, although they exist only in a small amount (0.03% – 0.05% in fresh leaves on a dry basis, 0.005% – 0.01% in green tea and 0.01% – 0.03% in black tea).

With regard to the chemical basis of tea tasting and tea color, it is based on the taste threshold value of chemical components in tea and the reaction of sensory organs to these components. The compounds which play the major role in taste are tea polyphenols, amino acids and polysaccharides. The freshness of green tea is a reflection of the amino acids and the fullness of green tea is a reflection of the suitable ratio of tea polyphenols and amino acids. The catechins and theaflavins are the most important compounds determining the taste of black tea. Strong black tea depends on the content of water extracts and the briskness and fullness of black tea mainly depend on the ratio of caffeine, theaflavins and amino acids. The conjugation of these compounds and caffeine creates the astringency feeling and strong taste. The color of tea and tea infusion are based on certain chemical compounds. The color of green tea is mainly determined by the chlorophyll and some flavone compounds, such as vitexin and isovetexin. Chlorophyll-a is a deep green in color and chlorophyll-b is yellow green in color. The different ratio of chlorophyll-a and chlorophyll-b results in a different grade of green color in green tea. The color of black tea is black in made tea and orange-red in an infusion. These colors are formed by the theaflavin and thearubigin. Theaflavin is yellow in color and thearubigin is red in color. Different ratios of these two compounds constitute the different degree of color. Semi-fermented Oolong tea is generally a green-brown color in made tea and a yellowish red color in an infusion.

1.5 Kinds of Tea and Manufacture

The fresh tea leaves plucked from the tea plant are manufactured into various kinds of tea including green tea, Oolong tea, black tea, etc., by means of different manufacturing methods. In terms of unfermented tea (green tea), the first step in manufacture is to treat the fresh leaves at high temperature and this is called de-enzyming or fixation. The nature of this step is to deactivate the polyphenol oxidase localized within the cells of leaves, so as to stop the fermentation process and maintain the original green color. On the other hand, when the fresh leaves begin withering and are not treated at high temperature, the tea polyphenols are oxidized by the polyphenol oxidase enzyme, thus producing fermented tea (black tea). When the enzymes in the tea leaves are not completely deactivated and the tea polyphenols are not fully oxidized, this produces intermediate tea products between black tea and green tea, termed semi-fermented tea (Oolong tea).

Black tea is the major kind of tea consumed in the world. Congou black tea is manufactured by the most traditional processes including withering, rolling, fermentation and drying. For the convenience of brewing, the tea cutter was developed first in India and the rolling process was changed to a rolling and wringing process; thus, the broken black tea product was produced. Subsequently, some alternative manufacturing machines and methods were developed successively in India, East African countries and China; they include the Rotovane process, CTC (crush, tear and curl) process, LTP (Laurie tea processor) process. The production of CTC black tea has increased rapidly in the past half century and occupied 61.7% of the total black tea production in the world in 2009.

The basic manufacturing processes of green tea include de-enzyming, rolling and drying. The de-enzyming process comprises pan de-enzyming and steam de-enzyming. Steamed green tea is the major tea product produced and consumed in Japan, while roasted green tea is the major green tea product in China.

The basic manufacturing processes of Oolong tea include sunlight withering, light rolling, de-enzyming, rolling and drying. The plucking standard for Oolong tea is different from other kinds of tea. It is recommended that shoots with one bud and three to four leaves be plucked as the raw material when the *Banjhi* is formed on the terminal of a shoot.

Besides the above mentioned three kinds of tea, there are many kinds of tea produced in the world, including dark tea (a kind of post-fermented tea), yellow tea and white tea, scented tea, fruit-flavored tea and instant tea, etc.

1.6 Tea and Human Health

Tea was adopted for medicinal use in ancient China and sustained over a long period of time in Chinese history. Tea was transferred from medicinal use to use as a beverage from the Tang Dynasty. Investigation into the medicinal function of tea developed intensely after the discovery of the inhibitory activity of tea polyphenols on cancer cells firstly reported by Fujiki in Japan (Yoshizawa *et al.*, 1987). Up to now, many miraculous medicinal functions have been proved using the investigations conducted throughout the world (Zhen *et al.*, 2002; Chen, 2009). The following major functions of tea were verified and described.

1.6.1 Antisenile Activity

Antisenile activity is based on the antioxidative activity and the free radical scavenging effect of tea components. Many investigations showed that tea polyphenols possessed a potent antioxidative activity. Among the individual catechin in green tea, the activity of scavenging the free radicals decreased as follows: EGCG >ECG >EGC>EC.

1.6.2 Improvement in Immunity

The active components in tea may contribute to potential immuno-modulating effects including the innate immune system and the acquired immune system. This is expressed in the increase in leucocyte and lymphocyte amounts and improves the role of interleutin in spleen cells.

1.6.3 Anticaries

The tea plant is a fluorine-accumulating plant. The caries-preventive effect of tea was first believed to be due to its fluoride content. Tea contains high fluoride, which may strengthen tooth enamel and improve dental health. Further investigations indicated tea polyphenol not only could kill the caries bacteria, it also inhibits the activity of glycan transferase, meaning the caries bacteria cannot adhere to the surface of teeth.

1.6.4 Reduction in Blood-Cholesterol and Prevention of Cardiovascular Disorders

Tea drinking could increase the antioxidative activity of blood and improve the blood parameters. Earlier cohort studies in the USA, Norway, India, Saudi Arabia, Japan and the Netherlands suggested an inverse relationship between tea consumption and the risk of death due to heart attack and stroke. Many epidemiological studies showed an inverse association between tea consumption and the risk of cardiovascular diseases. However, conflicting results also existed in human trials. The cardioprotective effects of tea may also be due to its hypolipidemic and hypocholesterolemic activities. Numerous studies with animals have shown that tea consumption can reduce serum and liver lipids and total serum cholesterol. The effect of tea polyphenols on obesity has been verified by many investigations.

1.6.5 Anticarcinogenic Activity

The inhibitory activity of tea extracts on the human cancer cell was firstly reported by Fujiki in 1987 in Japan (Yoshizawa *et al.*, 1987). Since then, around several thousand papers on the anticarcinogenic activity of tea have been published in the world. A number of epidemiological studies have shown that there exists an inverse correlation between tea consumption and the incidence of certain kinds of cancer in humans. The popular press heralds tea as a cancer preventive beverage because such an activity has been demonstrated in many animal models. These models include cancers of the skin, lung, oral cavity, esophagus, stomach, liver, small intestine, colon, pancreas, bladder, prostate and mammary glands. Many ecological, case-control and cohort studies have been conducted to investigate the effects of tea consumption on human cancer incidence. However, the results have been inconclusive. For example, studies in northern Italy have suggested a protective effect of tea against oral, pharyngeal and laryngeal cancer. In a case-control study in Shanghai, China, frequent consumption of green tea has been shown to be associated with a lower incidence of esophageal cancer, especially among non-smokers and non-alcohol drinkers. A protective effect against gastric cancer by drinking tea has also been suggested from studies in Kyushu (Japan), northern Turkey and central Sweden, but has not been seen in many other studies from different geographic areas. In studies in Saitama, Japan, women consuming more than 10 cups of tea daily have been shown to have a lower risk of cancer (all sites combined) and increased tea consumption was associated with a lower risk of breast cancer metastasis and recurrence. However, there were also some epidemiological studies showing no significant lower incidence of cancers after tea consumption. It appears that most reports showing positive cancer preventive effects were from studies on Asians who drink predominantly green tea, whereas studies of the black-tea drinking population infrequently observed protective effects. It is possible that the different results connecting tea and cancer are due to the different etiological factors involved as well as the types and quantities of tea consumed in different populations. Investigations conducted over a period of twenty years proved the mechanism of the anticarcinogenic activity of tea components. Many mechanisms have been proposed concerning the inhibitory action of tea against carcinogenesis including the following: antioxidative activity, modulation of the activity of the key enzyme related to carcinogenesis, the anti-proliferative effect, inhibition of cell transformation, inhibition of the transcription factor and the blocking of the signaling transduction pathways related to carcinogenesis, induction of apoptosis and the inhibition of angiogenesis (Chen, 2009).

References

Chen L, Zhao LP, Ma CL, Liu Z, Zhang YL, Yao MZ, Wang XC (2009) Recent progress in the molecular biology of tea (*Camellia sinensis*) based on the expressed sequence tag strategy: A review. Journal of Horticultural Science & Biotechnology, 84(5): 476-482.

Chen ZM (2009) Twenty years period in the investigation on the anticarcinogenic activity of tea. Journal of Tea Science, 29(3): 173-190 (in Chinese).

Chen ZM, Yu YM (1994) Tea, Encyclopedia of Agricultural Sciences. Beijing: Academic Press, Vol.4, pp.281-288 (in Chinese).

Chen ZM, Yang YJ (2011) China Tea Classics (2nd Edition). Shanghai: Shanghai Civil Publishers (in Chinese).

International Tea Committee (ITC) (2011) Annual Bulletin of Statistics, London.

Ming TL (2000) Monograph of the genus *Camellia*. Kunming: Yunnan Science and Technology Press (in Chinese).

Sealy JR (1958) A Review of the Genus *Camellia*. The Royal Horticultural Society, London, 233p.

Shi CY, Yang H, Wei CL, Yu O, Zhang ZZ, Jiang CJ, Sun J, Li YY, Chen Q, Xia T, Wan XC (2011) Deep sequencing of the *Camellia sinensis* transcriptome revealed candidate genes for major metabolic pathways of tea-specific compounds. BMC Genomics, 12: 131.

Yamanishi T (1995) Special Issue on Tea. Food Reviews International, 11(3): 371-546.

Yoshizawa S, Horiuchi T, Fujiki, Yoshida T, Okuda T, Sugimura T (1987) Antitumor promoting activity of epigallocatechin gallate, the main constituent of "Tannin" in green tea. Phytotherapy Research, 1(1): 44-47.

Zhen YS, Chen ZM, Cheng SJ, Chen ML (2002) Tea: Bioactivity and Therapeutic Potential. London: Taylor and Francis Publishers, 256p.

2

Tea Germplasm and Breeding in China

Ming-Zhe Yao, Liang Chen*

Tea Research Institute of the Chinese Academy of Agricultural Sciences; National Center for Tea Improvement, 9 South Meiling Road, Hangzhou, Zhejiang 310008, China
*E-mail: liangchen@mail.tricaas.com

Abstract: The tea plant is native to China, which was firstly found and made use of as a beverage thousands of years ago. The country has the greatest tea production and consumption in the world, especially green tea and Oolong tea. In 2009, China had 1,866.7 kilohectares of tea gardens and 1,344.4 kilotonnes of tea annual production. China has abundant tea genetic resources capable of providing diverse parent materials for tea breeding. Currently, it is estimated that more than 3,000 accessions have been collected and conserved in the national germplasm tea repository. Different methods, which included *ex-situ*, *in-situ* and *in vitro*, were employed to prevent the loss of tea germplasm and ensure the preservation of genetic diversity. The tea germplasms have been evaluated and appraised based on morphology, physiology, agronomy, cytology and molecular biology, through which the elite and unique tea accessions were screened and further utilized in tea breeding. Recently, the appraisal and evaluation of tea germplasms was further regulated and standardized to boost efficiency and informative management. An efficient tea breeding system, in which controlled hybridization and individual selection are the main breeding approaches, combined with molecular marker-assisted selection and micropropagation techniques, has now been established and gradually developed in China. Up to date, a total of 123 cultivars have been registered as national tea cultivars. More than 150 cultivars have been accredited for release in given provinces and 4 clones have been covered by Plant Varieties Protection (PVP) in China. A standardized tea cultivar propagation and extension system has been established and is being implemented in China. This chapter reviewed the current situation of collection, conservation, appraisal and evaluation of tea germplasms, the achievements of tea genetic improvement and breeding, and the establishment and development of tea breeding as well as extension systems. Finally, suggestions were made on possible areas of research emphasis for the genetic improvement of the tea plant in the near future in China.

2.1 General Introduction to the Chinese Tea Industry

The tea plant is believed to originate from Yunnan Province in southwestern China, where abundant and diverse wild tea plants were found growing in the primeval forest (Yu, 1986). Tea was first used for medicinal purposes, and later as a national and traditional beverage for thousands of years in China. Up to date, China still has the largest tea plantation, production and consumption in the world.

2.1.1 Historical Development of the Tea Industry

The utilization history of tea plant could be dated back to 2737 B.C. (Yamanishi, 1995). By the end of the 6th century, the Chinese began to regard tea as not only a medicinal drink, but as a refreshing beverage. Since then, drinking tea has become more and more popular among Chinese peoples.

The tea industry had been gradually developed from the Tang Dynasty (618 – 907 A.D.) to the Yuan Dynasty (1206 – 1368 A.D.), when fresh tea leaves were popularly processed to make commercial cake-like tea, named cake-tea. In those times, only noble people from the upper class enjoyed drinking tea. The complex and mysterious ceremony of tea drinking was formed and popularized in China. In 760 – 770 A.D., the first tea book, *Tea Classics* (*Cha Ching*), was written by Lu Yu. It is the first authentic literature on tea, which comprehensively described the origin of tea, cultivation, manufacture, drinking methods, history, culture, etc. At that time, tea seeds, cultivation, manufacture and drinking methods were introduced into Japan and Korea by Buddhists (Wu, 1987). Subsequently tea was introduced to the other parts of the world.

The cake-tea was replaced by diverse shapes and types of tea from the Song Dynasty (960 – 1279 A.D.) to the Qing Dynasty (1644 – 1910 A.D.), when more and more ordinary peoples embraced tea as a beverage and tea drinking became popular. In the 17th century, China began to supply tea products to Mongolia, Russia, Europe and North America. The monopoly of tea exports from China slowly came to an end in 1886 when 81% of exported tea (approximate 134.1 kilotonnes) in the world was supplied by this country (Wu, 1987). In 1834, either tea seeds or plants and processing technology were introduced to India by Gordon who was the Secretary of India Tea Committee. After that, the Indian tea industry was set up and developed gradually, followed by Ceylon (Sri Lanka). The year of 1887 was the turning point when Britain for the first time imported more tea from India and Sri Lanka than it did from China (Weatherstone, 1992). By the year

1900, India and Sri Lanka had become the biggest exporters of tea in the world. Subsequently, the amount of exported tea catastrophically dropped in China. After the People's Republic of China was established in 1949, the development of the modern tea industry was initiated. From 1961 to 2009, the harvest area increased from 355 kilohectares to 1,385 kilohectares, tea production from 97 kilotonnes to 1,344.4 kilotonnes, tea yield from 273.4 kg/ha to 971 kg/ha, respectively (Fig. 2.1). In the last ten years, growth in the tea industry has been facilitated by intensive research in tea breeding, cultivation, manufacture, functional chemical components and the popularization of tea culture in China.

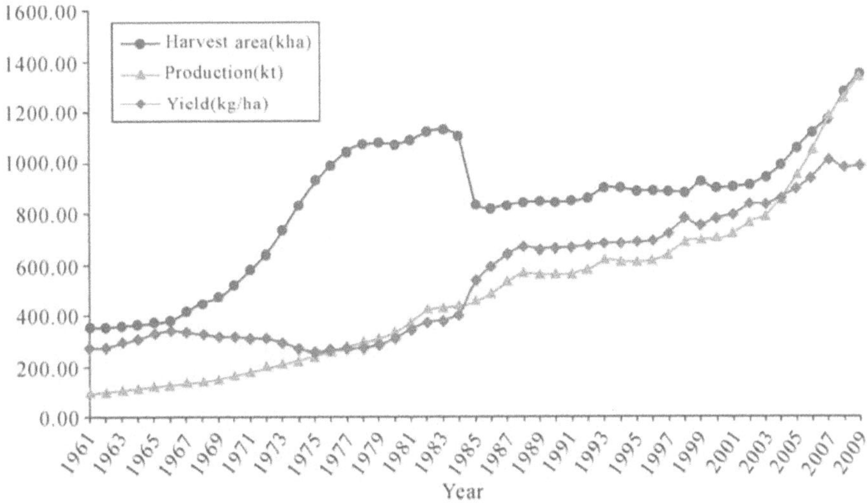

Fig. 2.1. The development of tea plantation, yield and production in China

2.1.2 *Location and Distribution of Tea Areas*

The tea-growing area is located between longitude E 94° (Milin, Tibet) and E122° (Taiwan Island), latitude N18°30′ (Yulin, Hainan) and N37°13′ (Rongcheng, Shandong). It covers more than 20 provinces and regions across China. However most of the tea-planting areas are distributed in the region of the southern Yangtze River. In 2009, tea plantation was estimated to be 1,866.7 kilohectares (Feng, 2010), of which 74.2% was harvest acreage (Table 2.1). More than 100 kilohectares of tea acreage are located in seven provinces: Yunnan, Sichuan, Fujian, Hubei, Zhejiang, Guizhou and Anhui. The largest tea-growing area (346.7 kilohectares) is in Yunnan Province, whereas the largest tea production (260.0 kilotonnes) comes from Fujian Province (Table 2.1). Recently, new tea gardens were developing rapidly in Guizhou Province, where the acreage of tea plantation increased by 108.5% from 69.8 kilohectares to 145.5 kilohectares, from 2007 to 2009.

Table 2.1 Regional level of the tea industry in mainland China (2009)

Region	Cultivated area (kilohectares)	Harvest area (kilohectares)	Clonal tea area (kilohectares)	Tea production (kilotonnes)
Fujian	193.3	163.3	183.7	260.0
Yunnan	346.7	233.3	117.3	180.0
Zhejiang	180.0	160.0	95.3	166.0
Sichuan	194.7	130.0	93.3	147.0
Hubei	190.0	140.0	35.3	140.0
Hunan	88.1	72.0	28.7	100.0
Anhui	130.0	114.7	16.0	78.0
Guangdong	37.3	32.7	25.0	48.2
Guizhou	145.5	60.1	93.1	41.0
Jiangxi	60.3	43.5	22.3	39.6
Guangxi	50.0	45.3	28.0	37.0
Chongqing	42.0	40.0	8.7	30.9
Henan	72.7	46.7	9.1	27.8
Shanxi	76.7	60.0	5.3	18.0
Jiangsu	30.0	24.3	8.8	15.3
Shandong	16.1	12.1	0.8	13.7
Gansu	12.3	5.7	0.3	1.1
Hainan	1.0	1.0	0.5	0.8
Total	1,866.7	1,384.7	771.5	1,344.4

Taking cognizance of environmental conditions and production modes, the tea growing region in China has been divided into four areas, namely the Jiangbei Area, Jiangnan Area, Southwest China Area and South China Area (Chen, 2000). The climatic and soil conditions in these four areas are presented in Table 2.2.

The tea area of Jiangbei is located to the north of the Yangtze River where tea plants are usually threatened by low temperature and drought. So the small-leaf type of cultivars with high tolerance to cold were usually selected and cultivated in this region. Green tea is the primary type of tea product.

Table 2.2 Climate and soil conditions in four tea areas

Tea areas	Temperature (°C)	Rainfall (mm)	Frostless period (day)	Soil type	Soil pH
Jiangbei	13 – 16	1,000	200 – 250	Yellow-brown	6.0 – 7.0
Jiangnan	15 – 18	1,100 – 1,600	230 – 280	Red and yellow	5.0 – 5.5
Southwest China	15 – 17	≥1,000	220 – 340	Yellow and lateritic red	5.5 – 6.5
South China	18 – 24	1,500	≥300	Laterite and yellow	4.5 – 5.5

The tea area of Jiangnan is located between the south of the Yangtze River and the north of Nanling Mountain, where warm temperatures and abundant rainfall facilitate tea growth. The middle-leaf types of cultivars are generally cultivated in this region, where green tea, Oolong tea and white tea are the dominant tea products.

The tea area of south China includes the south part of Fujian, Guangdong, Guangxi and Yunnan provinces, and the whole of Taiwan and Hainan Island, which belong to a subtropical and tropical monsoon climate. Most of the tea cultivars are arbor and semi-arbor trees with large leaves. Diverse tea products such as black tea, green tea, Oolong tea and dark tea are made and produced.

The tea area of southwest China is located in Guizhou, Sichuan, the middle and north of Yunnan, the southeast of Tibet, where the average annual temperature is about 20 °C, the rainfall more than 1,500 mm and the average altitude (1,000 – 1,500 m) is higher than other tea areas. Most of the cultivated tea comes from shrub trees with middle leaves. Green tea, dark tea and yellow tea are processed.

2.1.3 Tea Production and Its Global Standing

The primary Chinese tea products are traditionally categorized into six types, namely green tea, white tea, yellow tea, Oolong tea, black tea and dark tea, depending on their processing procedure, mainly the degree of fermentation. Green tea is a type of non-fermented tea, while white, yellow and Oolong teas belong to semi-fermented teas. Oolong tea has a distinct flower-like aroma due to a special processing method. Black tea belongs to fully fermented tea and dark tea is post-fermented tea. The basic flow chart of the processing procedure for different types of tea is shown in Fig. 2.2.

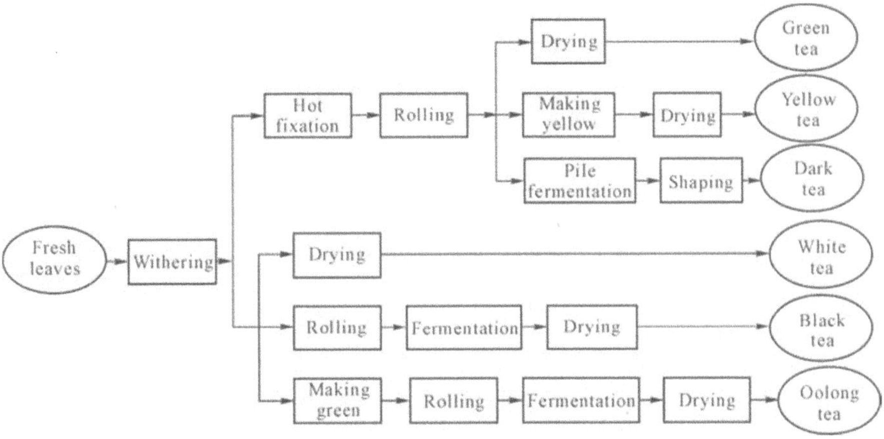

Fig. 2.2. Basic processing procedures for six types of tea products

Green tea is the major tea product in China. It is estimated that 926.6 kilotonnes of green tea was produced in 2008, occupying 73.7% of total tea production (Fig. 2.3). Oolong tea is the secondary tea product (144.1 kilotonnes) and 88.7% of this comes from Fujian Province. Only 6% of tea products are black tea (69.7 kilotonnes), of which 72.3% comes from Yunnan, Hunan and Hubei provinces. Other tea products such as dark tea, white tea and yellow tea are estimated to be 117.2 kilotonnes, which accounted for 9.0% of total products. In particular, a kind of dark tea, named as Pu'er tea, has become more and more popular in recent years. The production of Pu'er tea increased quickly with an increase of 25% – 50% from 2004 to 2007 (Ministry of Agriculture, 2008).

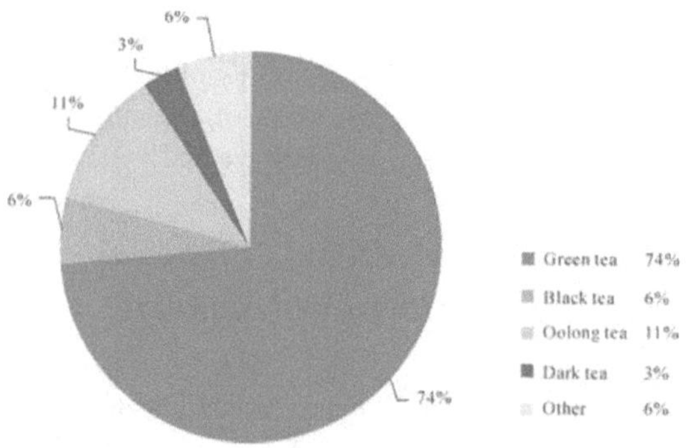

Fig. 2.3. Ratio of production of various tea types in China

According to statistical data from FAO (http://faostat.fao.org/) (2011), China became the largest tea producer and consumer once again in 2005 and 2006, respectively. The harvest tea acreage was 1,384.7 kilohectares, which accounted for 46.2% of whole world acreage in 2009 (Table 2.3). Total of tea production was 1,344.4 kilotonnes, of which 76.8% was used for domestic consumption.

Table 2.3 Tea area, production and yield of top ten countries

No.	Country§	Harvest area (kilohectares)	Production (kilotonnes)	Yield (kg/ha)
1	China*	1,384.7	1,344.4	970.9
2	India	474.0	805.2	1,698.6
3	Sri Lanka	221.9	290.0	1,306.4
4	Kenya	158.4	314.1	1,982.9
5	Vietnam	129.3	174.9	1,352.6
6	Indonesia	107.0	160.0	1,495.3
7	Turkey	75.9	198.6	2,618.3
8	Myanmar	74.5	26.5	355.7
9	Bangladesh	58.0	59.0	1,017.1
10	Japan	47.3	86.0	1,818.1
Sub-total		2,731.0	3,458.7	1,461.6
World total		2,995.5	3,885.3	1,297.0

§ The country order is sorted by harvest acreage;* Data source: http://faostat.fao.org, the data of China is from MOA

2.2 Collection, Conservation, Appraisal and Utilization of Tea Germplasm

The tea plant is originated in China, which has the most abundant and diverse tea genetic resources in the world. The genetic resource is presently considered as one of the most valuable fundamental materials for tea breeding and tea biotechnology, with valuable potential for the tea industry in the future. Success with tea collection, preservation, exploitation and utilization and present and long-term breeding programs depend largely on knowledge and understanding of the genetic diversity, relationship and identification of tea germplasms.

2.2.1 Collection and Conservation

The targeted and systematic investigation and collection of tea germplasms in China initiated during 1980s. The permanent facilities, China National Germplasm Hangzhou Repository and Menghai Branch were established in 1990 for the *ex-situ* conservation of tea germplasms in China. In additional, *in-situ* and *in vitro* conservation systems were established as useful supplement for the *ex-situ* conservation.

2.2.1.1 Collection

In modern China, the investigation of wild tea germplasms was initiated in the 1930s. However, the targeted investigation and collection of tea genetic resources commenced in the 1980s when the tea plant was listed in the national plans for crop germplasms investigation and collection. From 1980 – 1994, more than 1,500 tea accessions including wild types, landraces and cultivars from Yunnan, Sichuan, Guizhou, Guangxi, Guangdong, Fujian, Hainan and Shanxi, etc., had been collected and preserved in the China National Germplasm Tea Repository (CNGTR) (Chen *et al.*, 2004). Presently, the collection of tea germplasm still continues to be made spontaneously by individuals and public institutes. Modern equipments such as GPS (global positioning system), digital cameras, laptops, and so on, have been employed to facilitate the collection of tea germplasms.

2.2.1.2 Conservation

Permanent *ex-situ* conservation facilities, the China National Germplasm Tea Repository and Branch, have been established in Hangzhou, Zhejiang Province and Menghai, Yunnan province in 1990, to preserve all types of tea germplasm collection. The CNGTR in Hangzhou (Fig. 2.4) located at the Tea Research

Institute of the Chinese Academy of Agricultural Sciences (TRICAAS), which was used to preserve the national tea genetic resources of different regions, except Yunnan province. The Menghai Branch of the CNGTR, located at the Tea Research Institute of the Yunnan Academy of Agricultural Sciences, was uniquely used for conservation of local tea germplasm from Yunnan province. In the CNGTR, facilities such as the drainage and irrigation system, protective fences and greenhouses have been set up to safely preserve all types of tea plants.

Fig. 2.4. China National Germplasm Hangzhou Tea Repositories

By the end of 2010, about 3,000 accessions had been collected and preserved in the CNGTR, among which 2,800 accessions had been identified with their species and varieties according to Chen's taxonomic system (Chen L *et al.*, 2000). One hundred and eighty-two accessions could not be classified yet and 24 accessions belong to the related species of genus *Camellia* L. (Table 2.4). Most of the cultivated tea accessions were categorized as *C. sinensis* (L.) O. Kuntze and its varieties, and the majority of wild tea plants were classified as *C. taliensis* (W. W. Smith) Melchior, *C. tachangensis* F. C. Zhang, *C. crassicolumna* Chang, and *C. gymnogyna* Chang. Among all the preserved tea accessions, 59.7% are landraces, 11.1% wild types, 6.1% cultivars and 23.1% genetic materials including mutants, breeding strains and breeding stocks (Table 2.5). It is hoped that the amount of tea genetic resources will be increased gradually with persistent collection. Additionally, more than 5,000 accessions are estimated to be dispersedly conserved in local research and breeding institutions of tea. However, half of them might be duplicate accessions of those preserved in the CNGTR.

Table 2.4 Tea species and varieties conserved in the CNGTR

Species and varieties	Accessions	%
Camellia sinensis (L.) O. Kuntze	1,655	55.2
C. sinensis var. *assamica* (Masters) Kitamura	726	24.2
C. sinensis var. *pubilimba* Chang	177	5.9
C. taliensis (W. W. Smith) Melchior	138	4.6
C. crassicolumna Chang	48	1.6
C. tachangensis F. C. Zhang	24	0.8
C. gymnogyna Chang	26	0.9
Camellia sp.	182	6.1
Related species in *Camellia*	24	0.8
Total	3,000	100

Table 2.5 Types of tea collection preserved in the CNGTR

Type	Accessions
Wild types	334
Landraces	1,792
Cultivars and strains	182
Genetic materials	507
Other types	185
Total	3,000

For *ex-situ* conservation, it is required that at least 10 plants for each clonal accession and at least 20 individual plants for each seed-propagation accession and natural tea population are conserved to ensure safety and preservation of broadly genetic diversity. In addition to *ex-situ* conservation, *in-situ* conservation is also supported by national and local government. Plans are ongoing to conserve famous tea landraces and wild tea plants in Fujian, Zhejiang and Yunnan provinces. However, neither *ex-situ* nor *in-situ* conservation can prevent natural risks arising from extreme drought, extreme cold, hailstones, snow-storms and so on.

Therefore, an *in vitro* conservation system should be developed to utilize tissue culture and seed cryopreservation besides field conservation. Previous studies showed that tea plantlets could be developed from immature embryos and have been successively propagated for more than 20 generations with no significant genetic shift observed (Wang *et al.*, 1990). A length of 0.3 cm of nodal segments could be successfully induced to form a callus which subsequently grew into plantlets (Wang *et al.*, 1997). Cryopreservation may be the large-scale, long-term option for the conservation of species that are clones or have recalcitrant seeds (Li & Pritchard, 2009). The cryopreservation of tea seeds was studied under extra-low temperature (–196 °C) in liquid nitrogen with cryoprotectants (Wang *et al.*, 1999). However, innovational methods and technologies should be further studied and developed. Future studies should be concentrated on: (1) development of stable and effective cryoprotectants, (2) the rapid diagnosis and prediction of seed viability under cryopreservation; (3) cryopreservation of tea seed tissues.

2.2.2 *Appraisal and Utilization*

A standard descriptive system has been proposed to guide the appraisal and evaluation of tea accessions. The identification of tea germplasm was carried out by using multi-disciplinary approaches such as agronomy, morphology, biochemistry, cytobiology and molecular biology. A batch of elite tea accessions were screened and utilized for tea breeding and other research.

2.2.2.1 Descriptors of Tea Germplasm

In order to facilitate the evaluation, appraisal and digital management of tea germplasm, a total of 111 descriptors for tea germplasm were proposed (Chen L *et al.*, 2005a), of which 26 were passport data, 45 were morphological traits and biological characteristics, 29 were quality characteristics, 8 were abiotic tolerance and biotic resistance, and the other 3 were chromosome ploidy, DNA fingerprinting and remarks, respectively (Table 2.6).

Table 2.6 Descriptors of tea germplasm in China

No.	Descriptors	Explanation
Basic information/ Passport (26)		
1	Accession number	This number serves as a unique identifier for accessions and it is assigned when an accession is entered into the collection
2	Field genebank number	Number assigned by field genebank
3	Introduction number	Number assigned for alien accessions by introducer
4	Collecting number	Assigned by the collector(s) of the sample, normally composed of the name or initials of the place(s) followed by an Arabic number
5	Accession name	Either a registered or other formal designation given to the accession
6	Alien name	Native name in original place
7	Family	Latin name of family
8	Genus	Latin name of genus
9	Species	Latin name of species
10	Country of origin	Name of the country from which the sample was collected
11	Province of origin	Name of the primary administrative subdivision of the country from which the sample was collected
12	Origin	The original growing location at a village, a town

(To be continued)

(Table 2.6)

No.	Descriptors	Explanation
13	Altitude	Height of the location where the sample were collected
14	Longitude	Degrees and minutes followed by E (east) or W (west)
15	Latitude	Degrees and minutes followed by N (north) or S (south)
16	Sample source	Name of location where the sample was collected
17	Donor institution	Name of institution or individual responsible for donating the accessions
18	Donor accession number	Number assigned to an accession by the donor
19	Pedigree	Parentage for cultivars bred by crossing
20	Breeding institution	Name of institute breeding the cultivar
21	Releasing year	Year of cultivars authenticated to release
22	Breeding methods	Breeding methods such as selection, hybrid, mutation and others
23	Germplasm type	Refer to wild plant, landrace, cultivar, genetic materials and others
24	Propagating type	Refer to propagate by seed and cutting
25	Image filename	Filename of image related to the accessions
26	Experiment location	Location where assessment was carried out
Morphological and biological traits (45)		
27	Plant type	Described by shrub, semi-arbor and arbor
28	Growth habit	Described by upright, semi-upright and spreading
29	Sprouting density	Described by sparse, medium and dense
30	Date of 'one and a bud'	Time of 30% shoots sprouting 'one and a bud' in early spring
31	Date of 'two and a bud'	Time of 30% shoots sprouting 'two and a bud' in early spring
32	Young shoot color	Described by whitish, yellow green, light green, medium green and purple green
33	Young shoot pubescence	Described by absent, sparse, medium, dense and extremely dense
34	Length of 'three and a bud'	Average length of ten 'three and a bud' fresh shoots (cm)
35	Weight of 100 'three and a bud'	Weight of 100 'three and a bud' fresh shoots (g)
36	Leaf attitude	Described by upwards, semi-upwards, outwards and downwards

(To be continued)

(Table 2.6)

No.	Descriptors	Explanation
37	Leaf length	Average length from base to apex of ten mature leaves
38	Leaf width	Average width of maximal wide part of ten mature leaves
39	Leaf size	Described by small, medium, large and extremely large
40	Leaf shape	Described by near round, ovate, elliptic, oblong and lanceolate
41	Number of venation	Average number of vein pairs of ten mature leaves
42	Leaf color	Described by yellow green, light green, medium green and dark green
43	Leaf upper surface	Described by smooth, slightly rugose and rugose
44	Leaf cross section	Described by folded upwards, flat and recurved
45	Leaf texture	Soft degree described by soft, medium and hard
46	Sharpness of leaf serration	Described by sharp, medium and obtuse
47	Density of leaf serration	Described by sparse, medium and dense
48	Depth of leaf serration	Described by flat, medium and deep
49	Leaf base shape	Described by acute and obtuse
50	Leaf apex shape	Described by acute, attenuate, blunt and obtuse
51	Leaf margin undulation	Described by flat, slightly wavy and wavy
52	Time of full blooming	Time when 50% of flowers were full blooming
53	Number of calyxs	The average number of calyxs
54	Calyx color	Described by green and purple red
55	Calyx pubescence	Absence and presence of calyx pubescence
56	Flower diameter	Average size of ten flowers diameter
57	Petal colour	Described by white, greenish and pink
58	Petal texture	Thick degree of petal described by thin, medium and thick
59	Number of petals	Average number of petals
60	Ovary pubescence	Absence and presence of ovary pubescence
61	Style length	Average length of ten flower styles
62	Number of style splittings	Average number of style splittings of ten flowers
63	Position of style splittings	Relatively splitting position of flower style described by low, medium and high
64	Relative height between gynoecium and androecium	Gynoecium being lower, same height and higher compared to androecium
65	Fruit shape	Described by global, kidney-shaped, triangular, quadrangle, cinquefoil-shaped

(To be continued)

(Table 2.6)

No.	Descriptors	Explanation
66	Fruit diameter	Average size of ten fruits diameter (cm)
67	Thickness of percarp	Average thickness of percarp (mm)
68	Seed shape	Described by round, semi-round, cone-shaped, like kidney-shaped and irregular
69	Seed diameter	Average size of seed diameter (cm)
70	Seed color	Described by brown, brown grey and grey
71	Weight of 100 seeds	Weight of 100 seeds (g)
Quality traits (29)		
72	Processing suitability	Primary capability for making green tea, black tea, Oolong tea and others
73	Other processing suitability	Secondary capability for making green tea, black tea, Oolong tea and others
74	Total score of green tea	Total organoleptic evaluation score of green tea
75	Aroma score of green tea	Aroma score of green tea
76	Characteristics of green tea aroma	Description for the aroma of green tea
77	Taste score of green tea	Taste score of green tea
78	Characteristics of green tea taste	Description for the taste of green tea
79	Total score of black tea	Total organoleptic evaluation score of black tea
80	Aroma score of black tea	Aroma score of black tea
81	Characteristics of black tea aroma	Description for aroma of black tea
82	Taste score of black tea	Taste score of black tea
83	Characteristics of black tea taste	Description for taste of black tea
84	Total score of Oolong tea	Total organoleptic evaluation score of Oolong tea
85	Aroma score of Oolong tea	Aroma score of Oolong tea
86	Characteristics of Oolong tea aroma	Description for aroma of Oolong tea
87	Taste score of Oolong tea	Taste score of Oolong tea
88	Characteristics of Oolong tea taste	Description for taste of Oolong tea
89	Water extracts	Content of water extracts in dry tea (%)
90	Caffeine	Content of caffeine in dry tea (%)
91	Tea polyphenols	Content of polyphenols in dry tea (%)
92	Amino acids	Total content of free amino acids in dry tea (%)
93	Ratio of polyphenols to amino acids	Ratio between the content of polyphenols and free amino acids
94	Theanine	Content of theanine in dry tea (%)
95	Total catechins	Content of catechins in dry tea (%)
96	Epigallocatechin gallate	Content of epigallocatechin gallate in dry tea (%)
97	Epigallocatechin	Content of epigallocatechin in dry tea (%)

(To be continued)

(Table 2.6)

No.	Descriptors	Explanation
98	Epicatechin gallate	Content of epicatechin gallate in dry tea (%)
99	Epicatechin	Content of epicatechin in dry tea (%)
100	Gallocatechin	Content of gallocatechin in dry tea (%)
Resistant traits (8)		
101	Cold resistance	Described by strong, relatively strong, medium and weak
102	Drought resistance	Described by strong, relatively strong, medium and weak
103	Resistance to brown blight	Described by resistant, moderate resistant, susceptive and high susceptive
104	Resistance to tea anthracnose	Described by resistant, moderate resistant, susceptive and high susceptive
105	Resistance to blister blight	Described by resistant, moderate resistant, susceptive and high susceptive
106	Resistance to tea leafhopper	Described by resistant, moderate resistant, susceptive and high susceptive
107	Resistance to pink mite	Described by resistant, moderate resistant, susceptive and high susceptive
108	Resistance to tea red spider mite	Described by resistant, moderate resistant, susceptive and high susceptive
Other (3)		
109	Chromosome ploidy	Refer to the diploid, triploid, tetraploid, etc
110	Fingerprinting and molecular markers	DNA fingerprinting constructed by molecular markers
111	Remarks	Other characteristics cannot be included by above descriptors

2.2.2.2 Classification of Tea Germplasm

Different taxonomic systems for tea plants were proposed on the basis of morphological traits in the past 60 years. The characteristics of flower and fruit were considered as decisive descriptors for tea classification, and other traits such as tree type, leaf texture, size, shape and pubescence were also important indicators for tea classification (Sealy, 1958; Chang, 1981, 1984; Ming, 1992; Chen L *et al.*, 2000). Later, new parameters like the bio-chemical components (Du *et al.*, 1990), pollen morphology (Shu & Chen, 1996; Chen *et al.*, 1997), chromosome karyotype (Li, 1983; Li, *et al.*, 1986; Li & Liang, 1990) and DNA markers (Chen & Yamaguchi, 2002) were considered useful for tea classification.

Based on investigation and identification of wild tea plants in southwest China, Chang (1981) proposed that *Camellia* L. Sect *Thea* (L.) Dyer could be grouped into four series containing 17 species, 3 varieties on the basis of style splitting, ovary and leaf morphology. These series and species were named as Ser. *Quinqueloculāris* Chang (including *C. kwangsiensis* Chang, *C. quinquelocularis* Chang and *C. tetracocca* Chang) and Ser. *Pentastylae* Chang (including *C. crassicolumna* Chang, *C. pentastyla* Chang, *C. taliensis* (W. W. Smith) Melchior, *C. irrawadiensis* P. K. Barua and *C. crispula* Chang), Ser. *Gymnogynae* Chang (including *C. gymnogyna* Chang, *C. costata* Hu et Liang, *C. yungkiangensis* Chang and *C. leptophylla* S. Y. Liang) and Ser. *Sinensis* Chang (including *C. pubicosta* Merr, *C. angustifolia* Chang, *C. sinensis* (L.) O. Kuntze, *C. ptilophylla* Chang and *C. parvisepala* Chang). The former 2 series were characterized with 5 style arms and a glabrous ovary with 5 locules of flowers. While the latter 2 series have flowers with 3 style arms, and a pubescent ovary with 3 locules. In total, 44 species had been reported to date (Chang, 1984, 1990).

Later, Chang's taxonomic system was further revised according to the investigation and observation of wild tea plants and their specimens. More than 40 species were merged into 12 species and 6 varieties (Ming, 1992). They were named as *C. tachangensis* F. C. Zhang, *C. kwangsiensis* Chang (var. *kwangsiensis*, var. *kwangmanica* (Chang et Chen) Ming), *C. grandibracteata* Chang et Yu, *C. taliensis* (W. W. Smith) Melchior, *C. crassicolumna* Chang (var. *crassicolumna*, var. *multiplex* (Chang et Tang) Ming), *C. gymnogyna* Chang (var. *gymnogyna*, var. *remotiserrata* (Chang et Tan) Ming), *C. costata* Chang, *C. leptophylla* S. Y. Liang et Chang, *C. ptilophylla* Chang, *C. fangchengensis* S. Y. Liang, *C. purpurea* Chang et B. H. Chen and *C. sinensis* (L.) O. Kuntze (var. *sinensis*, var. *dehungensis* (Chang et Chen) Ming, var. *assamica* (Masters) Kitamura and var. *pubilimba* Chang). The former 5 species have flowers characterized with 5 style arms and ovary locules, while the latter 7 species have three style arms and ovary locules of flowers.

Based on previous taxonomic systems of the genus *Camellia* L., and long-term research of the wild tea plants and other genetic resources in China, the classification of section *Thea* (L.) Dyer was further simplified and revised into five species and two varieties (Chen L *et al.*, 2000). The tea plant characterized with 5 ovary locules and style splitting of the flower was classified into 3 species, *C. tachangensis* F. C. Zhang, *C. taliensis* (W. W. Smith) Melchior and *C. crassicolumna* Chang. Tea plants characterized with 3 ovary locules and style arms of the flower were classified into *C. gymnogyna* Chang and *C. sinensis* (L.) O. Kuntze (var. *sinensis*, var. *assamica* (Masters) Kitamura, var. *pubilimba* Chang). Compared with Chang's and Ming's systems, this proposal seems more concise and functional, and was validated by molecular marker analysis (Chen & Yamaguchi, 2002).

In addition to *C. sinensis* (L.) O. Kuntze and its varieties are widely distributed and cultivated in China and other tea growing countries, other species mostly were wild types distributed in southwest China. As a kind of perennial cross-pollination plants, it is difficult to clearly discriminate the species and varieties of tea plants due to overlapping morphological traits resulting from natural hybridization of different tea species, which lead to keep being under discussion for tea classification today. So it is still necessary to improve the classification system of the tea plant using modern molecular technologies.

2.2.2.3 Appraisal and Evaluation of Tea Germplasm

Morphological and biological traits, chemical components, cup tea quality, tolerance and resistance to biotic and abiotic stresses of more than 1,500 accessions of tea genetic resources have been identified and appraised using multi-disciplinary approaches in the past 20 years. Accessions characterized with special traits such as extremely early and late sprouting time in the spring, extremely large and small leaf blades, high tea polyphenols content, high and low caffeine content, high amino acids (theanine) content, high tolerance to low temperature, have been screened (Lv & Lou, 1991; Lv *et al.*, 1990; Yu *et al.*, 1992; Zhu, 1992; Lu *et al.*, 1995; Xu *et al.*, 1997; Yang *et al.*, 2003c; Chen *et al.*, 2004; Chen & Zhou, 2005; Yao *et al.*, 2008b). The book *Descriptors and Data Standard for Tea* (*Camellia* spp.) was published (Chen L *et al.*, 2005a) and the agricultural technique standards of the Ministry of Agriculture *Technical Code for Evaluating Crop Germplasm - Tea Plant* (*Camellia sinensis*), *Evaluating Standards for Elite and Rare Germplasm Resources-Tea Plant (Camellia sinensis* (L.) O. Kuntze*)* were released to the public (Chen *et al.*, 2007b, 2011). These publications will further regulate and standardize the appraisal and evaluation of tea germplasms. To enhance the efficiency of evaluation and appraisal, the primary core collection of Chinese tea germplasms containing 532 tea accessions was constructed with optimum sampling strategies, which represents 99.7% of phenotypic variation of the whole collection (Wang *et al.*, 2009).

2.2.2.3.1 Evaluation of Morphological and Biological Traits

The arbor, semi-arbor and shrub type of tea plants can be found widely among wild and cultivated tea. The height of tea plants varied from 1.2 m to 26.5 m (Fig. 2.5). Some wild tea plants such as 'Lancang Dachashu', 'Qianjiazhai Dachashu' and 'Bada Dachashu' were higher than 20 m. It was observed that wild tea generally belongs to the type of arbor and semi-arbor plants, and cultivated tea belongs to the type of semi-arbor and shrub plants (Yu, 1986). Two hundred tea accessions originating from 14 regions were evaluated for their agronomic traits (Yu *et al.*, 1991). The results showed that 91.7% of arbor plants were characterized with large and extra large leaf size, while only 3.3% of shrub plants had a large leaf size. The sprouting time of fresh shoots with one to three buds varied among different accessions. Early types of tea plants may be earlier by 13 to 24 days than later types. The length and weight of 'three and a buds' were observed to correlate with plant type and leaf size. The morphological investigation of flowers organ among 200 tea accessions indicated that 174 (89.7%) were present at ovary pubescence, only 20 (10.3%) were absent; 189 (97.4%) had three style arms, only

5 (2.6%) had five style arms (Yu *et al.*, 1991). Greatly morphological variation was observed within species and varieties of the genus *Camellia* L. Most of the quantitative traits of leaves and flowers showed significant differences among different species; however fruit characteristics did not differ significantly.

(a) (b)

Fig. 2.5. Wild tea plant (a) and cultivated tea plant (b)

2.2.2.3.2 Evaluation of Bio-chemical Components

The bio-chemical components such as tea polyphenols, caffeine and amino acids are determinant factors for cup-tea quality. The analysis of tea germplasm from Yunnan revealed that the coefficient of variation (*CV*) ranged from 13.4% to 22.7% among the contents of caffeine, polyphenols, free amino acids and water extract solid. *C. taliensis* (W. W. Smith) Melchior and *C. sinensis* var. *assamica* (Masters) Kitamura had a significantly higher content of polyphenols and water extract content than *C. sinensis* var. *sinensis* (L.) O. Kuntze, *C. sinensis* var. *dehungensis* (Chang et Chen) Ming and *C. crassicolumna* Chang. However, no significant difference was detected for caffeine and free amino acids contents among different species and varieties. Based on allozyme analysis, it was observed that the percentage of polymorphic locus (*P*) varied from 21.4% to 50.0%, and the mean of heterozygosity per locus (H_e) varied 0.114 – 0.218 among species and varieties (Chen J *et al.*, 2005). Further assessment of polyphenols, catechins, amino acids, caffeine and water extract were carried out among 596 accessions (Chen & Zhou, 2005). The content of tea polyphenols (TP) varied from

13.6% to 47.8%, averaging 28.4%. The TP increases gradually from the northern and eastern regions to the southern region, and the highest TP of tea accession was found in Yunnan. The content of total catechins ranged from 81.9 g/kg to 262.7 g/kg, with an average of 144.6 g/kg. The amino acids varied from 1.1% to 6.5%, with an average of 3.3%. The average caffeine content was 4.2%, varying from 1.2% to 5.9%. Most accessions of high caffeine content have widespread distribution in Yunnan Province, followed by Fujian Province. The average water extract content was 44.7%, varying from 24.4% to 57.0%. Based on previous studies, a few tea genetic resources were identified and screened with trans-normal bio-chemical components, like low caffeine, high polyphenols, high total catechins, high epigallocatechin gallate (EGCG), high amino acids, etc. (Yu *et al.*, 1992; Yang *et al.*, 2003c; Chen & Zhou, 2005; Lin *et al.*, 2005). These could be used directly or indirectly for commercial functional components extraction, breeding and development of new tea cultivars.

2.2.2.3.3 Sensory Evaluation of Cup Tea

Appearance, color, flavor and taste are crucial parameters in appraising the sensory quality of cup tea. It was observed that the color of tea infusion significantly correlated with aroma ($r = 0.98$) (Yu *et al.*, 1991). Different accessions showed varying suitability in the making of different types of tea. The sensory quality of cup tea was evaluated among 400 accessions, of which 4 were identified to be very suitable for making green tea, 17 for black tea and 2 for Oolong tea (Yu *et al.*, 1992; Yang *et al.*, 2003c). Several accessions were characterized with special aromas such as sweet aroma, chestnut-like and fruit-like smells, because their fragrance components like geraniol, linalool, phenylacetaldehyde, etc., were different among these accessions (Yu *et al.*, 1992).

2.2.2.3.4 Tolerance to Biotic and Abiotic Stresses

Resistance to pests and diseases was evaluated among tea genetic resources from different regions in previous studies. Several highly resistant accessions were screened for tea leafhopper (*Empoasca vitis* Gothe), tea geometrid (*Ectropis oblique* Prout), pink mite (*Acaphylla theae* Watt), tea mite (*Oligonychus coffeae* Nietner), broad mite (*Polyphagotarsonemus latus* Banks), brown blight (*Guighardia camelliae* (Cooke) Butler) and gray blight (*Pestalotiopsis theae* (Sawada) Steyaert) (Lv *et al.*, 1990, 1991; Peng, 1990; Lv & Lou, 1991; Zhu, 1992; Yao *et al.*, 2008b). However, few accessions were observed to exhibit horizontal resistance to a variety of pests and diseases. A tolerant level of resistance to cold and drought varied among different tea germplasms, and few strong and tolerant genotypes were identified and screened (Yu *et al.*, 1991; Lu *et al.*, 1995).

2.2.2.3.5 Cytological and Chromosome Characterization

Pollen morphology was regarded as an important indicator for studies on tea classification and evolution. Tea pollen was viewed as a prolate and trilobate sphere across equatorial and polar sections, respectively (Chen *et al.*, 1997). Three colporate grains were covered with reticulate and obscurely reticulate ornamentations on the pollen coat. A broad variation could be observed in the pollen size, the ornamentations on the pollen surface among and within varieties of tea. Studies indicated that pollens might evolve from a large to small size, from smooth to coarse reticulate ornamentation on the surface (Shu & Chen, 1996; Chen *et al.*, 1997).

Tea is generally considered as a kind of diploid plant which has 15 pairs of chromosomes ($2n = 2x = 30$) (Li, 1983; Li & Liang, 1990). However, it was observed natural tri- and tetra-ploidy accessions such as 'Wuyi Shuixian', 'Fuding Dahaocha', etc. (Liang & Liu, 1988; Zhan *et al.*, 1993; Li, 1996). Most karyotypes of tea were classified as type 2A. However, variations occurred in chromosome length, number and position of the SAT-chromosome, the ratio between the long arm and short arm among and within species (Li, 1983; Li *et al.*, 1986).

2.2.2.3.6 Characterization Based on Molecular Markers

The tea plant is a heterogenous plant with many overlapping morphological, biochemical and physiological traits. It has been argued that phenotypic descriptors may not reflect the true level of genetic differentiation, because most of them shows continuous variation and a high degree of plasticity when they are subject to environmental and developmental influences. Molecular markers, such as random amplified polymorphic DNA (RAPD), amplified fragment length polymorphism (AFLP) and inter-simple sequence repeat (ISSR), are popular tools used in research of identification, genetic diversity and the genetic relationship of tea germplasms (Ni *et al.*, 2008). Recently, microsatellite markers, genomic SSRs (gSSRs) and expressed sequence tag based SSRs (EST-SSRs), have been developed and applied to tea plant characterization (Jin *et al.*, 2006; Zhao *et al.*, 2008a; Yang *et al.*, 2009; Ma JQ *et al.*, 2010b), which hopefully provide stronger and more effective DNA markers for the tea plant.

Molecular markers had been proved to be useful and sufficient to characterize and discriminate tea varieties and cultivars, even those that could not be distinguished on the basis of morphological and phenotypic traits. Twenty-four wild tea germplasms, composed of different species and varieties originating from China, could easily be discriminated by unique RAPD markers, specific band patterns, a combination of the band patterns and DNA fingerprinting provided by different primers (Chen & Yamaguchi, 2005). Recent attention has been focused on the potential use of different molecular descriptors for estimation of the genetic relationship and determination of the genetic diversity of tea germplasm. The genetic diversity and phylogeny of wild tea plants were investigated by RAPD analysis. The relationship was determined among *C. sinensis* (L.) O. Kuntze and its related 23 species and varieties. The genetic diversity of tea landraces in different regions were also analyzed and compared by using RAPD (Chen *et al.*, 1998; Luo *et al.*, 2002), AFLP (Huang *et al.*, 2004), ISSR (Yao *et al.*, 2005, 2007; Hou *et al.*, 2007; Guo *et al.*, 2008) and EST-SSR (Jin *et al.*, 2007; Liu Z *et al.*, 2008; Yao *et al.*, 2009; Qiao *et al.*, 2010). A higher level of genetic diversity was revealed from intra-groups (populations) rather than from that of inter-groups (populations) (Shen *et al.*, 2007; Yao *et al.*, 2007, 2008a, 2009), and high gene flow (N_m) indicated that frequently gene introgression happened between populations (Yao *et al.*, 2008b, 2009). The high N_m may gradually lead to a similar frequency of gene and genotype among different populations for cross-pollinated crops like tea, resulting in low genetic differentiation and close genetic distance among populations.

The genetic diversity and population structure were studied based on the 272 accessions from the primary tea collection in China using EST-SSR markers (Yao, 2009). Results indicated that the level of genetic diversity showed a decreasing tendency from the original center to the north and east regions, and a higher level of genetic diversity existed in the coastal region than in the inland region. Meanwhile, a higher level of genetic diversity was observed in the wild tea plants rather than traditional landraces and bred cultivars, which indicated that the genetic diversity was influenced by long-standing domestication and artificial selection. A total of 272 accessions were grouped into five populations by structure analysis based on a mathematical simulation model (Fig. 2.6), which was confirmed by N-J (Neighbor-joining) methods based on Nei's genetic distance (Fig. 2.7). It was confirmed that the population structure was not only related to geographic origin, but accession types. Though some accessions from similar region were usually clustered together in small sub-populations, most of the tested accessions within the same origin and the same type were dispersed in different populations. It indicated that there is a broadly genetic variation among the Chinese tea collection.

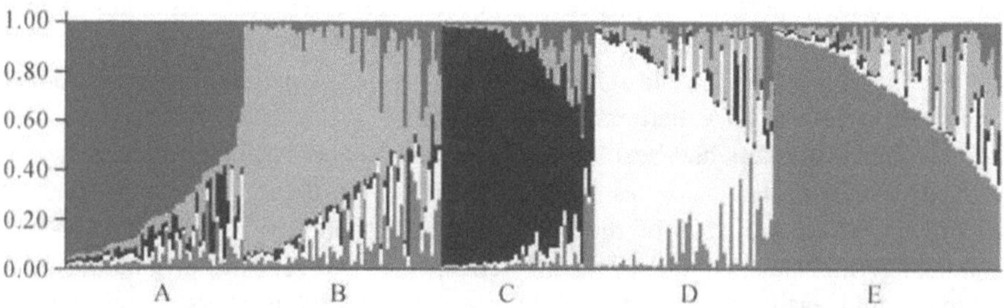

Fig. 2.6. The population structure of 272 Chinese primary core collection of tea germplasms based on EST-SSR markers (A-E represent different populations)

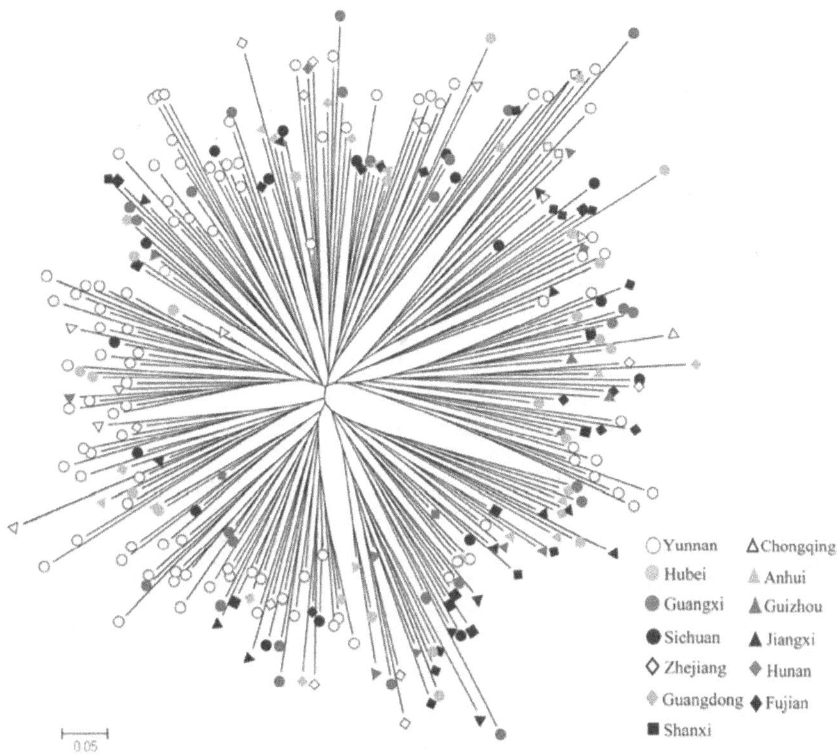

Fig. 2.7. The N-J unrooted tree of Nei's genetic distance for 272 tea accessions based on EST-SSR markers

2.2.2.3.7 Identification and Expression Analysis of Functional Gene

In the past ten years, many functional genes controlling plant growth and development, tea quality and resistance, such as S-adenosylmethionine synthase, polyphenol oxidase, β-1, 3-glucanase and so on, have been isolated, cloned and expressed (Table 2.7) (Li *et al.*, 2004; Chen *et al.*, 2006; Yu *et al.*, 2008a, 2008b; Zhu *et al.*, 2008a, 2008b, 2008c; Qiao *et al.*, 2009; Tan *et al.*, 2009; Ma & Chen, 2007, 2009; Ma CL *et al.*, 2010). The differential expressed genes have been identified and detected under treatment of cold stress (Mei *et al.*, 2007; Zou *et al.*, 2008; Fang *et al.*, 2009), pest attacks (Wei *et al.*, 2007), at different growth stages (Zhao *et al.*, 2006a; Wang *et al.*, 2008). The cDNA libraries from both tender shoots and young roots of tea plants have been constructed and sequenced. More than 6,000 ESTs were obtained through sequencing of cDNA clones. The blasting results showed that 45% of ESTs were partial or full sequences of novel genes (Chen L *et al.*, 2005b, 2005c; Zhao *et al.*, 2008b). These ESTs were identified with their function of metabolism (secondary metabolism), energy, cell growth and division, transcription, protein synthesis, protein destination and storage, transporters, cell structure, signal transduction, disease and defense (Chen L *et al.*, 2005b). A total of 1,680 genes obtained in the EST project were selected from the cDNA library to develop the first cDNA microarray of the tea plant. The cDNA microarray contained 6,912 dots, including 6,720 ESTs, 160 positive controls and 32 negative controls. The density of the microarray was 1,037 dots per cm^2 (Zhao *et al.*, 2006b). The microarray was further used to identify and reveal the different expression profiles between clones with high and low content of tea polyphenols based on hybridization experiments. The cDNA microarray could be applied in various research fields to develop high-throughput detection for gene expression profiling of the tea plant (Ma *et al.*, 2011). Furthermore, the initiation and primary progress of functional genomics of the tea plant provide a novel and robust approach to understand the mechanisms of growth, development, differentiation, metabolism, quality of cup tea and stress resistance on the whole genome level.

Table 2.7 A partial list of cloned genes of the tea plant published in China

Gene	GenBank accession	Length (bp)	Amino acid residues
Violaxanthin de-epoxidase (VDE)	AF462269	1,632	473
S-adenosylmethionine synthase (SAM)	AB041534	1,303	394
Polyphenol oxidase	AF269192	1,006	335
β-1,3-glucanase	AF399920	1,329	495
ACC synthase	EF205149	1,579	478
ACC oxidase	DQ904328	1,232	320
Cyclophilin	DQ904327	949	164
Calcium-dependent protein kinase	EU732607	2,281	760
Chalcone isomerase	DQ904329	1,163	240
C-repeat binding factor	EU857638	981	275
14-3-3 protein	DQ444463	1,072	260
Cystatin	/	627	101
Pollen specific protein (CsPSP1)	DQ887753	2,079	567
Photosystem II protein D1	AY665295	1,678	353
5.8S ribosomal RNA	AF315492	633	/
Histone H1-like protein	EU716314	879	207
Alpha tubulin 1	DQ340766	1,537	450
Flavone synthase II	FJ169499	1,824	534
ATP sulfurylase (APS1, APS2)	EF218618 EF218619	1,415 1,706	362 465
Selenocysteine methyltransferase (SMT)	DQ480337	1,386	351
Flavonol synthase (FLS)	EF205150	1,317	331
Leucoanthocyantin reducase(LAR)	EF205148	1,301	342
β-glucosidase	AF537127	1,475	450
Photosystem II CP43 protein	AY741479	1,520	433
Magnesium chelatase H subunit	HQ660368	4,417	1382

(To be continued)

(Table 2.7)

Gene	GenBank accession	Length (bp)	Amino acid residues
Chlorophyllide a oxygenase	HQ660369	2,146	536
Chlorophyll synthase	HQ660370	1,463	374
Glutamyl-tRNA reductase	HQ660371	2,165	554
Chloroplast ferredoxin Ⅰ	HQ660372	583	147
R2R3-MYB transcription factor 1	HQ660373	1,132	292
R2R3-MYB transcription factor 2	HQ660374	1,020	224
MYB transcription factor 1	HQ660375	919	271
bHLH transcription factor 1	HQ660376	1,103	235
Auxin-repressed protein	HQ225758	711	118
Actin	HQ235647	1,470	377
Vacuolar ATP synthase subunit E	HQ235648	1,021	229
Ubiquinol-cytochrome C reductase	HQ235649	504	124
ABC transporter family protein	HQ423146	1,470	287
NADH-plastoquinone oxidoreductase subunit J protein	HQ423147	1,041	159
Unknown gene	HQ423148	1,021	747
Cytoplasmic ribosomal protein	HQ423149	758	152
Late embryogenesis abundant protein	HQ852396	889	153
Auxin induced protein	HQ852397	1,568	460
Brassinosteroid-regulated protein	HQ852398	1,306	286
Cyclin-dependent kinase	JF449383	1,245	307
Germin-like protein	JF449384	926	212
Oxygen-evolving enhancer protein	JF449385	1,258	333
RAV transcription factor	GQ227992	1,329	362
Dehydrin	GQ228834	1,091	251
Inducer of CBF expression (ICE)	GQ229032	1,964	518
ERF transcription factor	GU393024	977	212

2.2.2.4 Screening and Utilization of Elite Tea Genetic Resources

The screening of elite and unique tea germplasms was important for tea breeding and its exploitation. Several tea accessions were identified with specific morphological traits, like whitish, yellow and purple buds (Fig. 2.8), extra-small or extra-large size of leaves, zigzag branches (Fig. 2.9), etc. An excellent example is a temperature-sensitive mutant, 'Anji Baicha', its tender shoots will change to be whitish in spring at 19 – 22 °C and turn green again when the temperature is higher than 25 °C (Li *et al.*, 1997), with very high amino acids content (6.2%) and low tea polyphenols (10.7%). It has been extended and popularized in northern Zhejiang Province because the tea product exhibits a fascinating appearance and good taste. But the vigor and yield need to be further improved. Recently, more and more similar tea clones with fresh white or yellow shoots have been reported.

(a) (b) (c)

Fig. 2.8. Tea collection characterized with whitish(a), yellow(b) and purple(c) buds

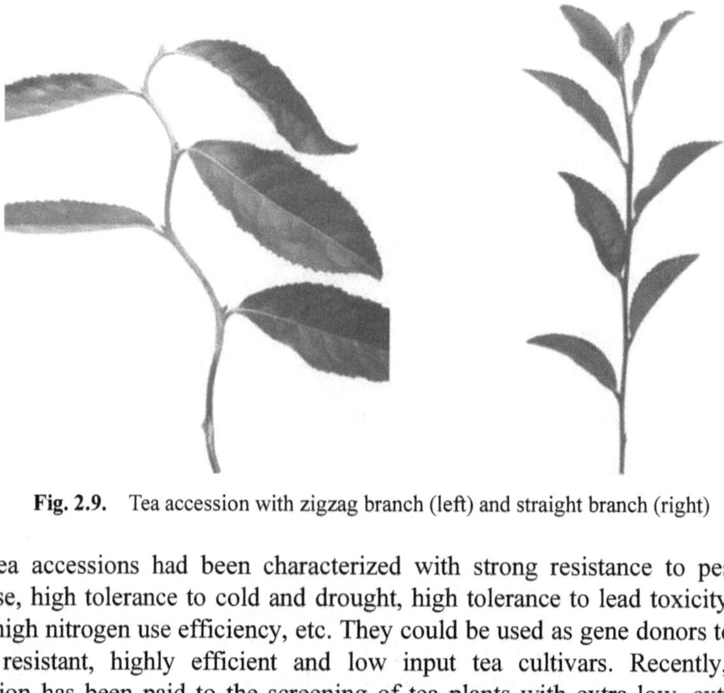

Fig. 2.9. Tea accession with zigzag branch (left) and straight branch (right)

Tea accessions had been characterized with strong resistance to pests and disease, high tolerance to cold and drought, high tolerance to lead toxicity in the soil, high nitrogen use efficiency, etc. They could be used as gene donors to breed high resistant, highly efficient and low input tea cultivars. Recently, more attention has been paid to the screening of tea plants with extra-low, extra-high bio-chemical components and special flavors. Tea genetic resources with low caffeine, low fluoride content, high EGCG, high EGCG-3-3-*O*-methyl, high amino acids and high theanine have been identified and screened. They will satisfy the diverse requirements for deep exploitation of tea.

The elite landraces play great roles in tea breeding in China. Forty-seven national and 62 provincial registered improved clones have been bred by hybridization and systematic selection using these landraces. Landrace 'Longjing Qunti' was well-known as being used to make 'Longjing Tea' (Dragon Well Tea)—a famous Chinese green tea. Three famous clones, 'Longjing 43', 'Longjing Changye' and

'Zhongcha 102', were bred through individual selection from 'Longjing Qunti'. They became the most popular clones which have been extended to more than 10 provinces in China. An excellent black tea clone, 'Yunkang 10', which was selected from genetic resources in Nannuoshan, Xishuan Banna, Yunnan Province, is now a predominant clone in Yunnan Province. More than 20 clones were bred from the progenies of 'Fuding Dabaicha' (Fig. 2.10), which is an elite landrace from Fujian Province and usually used as one of the parents in tea breeding. A landrace, 'Tieguanyin', is famous for its excellent quality in processing Oolong tea, which was used as hybrid parent to breed and develop 3 national and 4 provincial clones, respectively (Ye *et al.*, 2004).

Fig. 2.10. 'Fuding Dabaicha', a core parent for tea breeding in China

2.3 Tea Breeding and Selection Techniques

There are more than 13 national and local tea research institutes involved in programs of tea breeding and extension of new clones. And more than 20 universities and colleges are concerned with education and research of tea breeding. Most of the tea breeding programs are carried out by public institutions sponsored by government. However, more and more private breeders have thrown themselves into tea breeding in recent years.

Knowledge and selection of tea cultivars dates back thousands of years in ancient China. The first authentic literature on tea, *Tea Classics* (*Cha Ching*), written by Chinese tea scientist Lu Yu in 760 – 770 A.D. had a clear record about the leaf color of different cultivars (Wu, 1987). Later, a tea farmer in Fujian

Province, China, successfully developed the vegetative propagation cutting method in the 1780s. Two famous Oolong and green tea clones 'Tieguanyin' and 'Fuding Dabaicha' were bred successfully in the 1780s and 1857, respectively. Currently, they are both still predominant tea cultivars in Chinese tea gardens (China Tea Varieties Compilation Committee, 2001).

In addition to high yield and excellent quality, higher standards and requirements for breeding targets are proposed to breed outstanding cultivars. Today, multiple targets such as good flavor, high resistance to diseases and pests, strong tolerance to cold and drought, high content of functional components, high usage efficiency of soil nutrition, and so on, should be given priority in tea breeding programs. In China, the earlier sprouting time in spring is also an important objective in tea cultivars selection for making premium green tea.

Traditionally, it requires at least 20 – 25 years to finish the whole procedure of tea breeding, from individual selection to local adaptability testing and final release to the public (Fig. 2.11). But a highly efficient tea breeding system, in which controlled hybridization and individual selection are the major breeding approaches, combined with molecular marker-assisted selection and micropropagation techniques, has now been established and gradually developed in China (Chen *et al.*, 2007a).

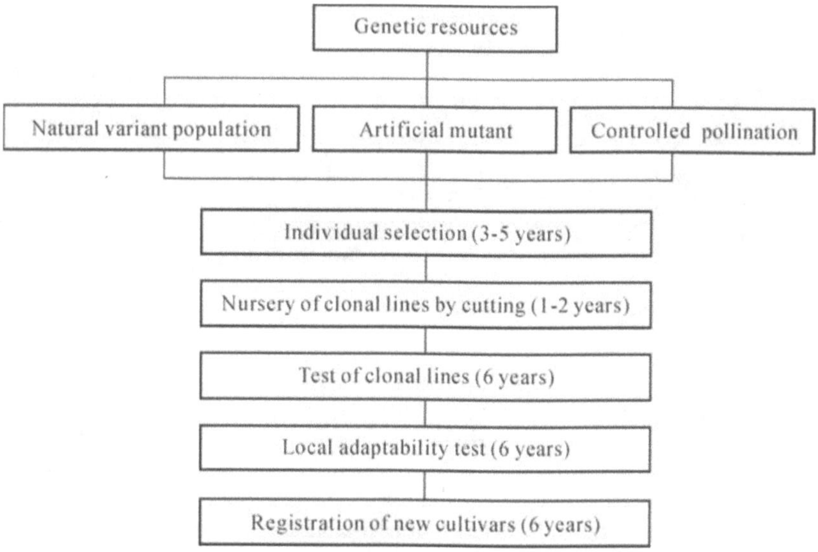

Fig. 2.11. Simplified procedure of tea breeding in China

2.3.1 Breeding Techniques

Breeding techniques include the creating and characterizing the useful variations.

According to the resources of variants, tea breeding methods could be classified into individual selection, controlled hybridization, mutation breeding. Most recently, the marker-assisted selection and transgenic technology are becoming more and more applicable for tea breeding.

2.3.1.1 Individual Selection

Individual selection from naturally existing variations among local tea populations (*jats*) and open-pollinated progenies is still considered as an effective method for tea breeding. There have been more than 200 tea clones released in China, in which 76% were bred through individual selection. The elite clones, either 'Longjing 43' (Fig. 2.12) or 'Longjing Changye'—both screened from the landrace 'Longjing Qunti'—are the most popular clones which have been extended to more than 10 provinces. Three clones named 'Anhui 1', 'Anhui 3' and 'Anhui 7' were all selected from the local population 'Qimen Qunti' in Qimen, Anhui Province. An excellent black tea clone, 'Yunkang 10', which was selected from genetic resources in Nannuoshan, Xishuan Banna, Yunnan Province, is now becoming a predominant clone in Yunnan Province. In total, 15 clones, such as 'Yingshuang', 'Zhenong 139' and 'Zhenong 117' etc., all have been selected from open-pollinated progenies between two elite landraces like 'Fuding Dabaicha' and 'Yunnan Dayecha'. Many of them have been popular clones cultivated in tea gardens. Nevertheless, the ratio of bred clones through individual selection decreased gradually, from 85% for the first time (1987) to 66.7% for the second (1994) and the third (2002) time of national registration, respectively (Chen *et al*., 2007a).

Fig. 2.12. 'Longjing 43', a popular clone for making green tea in China

2.3.1.2 Controlled Hybridization

Controlled hybridization has gradually become a predominant method for tea breeding. In China, the elite landraces such as 'Fuding Dabaicha', 'Yunnan Dayecha', 'Tieguanyin' etc., were the core parents used in hybrid breeding programs (Guo *et al.*, 2003; Liu *et al.*, 2005; Ye *et al.*, 2004). Eleven nationally released clones, corresponding to 73.3% of all clones through hybridization, were bred from the controlled hybridization of 'Yunnan Dayecha' (including 'Fengqing Dayecha') used as parent. The percentage of registered clones bred using the hybridization method increased significantly over the years, from 9.1% in 1987 to 22.2% in 2002 (Chen *et al.*, 2007a). Hybridization among few limited parents may lead to a narrow genetic base of tea clones. So it is important to select breeding parents with a distant relationship to sustain the genetic diversity of tea cultivars, and this may help in obtaining progenies with higher hybrid vigor. Distant hybridization is considered as a powerful strategy for broadening the genetic base of new cultivars. However, the fruit set success and seed viability was so low that it is difficult to get enough progenies. So young embryo rescuing tissue culture strategy is now being studied, and developed to improve the opportunities for successful distant hybridization.

2.3.1.3 Mutation Breeding

Mutation breeding is also an important method in tea breeding. Physical and chemical methods have been used to produce mutants, from which to screen valuable genetic materials for further breeding. The γ-rays from Co^{60} radiation have been widely applied to mutation breeding. The biological harm to the tea plant from the combined effects of γ-rays and chemical mutagens was analyzed. Results showed that radical damage was correlated with the moisture content, the chemical components of tea seeds. A mathematical model has been proposed to select the optimal radical dose and concentration of chemical mutagens (Yang & Lin, 1992). One excellent new strain, which was characterized with very early sprouting time, high cup-tea quality, high resistance to disease and suitability for making Longjing tea, has been selected from the cuttings of 'Longjing 43' under Co^{60} γ-rays radiation (Yang *et al.*, 2003b). It is now the national registered clone, named 'Zhongcha 108'. Using Co^{60} γ-rays as radical source, two other clones bred by the Tea Research Institute of Hunan Province and Anhui Agricultural University had registered a provincial and national clone in 1997 and 2002, respectively. The implantation of N^+ (nitrogen ions) had also been applied in tea mutation breeding, through which two elite lines named 'Chanong 1' and 'Chanong 8' were screened (Wang *et al.*, 2006a, 2006b). Chemical mutagens such as colchicines and ethyl methylsulfonate could be used to treat tea buds, shoots, seeds and seedlings. But successful applications were still not reported for tea plants in China.

2.3.1.4 Marker-Assisted Selection (MAS)

Molecular markers provide an effective tool for the discrimination of cultivars, identification of hybrid parents and characterization of genetic relatedness. The RAPD and ISSR fingerprints were constructed (Chen *et al.*, 1998; Yao *et al.*, 2005) to discriminate different tea cultivars. And the parentage of some hybrid progenies was identified by RAPD (Li *et al.*, 2001) and ISSR (Hou *et al.*, 2006; Liu BY *et al.*, 2008). These studies showed that DNA markers were useful to authenticate the released cultivars and their parents, and helpful to protect the cultivars for intellectual property rights. The genetic similarity and relationship between tea cultivars were revealed by molecular markers, which could have necessary implications for parental selection in tea breeding programs (Huang JA *et al.*, 2006; Shen *et al.*, 2007; Yao *et al.*, 2008a; Yu *et al.*, 2009).

The findings of markers linkage/associated with phenotypic traits are very important for individual selection. The important traits such as yield, quality, biotic and abiotic resistance are controlled by quantitative trait loci (QTLs). The detection and location of QTLs depend on construction of a molecular linkage map with a high density of DNA markers. The linkage map of a tea plant based on AFLP markers was first reported in 2005 in China. This map of a female parent included 17 linkage groups, on which 208 markers were located and covered a total length of 2,457.7 cM, with 11.9 cM of average distance between markers. Another map from a male parent had 16 linkage groups, on which 200 markers were located and covered a total length of 2,545.3 cM, with 12.8 cM of average distance between markers (Huang *et al.*, 2005). Another partial linkage map of tea was constructed by RAPD and ISSR markers based on backcross generation (Huang FP *et al.*, 2006). However, in the previous studies, the individual number applied for mapping was too small to meet the demands of precise mapping. So these constructed maps had limited ability to locate QTLs linked with some important traits, due to their low distribution of molecular markers. Although a practical genetic map is not available for the tea plant right now, the potential advantage of marker-assisted selection is clear and strong.

In addition to linkage analysis based on mapping populations, an alternative approach which makes use of linkage disequilibrium (LD)—based on association analysis was proposed to study the genetics of complex traits in the natural population. Based on analysis of LD and the population structure of 112 tea cultivars, five EST-SSR markers associated with phenotypic traits were screened out and identified at the level of $p<0.01$. Among them, three markers CS298, CS317 and CS84 were associated with the weight of 100 fresh shoots with 'two and a bud', leaf length and content of tea polyphenols. And they could explain phenotypic variation of 0.11, 0.15 and 0.35, respectively. Another two markers CS330 and CS338 were found to associate with caffeine content, and they could explain phenotypic variation with 0.14 and 0.15, respectively (Yao *et al.*, 2010). The identification and screening of alleles associated with phenotypic traits may help the development of functional markers for applying MAS in tea breeding.

2.3.1.5 Transgenic Technology

Genetic transformation is becoming a strong tool in tea breeding. The isolation and cloning of functional genes provide an opportunity to genetically manipulate and control the tea plant. Transgenic protocols were developed via *agrobacterium* and particle bombardment mediated transformation of somatic embryos. Based on *β*-glucuronidase (GUS) transient expression, several factors that affect the efficiency of agrobacterium mediated transformation (AGR) were compared, and then the appropriate AGR system was proposed in tea plants (Zhao *et al*., 2001). The *Bt* gene was transformed into tea callus by the AGR method and verified by reverse transcription polymerase chain reaction (RT-PCR) and the GUS reporter gene (Luo & Liang, 2000; Zhang *et al*., 2006a). The micro-projection protocol was developed for transformation of the GUS reporter gene into tea callus using a particle delivery system (Wu *et al*., 2005). The study was carried out to compare the effect and efficiency of three transformation methods, agrobacterium mediated transformation (AGR), particle bombardment (BOM), and a combination of the above two. Results indicated that the method combining AGR and BOM was the most effective to apply in tea callus transformation (Wu *et al*., 2005). In order to breed low caffeine tea cultivars, the RNA*i* vector of the caffeine synthase gene (*TCS*) was constructed successfully and it was tried to be transformed to tea callus. It inhibited expression of the *TCS* gene (Zhang *et al*., 2006b). However, transgenic plantlets of tea have not still been successfully obtained to date in China.

2.3.2 Selection Techniques Used for Yield, Quality and Resistance

The selection of new cultivars has usually been focused on target traits such as excellent quality, high production, strong resistance and early budding time. It is essential to use the correct selection criteria to improve the efficiency of tea breeding.

2.3.2.1 Selection of Early Sprouting Tea Clones

In China, tea farmers commonly prefer tea clones with an early sprouting time for producing premium green tea. So screening of early sprouting clones is an important target when breeding green tea cultivars. The sprouting times of 'one and a bud' and 'two and a bud' are observed in the spring season in three successive years. Compared with the controlled standard green tea cultivars, 'Fuding Dabaicha (FD)', if the sprouting time (ST) of tested clones is on average earlier by at least 5 days, they are classified as early sprouting clones; if the ST of the tested clones is similar to FD, they are middle sprouting clones; if the ST is more than 5 days later, they are late-sprouting clones. The genetic analysis showed that the average time of bud sprouting among F_1 progenies was close to the mean of the hybrid parents, and the transgressive inheritance was observed in hybrid progenies between the two early types (Guo *et al*., 2004).

2.3.2.2 Selection of High Yielding Tea Clones

Studies showed that tea yield correlated with many traits such as tree vigor, tender shoot length, tender shoot weight, plucking density, leaf size, chlorophyll content and photosynthetic rate, etc. The numbers and weight of new shoots were observed to significantly correlate with yield ($r = 0.91$, $r = 0.60$, respectively), hence they could be considered as the primary criteria for screening the high yield of tea clones (Liu & Zhou, 1994). The tea yield also had significant correlation with other criteria, such as leaf area index, thickness and density of palisade tissue, root vigor, dry weight of the root, etc. (Liu & Zhou, 1994). The photosynthetic ability of tea leaves was determined by the content of chlorophyll which varied in different tea genotypes. The genetic variation coefficient (*GVC*) and broad heritability (h_B) of chlorophyll content were estimated on average as 17.4% and 55.6%, respectively (Zhou *et al.*, 1993). The net photosynthetic rate also varied among F_1 progenies, and its *GVC* and h_B were estimated to be 22.9% and 59.4%, respectively (Ye *et al.*, 1990). For Oolong tea clones, the *GVC* and h_B of yield of fresh leaves were estimated to be in the range of 26.9% – 40.5%, 67.2% – 83.1%, respectively (Guo *et al.*, 1992).

2.3.2.3 Selection of High Quality Tea Clones

Quality is an inherent polygenic trait determined by many factors such as flavor, aroma and color of the tea infusion. Quality-related morphological and biochemical parameters could be developed to be applied in the selection of high quality tea. It was observed that tea quality significantly correlated with morphological characteristics such as leaf size, color and pubescence. Abundant amino acids were identified in the pubescence on young buds. Studies indicated that the density and length of pubescence would have an influence on the appearance of green tea, on the flavor and aroma of black tea (Xiao & Wang, 1991). The statistic model for tea quality was proposed based on investigation of bio-chemical components like tea polyphenols (TP), amino acids (AA), caffeine, catechins and the ratio of TP to AA (RTA). They were used for the purpose of predicting the quality of green and black tea (Yang & Ying, 1990, 1991). The value of RTA was proposed as a decisive indicator to determine the processing suitability of tea. When the $RTA<10$, the tea clones are suitable to make green tea, while $10 \le RTA<20$, the clones can be fit for making either green tea or black tea, but if $20 \le RTA<35$, the clones may be better for black tea (Yang & Ying, 1990).

2.3.2.4 Selection for High Resistance in Tea Clones

It is still very difficult, costly and unrealistic to characterize and evaluate biotic and abiotic resistance. So it is essential to identify morphological and biochemical descriptors as alternative indicators in selecting high resistance clones under biotic and abiotic stress. Previous studies indicated that the resistance to pink mite (*Acaphylla theae* Watt) was determined by a high density of pubescence and low density of stomata of young leaves (Chen *et al.*, 1996; Chen HC *et al.*, 2000). The content of amino acids and caffeine in high resistance cultivars was remarkably higher than in cultivars susceptible to pink mite (Chen *et al.*, 1996; Chen HC *et al.*, 2000). Several morphological, anatomical and biochemical parameters including pubescence length, leaf pose, thickness of leaf surface, stomata density, amino acid and soluble sugar were proposed to screen resistant cultivars to the *Polyphagotarsonemus latus* Banks (Liu *et al.*, 1994). Tea leafhopper (*Empoasca vitis* Gothe) resistance was observed to vary among different genotypes (Zhang *et al.*, 1994a; Zeng *et al.*, 2001), as well as for tea weevil (*Myllocerinus aurolineatus* Voss) (Zhang *et al.*, 1994b) and tea geometrid (*Ectropis oblique* Prout) (Lv *et al.*, 1990). It had been observed that the resistant cultivars were generally characterized by late sprouting time, thick horny cells and palisade tissue of young leaves (Zhu, 1992). Another study demonstrated that yellow-green shoots could be more attractive to leafhoppers than dark-green shoots, and the tender shoots more attractive than mature shoots (Zhang *et al.*, 1994). The population density of leafhoppers was negatively correlated with anatomical structures such as the thickness of palisade tissue, spongy tissue, collenchyma and epidermis (Huang *et al.*, 1998; Zhang *et al.*, 1994). A high density of pubescence and abundant caffeine content could inhibit the population density of leafhopper (Zhang *et al.*, 1994), but these factors did not impact on the harm of the tea weevil (Huang *et al.*, 1994). The growth and development of tea geometrid was inhibited when fed fresh leaves with high polyphenols, which indicated that *Ectropis oblique* Prout was sensitive to tea polyphenols (Gao *et al.*, 1997). Pathogens, like *Pestalotia theae* Sawada and *Exobasidium vexans* Massee, were isolated and inoculated *in vitro* to identify disease resistance among tea accessions (Zhong *et al.*, 1997; Ran *et al.*, 2008).

The drought resistance of the tea plant was comprehensively evaluated with morphological and physiological parameters including the water loss rate in fresh leaves, relative permeability of plasma membrane, superoxide dismutase (SOD) activity, catalase (CAT) activity, root length, ratio of root to shoot, and so on (Lu *et al.*, 1995). Studies indicated that drought tolerance was significantly correlated to activities of SOD and CAT. However, it was observed that there was no correlation with peroxidase (POD) activity (Tong *et al.*, 1992). There had been remarkable differences for CAT activity among tea cultivars under drought stress, and activity of stress-induced CAT was significantly correlated with relative permeability of the membrane (Lu *et al.*, 1992). The method based on triphenyltetrazolium chloride (TTC) and electric resistance had been used to evaluate the degree of cold tolerance (Yang & Lin, 1987; Yang, 1989). Under cold stress, the resistant tea cultivars showed high activity of SOD and CAT (Huang *et al.*, 1986; Huang, 1990; Luo *et al.*, 2001). Criteria were proposed to evaluate and screen high resistance tea plants via the anatomical structure of mature leaves. Tea plants could be identified as being tolerant to cold when they were characterized with more than 20 μm thickness of leaf epicuticle, more than two layers of palisade cell, more than a 0.6 ratio between thickness of the palisade and spongy tissue (Shu 1995; Su & Zhang, 1997). In previous studies, the cold tolerance of tea plants was also proved to be related to high content of chlorophyll and amino acid (Chen, 1980).

2.3.3 Local Adaptability Tests and Registration

In China, the Ministry of Agriculture (MOA) accredited the National Authentication Committee of Tea Cultivars (NACTC), set up in 1979, to take charge of the examination and approval of new cultivars applied for release in the whole country. In most of the tea producing regions, local authentication committees for tea cultivars were accredited by the agricultural administration of the government. They organized field tests of new tea cultivars and approved their release in a restricted region. Only when they had passed adaptability tests, could they be registered as national and provincial cultivars and then permission was granted to release them to the public. From 1984, the NACTC has organized 4 national-level adaptability tests for tea cultivars, in 1984, 1998, 2002 and 2010, respectively. To date, 123 tea cultivars have been registered as national tea cultivars(Table 2.8), and more than 150 cultivars have been accredited as provincial tea cultivars in China (Table 2.9).

Table 2.8 List of national registered tea cultivars in China

No.	Cultivar	Registered year	Breeding method	Origin
1	Fuding Dabaicha	1985	Landrace	Fujian
2	Fuding Dahaocha	1985	Landrace	Fujian
3	Fu'an Dabaicha	1985	Landrace	Fujian
4	Meizhan	1985	Landrace	Fujian
5	Zhenghe Dabaicha	1985	Landrace	Fujian
6	Maoxie	1985	Landrace	Fujian
7	Tieguanyin	1985	Landrace	Fujian
8	Huangdan	1985	Landrace	Fujian
9	Fujian Shuixian	1985	Landrace	Fujian
10	Benshan	1985	Landrace	Fujian
11	Daye Wulong	1985	Landrace	Fujian
12	Mengku Dayecha	1985	Landrace	Yunnan
13	Fengqing Dayecha	1985	Landrace	Yunnan
14	Menghai Dayecha	1985	Landrace	Yunnan
15	Lechang Baimaocha	1985	Landrace	Guangdong
16	Hainan Daye	1985	Landrace	Hainan
17	Fenghuang Shuixian	1985	Landrace	Guangdong
18	Damianbai	1985	Landrace	Jiangxi
19	Shangmeizhouzhong	1985	Landrace	Jiangxi
20	Ningzhouzhong	1985	Landrace	Jiangxi
21	Huangshanzhong	1985	Landrace	Anhui
22	Qimenzhong	1985	Landrace	Anhui
23	Jiukengzhong	1985	Landrace	Zhejiang

(To be continued)

(Table 2.8)

No.	Cultivar	Registered year	Breeding method	Origin
24	Yuntaishanzhong	1985	Landrace	Hunan
25	Meitan Taicha	1985	Landrace	Guizhou
26	Lingyun Baimaocha	1985	Landrace	Guangxi
27	Ziyangzhong	1985	Landrace	Shanxi
28	Zaobaijian	1985	Landrace	Sichuan
29	Yichang Dayecha	1985	Landrace	Hubei
30	Yixingzhong	1985	Landrace	Jiangsu
31	Qianmei 419	1987	OP	Guizhou
32	Qianmei 502	1987	HP	Guizhou
33	Fuyun 6	1987	OP	Fujian
34	Fuyun 7	1987	OP	Fujian
35	Fuyun 10	1987	OP	Fujian
36	Zhuyeqi	1987	Field clone	Hunan
37	Longjing 43	1987	Field clone	Zhejiang
38	Anhui 1	1987	Field clone	Anhui
39	Anhui 3	1987	Field clone	Anhui
40	Anhui 7	1987	Field clone	Anhui
41	Yingshuang	1987	OP	Zhejiang
42	Cuifeng	1987	OP	Zhejiang
43	Jingfeng	1987	OP	Zhejiang
44	Biyun	1987	OP	Zhejiang
45	Zhenong 12	1987	OP	Zhejiang
46	Shuyong 1	1987	HP	Chongqing
47	Yinghong 1	1987	Field clone	Chongqing
48	Shuyong 2	1987	HP	Chongqing
49	Ningzhou 2	1987	Field clone	Jiangxi
50	Yunkang 10	1987	Field clone	Yunnan
51	Yunkang 14	1987	Field clone	Yunnan
52	Juhuachun	1987	OP	Zhejiang
53	Guihong 3	1994	Field clone	Guangxi
54	Guihong 4	1994	Field clone	Guangxi
55	Yangshulin 783	1994	Field clone	Anhui
56	Wannong 95	1994	Field clone	Anhui
57	Xicha 5	1994	Field clone	Jiangsu
58	Xicha 11	1994	Field clone	Jiangsu
59	Hanlü	1994	Field clone	Zhejiang
60	Longjing Changye	1994	Field clone	Zhejiang
61	Zhenong 113	1994	OP	Zhejiang
62	Qingfeng	1994	Field clone	Zhejiang
63	Xinyang 10	1994	Field clone	Henan
64	Baxiancha	1994	Field clone	Fujian
65	Qianmei 601	1994	HP	Guizhou
66	Qianmei 701	1994	HP	Guizhou
67	Gaoyaqi	1994	Field clone	Hunan
68	Zhuyeqi 12	1994	Field clone	Hunan
69	Baihaozao	1994	Field clone	Hunan
70	Jianbohuang 13	1994	Field clone	Hunan
71	Shuyong 703	1994	HP	Chongqing
72	Shuyong 808	1994	HP	Chongqing

(To be continued)

(Table 2.8)

No.	Cultivar	Registered year	Breeding method	Origin
73	Shuyong 307	1994	HP	Chongqing
74	Shuyong 401	1994	HP	Chongqing
75	Shuyong 3	1994	HP	Chongqing
76	Shuyong 906	1994	HP	Chongqing
77	Yihongzao	1998	Field clone	Hubei
78	Fuzao 2	2002	Field clone	Anhui
79	Lingtou Dancong	2002	Field clone	Guangdong
80	Xiuhong	2002	Field clone	Guangdong
81	Wulinghong	2002	Field clone	Guangdong
82	Yunda Danlü	2002	Field clone	Guangdong
83	Gancha 2	2002	OP	Jiangxi
84	Shuyong 808	2002	HP	Chongqing
85	Shuchazao	2002	Field clone	Anhui
86	Wannong 111	2002	Mutation	Anhui
87	Zaobaijian 5	2002	Field clone	Chongqing
88	Nanjiang 2	2002	Field clone	Chongqing
89	Zhenong 21	2002	Field clone	Zhejiang
90	E'cha 1	2002	HP	Hubei
91	Zhongcha 102	2002	Field clone	Zhejiang
92	Mingke 2	2002	HP	Fujian
93	Yuemingxiang	2002	Field clone	Fujian
94	Mingke 1	2002	HP	Fujian
95	Huangqi	2002	OP	Fujian
96	Guilü 1	2003	Field clone	Guangxi
97	Mingshan Baihao	2005	Field clone	Sichuan
98	Xiapu Chunbolü	2010	Field clone	Fujian
99	Chunyu 1	2010	Field clone	Zhejiang
100	Chunyu 2	2010	Field clone	Zhejiang
101	Maolü	2010	Field clone	Zhejiang
102	Nanjiang 1	2010	Field clone	Chongqing
103	Shifocui	2010	Field clone	Anhui
104	Wancha 91	2010	Field clone	Anhui
105	Yaoshan Xiulü	2010	Field clone	Guangxi
106	Guixiang 18	2010	Field clone	Guangxi
107	Yulü	2010	HP	Hunan
108	Zhenong 139	2010	OP	Zhejiang
109	Zhenong 117	2010	OP	Zhejiang
110	Zhongcha 108	2010	Mutation	Zhejiang
111	Zhongcha 302	2010	HP	Zhejiang
112	Dangui	2010	Field clone	Fujian
113	Chunlan	2010	Field clone	Fujian
114	Ruixiang	2010	Field clone	Fujian
115	E'cha 5	2010	Field clone	Hubei
116	Hongyan 9	2010	Field clone	Guangdong
117	Hongyan 12	2010	Field clone	Guangdong
118	Hongyan 7	2010	Field clone	Guangdong
119	Hongyan 1	2010	Field clone	Guangdong
120	Baimao 2	2010	Field clone	Guangdong
121	Jinmudan	2010	HP	Fujian

(To be continued)

(Table 2.8)

No.	Cultivar	Registered year	Breeding method	Origin
122	Huangmeigui	2010	HP	Fujian
123	Zimudan	2010	Field clone	Fujian

* Breeding methods include Landrace: traditional cultivars, including *jat* cultivars and clones; Field clone: individual selected clones from seedlings/*jats*; OP: clones selected from open-pollinated progenies; HP: clones selected from controlling-pollinated progenies

Table 2.9 An incomplete list of provincial registered tea cultivars in China

No.	Province	Cultivar	Registered year	Breeding method
1	Anhui	Qingyang Tianyuncha	1982	Landrace
2	Anhui	Songluozhong	1982	Landrace
3	Anhui	Shidacha	1982	Landrace
4	Anhui	Xuancheng Jianyezhong	1982	Landrace
5	Anhui	Yongxi Liuyezhong	1982	Landrace
6	Anhui	Huoshan Jinjizhong	1982	Landrace
7	Anhui	Yangshulin 781	1987	Field clone
8	Anhui	Bohao	1987	Field clone
9	Anhui	Huangshan Zaoya	1987	Field clone
10	Anhui	Huangjingcha	1987	Field clone
11	Anhui	Mingzhou 12	1995	Field clone
12	Anhui	Shifoxiang	2000	Field clone
13	Anhui	Xianyuzao	2000	Field clone
14	Chongqing	Chongpi 71-1	1995	Field clone
15	Chongqing	Yucha 1	2001	HP
16	Chongqing	Yucha 2	2001	HP
17	Chongqing	Bayu Tezao	2005	Field clone
18	Fujian	Zaofengchun	1985	Field clone
19	Fujian	Rougui	1985	Landrace
20	Fujian	Foshou	1985	Landrace
21	Fujian	Fuyun 595	1988	OP
22	Fujian	Zhaoyang	1994	OP
23	Fujian	Baiya Qilan	1996	Field clone
24	Fujian	Jiulong Dabaicha	1998	Landrace
25	Fujian	Fengyuanchun	1999	Field clone
26	Fujian	Xinrencha	1999	Landrace
27	Fujian	Xiapu Yuanxiaolü	1999	Field clone
28	Fujian	Jiulongpao	2000	Field clone
29	Fujian	Zaochunhao	2003	OP
30	Fujian	Fuyun 20	2005	HP
31	Fujian	Zimeigui	2005	HP
32	Guangdong	Fenghuang Dancong	1988	Field clone
33	Guangdong	Lechang Baimao 1	1988	Field clone
34	Guangdong	Liannan Dayecha	1988	Landrace
35	Guangdong	Yinghong 9	1988	Field clone
36	Guangdong	Huangye Shuixian	1988	Field clone
37	Guangdong	Heiye Shuixian	1988	Field clone
38	Guangdong	Fenghuang Huangzhixiang Dancong	2000	Landrace
39	Guangdong	Danxia 1	2010	Field clone
40	Guangdong	Danxia 2	2010	Field clone

(To be continued)

(Table 2.9)

No.	Province	Cultivar	Registered year	Breeding method
41	Guangxi	Guire 1	2006	Field clone
42	Guangxi	Guire 2	2006	Field clone
43	Hubei	E'cha 2	1995	Field clone
44	Hubei	E'cha 3	1995	Field clone
45	Hubei	E'cha 6	2000	Field clone
46	Hubei	E'cha 7	2004	Field clone
47	Hubei	E'cha 8	2005	Field clone
48	Hubei	E'cha 9	2006	Field clone
49	Hubei	E'cha 10	2009	Field clone
50	Hubei	E'cha 11	2009	Field clone
51	Hubei	E'cha 12	2009	Field clone
52	Hubei	Jinfeng	2010	Field clone
53	Hubei	Jinxiang	2010	Field clone
54	Hunan	Dajianye	1987	Field clone
55	Hunan	Jianghua Kucha	1987	Landrace
56	Hunan	Donghuzao	1987	Field clone
57	Hunan	Rucheng Baimaocha	1987	Landrace
58	Hunan	Jianbohuang	1987	Field clone
59	Hunan	Chengbu Dongcha	1987	Landrace
60	Hunan	Gaoqiaozao	1987	Field clone
61	Hunan	Xiangbolü	1987	Field clone
62	Hunan	Taoyuan Daye	1992	Field clone
63	Hunan	Mingfeng	1993	HP
64	Hunan	Bixiangzao	1993	HP
65	Hunan	Fuhao	1996	HP
66	Hunan	Anmingzao	1997	Field clone
67	Hunan	Fufeng	1997	Mutation
68	Hunan	Xianghongcha 1	1998	HP
69	Hunan	Xianghongcha 2	2003	HP
70	Hunan	Xiangfeicui	2003	Field clone
71	Hunan	Yusun	2009	Field clone
72	Hunan	Huangjincha 1	2010	Field clone
73	Jiangsu	Xicha 10	1987	Field clone
74	Jiangsu	Sucha 120	2010	Field clone
75	Jiangsu	Dongtingchun	2010	Mutation
76	Jiangxi	Gancha 1	1992	Field clone
77	Shandong	Luohan 1	2006	Field clone
78	Sichuan	Gulin Niupicha	1985	Landrace
79	Sichuan	Nanjiang Dayecha	1985	Landrace
80	Sichuan	Chongqing Pipacha	1985	Landrace
81	Sichuan	Beichuan Zhongyezhong	1989	Landrace
82	Sichuan	Mengshan 9	1989	Field clone
83	Sichuan	Mengshan 11	1989	Field clone
84	Sichuan	Mengshan 16	1989	Field clone
85	Sichuan	Mengshan 23	1989	Field clone
86	Sichuan	Mingshanzao	1997	Field clone
87	Sichuan	Mingshan Tezao 213	1997	Field clone
88	Sichuan	Huaqiu 1	2003	Field clone
89	Sichuan	Tianfu 28	2003	Field clone

(To be continued)

(Table 2.9)

No.	Province	Cultivar	Registered year	Breeding method
90	Sichuan	Tianfu 11	2003	Field clone
91	Sichuan	Chuannong Huangyazao	2009	Field clone
92	Sichuan	Chuanmu 28	2010	Field clone
93	Sichuan	Mabianlü 1	2010	Field clone
94	Taiwan	Taicha 1	1969	HP
95	Taiwan	Taicha 2	1969	HP
96	Taiwan	Taicha 3	1969	HP
97	Taiwan	Taicha 4	1969	HP
98	Taiwan	Taicha 5	1974	Field clone
99	Taiwan	Taicha 6	1974	Field clone
100	Taiwan	Taicha 7	1974	Field clone
101	Taiwan	Taicha 8	1974	Field clone
102	Taiwan	Taicha 9	1975	HP
103	Taiwan	Taicha 10	1975	HP
104	Taiwan	Taicha 11	1975	HP
105	Taiwan	Taicha 12	1981	HP
106	Taiwan	Taicha 13	1981	HP
107	Taiwan	Taicha 14	1983	HP
108	Taiwan	Taicha 15	1983	HP
109	Taiwan	Taicha 16	1983	HP
110	Taiwan	Taicha 17	1983	HP
111	Taiwan	Taicha 18	1999	HP
112	Taiwan	Taicha 19	2004	HP
113	Taiwan	Taicha 20	2004	HP
114	Taiwan	Taicha 21	2008	HP
115	Yunnan	Yunkang 43	1985	Field clone
116	Yunnan	Changye Baihao	1985	Field clone
117	Yunnan	Yunmei	1992	Field clone
118	Yunnan	Yungui	1992	Field clone
119	Yunnan	Aifeng	1992	Field clone
120	Yunnan	Yunkang 27	1995	Field clone
121	Yunnan	Yunkang 37	1995	Field clone
122	Yunnan	Yunxuan 9	1995	Field clone
123	Yunnan	73 – 8	1999	Field clone
124	Yunnan	73 – 11	1999	Field clone
125	Yunnan	76 – 38	2000	Field clone
126	Yunnan	Foxiangcha 1	2003	HP
127	Yunnan	Foxiangcha 2	2003	HP
128	Yunnan	Foxiangcha 3	2003	HP
129	Yunnan	Foxiangcha 4	2003	HP
130	Yunnan	Foxiangcha 5	2003	HP
131	Yunnan	Yuncha 1	2005	Field clone
132	Yunnan	Yunkang 48	2005	Field clone
133	Yunnan	Yunkang 50	2005	Field clone
134	Yunnan	Yunkang 12	2010	Field clone
135	Yunnan	Yunkang 47	2010	Field clone
136	Yunnan	Yuncha Chunyun	2010	HP
137	Yunnan	Yuncha Chunhao	2010	HP
138	Zhejiang	Muhezhong	1988	Landrace

(To be continued)

(Table 2.9)

No.	Province	Cultivar	Registered year	Breeding method
139	Zhejiang	Shuigucha	1988	Landrace
140	Zhejiang	Jiaming 1	1988	Landrace
141	Zhejiang	Pingyun	1988	OP
142	Zhejiang	Zhenong 121	1988	OP
143	Zhejiang	Bifeng	1988	Field clone
144	Zhejiang	Tengcha	1988	Field clone
145	Zhejiang	Longjingzhong	1992	Landrace
146	Zhejiang	Zhenong 25	1992	Field clone
147	Zhejiang	Huangyezao	1992	Field clone
148	Zhejiang	Rui'an Baimaocha	1992	Field clone
149	Zhejiang	Rui'an Qingmingzao	1992	Field clone
150	Zhejiang	Taixiangzi	1994	Field clone
151	Zhejiang	Meifeng	1995	Field clone
152	Zhejiang	Shuangfeng	1995	OP
153	Zhejiang	Baiye 1	1998	Field clone
154	Zhejiang	Pingyang Tezaocha	1998	Field clone
155	Zhejiang	Yinhoucha	2001	Field clone
156	Zhejiang	Xiangshanzao 1	2004	Field clone
157	Zhejiang	Huangjinya	2008	Field clone
158	Zhejiang	Qiannianxue	2008	Field clone

There are ten national adaptability experimental stations across tea producing regions in China so far. These stations are located at Xinyang in Henan Province, Hangzhou in Zhejiang Province, Fu'an and An'xi in Fujian Province, Guilin in Guangxi Province, Yingde in Guangdong Province, Chengdu in Sichuan Province, Wuhan in Hubei Province, Changsha in Hunan Province and Pu'er in Yunnan Province, respectively. They represent different ecological and climatic conditions for growing across the country. According to the regulations, it takes six years to finish the whole procedure of an adaptability test. The first three years are for cultivating tea plants and the last three years for repeating tests on agronomic traits, cup tea quality, yield and resistance to biotic and abiotic stresses. Any new cultivars undergoing adaptability tests are required to be tested at three locations representing different ecological and climatic conditions.

2.4 Propagation and Extension System of New Cultivars

Tea cultivars are popularly propagated by cuttings in China. However, new technologies of micropropagation have been studied and developed to improve the propagation efficiency of tea clones. Tea farmers are encouraged to replace the seedling population with elite clones. Currently, the clonal tea gardens are estimated to be 973 kilohectares, which accounts for 46% of the whole plantation area in 2010. Distinctness, Uniformity and Stability (DUS) Test Guidelines for the tea plant (TG/238/1) has been released (UPOV, 2008), and all the tea plants of

Camellia L. Sect. *Thea* (L.) Dyer had been listed in the protective species of new plant cultivars by the MOA of the People's Republic of China. Thus, new tea clones can be protected by intellectual property rights, which will further encourage and promote tea breeding in China.

2.4.1 Propagation Techniques

Tea plant can be propagated by both sexual and asexual methods. The asexual methods include cutting, grafting and micropropagation, etc. Cutting is the most popular vegetative propagation technique.

2.4.1.1 Propagation Based on Cutting

To avoid extensive genetic non-uniformity resulting from seed propagation and, consequently, dilution of a quality product from a particular clone, tea clones are traditionally multiplied by cuttings. Propagation by cutting was first developed by tea farmers, and this technique has been used for more than 200 years in China. Now it is still the most popular technique of propagation in the world. In China, the lignified young shoots with one mature leaf and a healthy auxiliary bud or young shoot are generally used as cuttings for propagation. Studies indicated that semi-lignified shoots have a higher efficiency in regenerating roots than fully-lignified shoots used as cuttings (Dong & He, 1991). And the better effect had been observed on cutting of mature leaves with a short shoot than with a bud (Rao & Cheng, 1998). The optimal cutting density is about 500 plantlets per m^2 on a nursery bed for middle-small leaf cultivars, which would improve the survival rate and seedling quality (Yang & Ying, 1993). The method and technology, such as soil sterilization, hormonal treatment, covering by plastic film and nutrient application were studied and applied to cutting propagation to improve the survival rates and seedling quality (Shen *et al.*, 2001; Qin *et al.*, 2004; Xu *et al.*, 2005). Genetic stability of clones is vital for tea germplasm preservation, breeding and production. RAPD analysis showed no variation occurring at the DNA level between Chinese elite cultivars and their cutting offspring (Chen *et al.*, 1999).

2.4.1.2 Propagation Based on Grafting

Scion grafting was usually also used to replant old tea bushes, particularly in the southern tea growing area, such as Guangdong and Yunnan provinces. Studies showed that the grafted tea garden could be ready 2 – 3 years earlier for harvest compared with replanting a tea garden with cuttings. It is important for scion selection to have a high yield and high quality of tea. The survival rates varied from 78% to 89% using scion from different clonal cultivars (Luo *et al.*, 2000), but elite tea clones are preferred to use of the grafted scion. Grafting is more suitable in either April or October when the temperature is 19 – 20 °C, for the recovery and growth of grafted plants. A higher or lower temperature could decrease the survival rate of a grafted plant (Zhang *et al.*, 2008). Both light and water were also essential factors; the survival rate could be obviously increased when grafted plants were kept under shade and in a certain humidity (Cao *et al.*, 2001). Previous studies demonstrated that the survival rate of grafted plants could be improved significantly by treatment with hormones such as NAA (1-naphthyl acetic acid), IAA (indole-3-acetic acid) etc. (Cao *et al.*, 2001; Nian 2002). The biochemical components, branch habit and photosynthetic ability among the mother plants, scion clones and grafted plants were compared (Luo *et al.*, 2000; Wu & Luo 2001; Jin *et al.*, 2003).

2.4.1.3 Micropropagation

It might take 8 – 10 years to multiply 4,000 – 5,000 plantlets using the conventional single node cutting method when very few cuttings were available. However, two years might be sufficient to obtain the same quantity of plantlets using the micropropagation method. The optimal culture conditions [MS (Murashige and Skoog) medium + BA (butyl acetate) 2.0 mg/L + NAA 0.1 mg/L + GA_3 (gibberellic acid) 3.0 mg/L] were screened, on which 2.75 folds of the proliferation rate of cultured buds were observed after 30 days incubation, and more than 20% of those buds developed into young plantlets with a height of more than 5 cm (Zhou *et al.*, 2005). Young plantlets growing in a culture medium were directly transplanted into greenhouses. They grew to 20.4 cm in height after 5 months under the control of light, temperature, water and fertilizers (Cheng *et al.*, 2007). The technique combining micropropagation with an industrial approach to multiply and raise plantlets gradually became refined and available (Fig. 2.13). It is possible to obtain a large number of plantlets through micropropagation in a relatively short time. So the technique of micropropagation is helping to shorten the gestation period for new materials and new clones, and to promote the efficiency of breeding as well as the extension of new clones.

Fig. 2.13. Micropropagation based on tissue culture and incubation in the greenhouse

2.4.2 Extension Systems

The MOA takes responsibility for supervising the extension of new tea cultivars in the country, while local agricultural administrations take charge of the release and propagation of local tea clones. In China, 14 national-level stock plantations have been set up for propagation and release of improved tea cultivars. In tea producing areas, regional tea stock plantations were also established by local governments to propagate and sell tea clones. Additionally, privately registered tea farms were also encouraged to propagate and sell tea clones under the supervision of the local agricultural administrations. A mandatory national standard (GB 11767) for tea seedlings was issued in 1989, updated and revised in 2003 to guide the propagation of tea clones and supervise the quality of tea plantlets (Yang *et al.*, 2003a).

2.4.3 Strategy for Promotion of New Cultivars

Tea farmers were encouraged by central government to replace traditional populations with elite tea clones. Those who transplanted elite tea clones could get financial aid from local government in some tea producing regions. The promotion of new cultivars was listed in the development plan in tea producing provinces such as Zhejiang Province, in which the cultivated area of elite tea clones reached 93,000 ha and 60% of the tea garden consisted of clonal tea cultivars by 2010. The clonal tea garden has rapidly increased in recent years, especially in the 21st century. It is estimated that approximately 46% of the tea garden (932 kilohectares) consisted of elite tea clones in 2010. However, the size of the clonal tea garden varied from 2.2% to 95% in different provinces and regions (Table 2.1). The largest clonal tea garden (183.7 kilohectares) can be found in Fujian Province where 95% of the tea area is planted with clones, while the smallest clonal tea garden (300 hectares) is located in Gansu Province where only 2.2% of the tea area is planted with clones.

2.4.4 The Intellectual Property Protection of New Cultivars

With the rapid progress in science and technology and the globalization of the world economy, the importance of intellectual property rights in economic and social activities has increased dramatically. The protection of intellectual property rights has gained wide and serious attention in the world (State Council Information Office of the People's Republic of China, 2005). Plant Varieties Protection (PVP), or Plant Breeder's Rights (PBR), are the rights granted to the breeders of new plant varieties to use the new varieties exclusively. As with industrial patents, copyrights, trademarks and industrial design, PVP is a very important part of the protection of intellectual property rights. The International Union for the Protection of New Varieties of Plants (UPOV) is an intergovernmental organization with headquarters in Geneva, Switzerland. It was established by the International Convention for the Protection of New Varieties of Plants. The objective of the Convention is the protection of new varieties of plants by an intellectual property right. The mission of UPOV is "to provide and promote an effective system of plant variety protection, with the aim of encouraging the development of new varieties of plants, for the benefit of society" (http://www.upov.int/). China acceded to UPOV in line with the 1978 Act and became the 39th member of UPOV on April 23, 1999. Distinctness, Uniformity and Stability (DUS) testing is the technical base of PVP and the scientific basis for the approval of PBR. DUS Test Guidelines are not only the technical manuals for the DUS testing authorities to conduct the testing, but also the technical standards for the competent authorities to examine the DUS of new varieties of plants. UPOV-DUS Test Guidelines for the tea plant (TG/238/1) were prepared by Dr. Liang Chen as the leading expert and adopted by the UPOV Technical Committee on April 8, 2008, which is the first DUS test guidelines prepared by Chinese experts for the UPOV (Chen *et al.*, 2008). In 2008, the tea plants including all the species and varieties of *Camellia* L. Sect. *Thea* (L.) Dyer were listed as a protective species of new plant cultivars by the MOA of the People's Republic of China. Thus, new cultivars of tea can be protected by intellectual property rights, which will further encourage and promote tea breeding in China.

So far, 4 tea cultivars, namely 'Zijuan', 'Yuncha 1', 'Kekecha 1' and 'Kekecha 2', are under PVP protection in China (http://www.cnpvp.net/), and more new tea cultivars are having DUS testing in the TRICAAS (http://www.cnpvp.cn/).

2.5 Developments and Prospects for Tea Breeding

Significant advances in tea breeding have been achieved in China. Nevertheless, to match and satisfy the diverse requirements of domestic and international markets, it is necessary to improve breeding efficiency and breed more excellent cultivars. In the near future, more attention should be paid to precise identification and screening of genetic resources, development of molecular breeding technologies, a deep understanding of the genetics of important agronomic traits, quality and resistance.

2.5.1 Precise Appraisal of Tea Germplasms

Except for functional components and cup-tea quality, the evaluation of resistance to abiotic and biotic stress is becoming more and more important due to climate change as well as the requirement to reduce the use of chemical pesticides.

However, rapid and effective technology is still not well developed to assess the resistance level to cold, drought, pests and diseases. Though the efficiency of evaluation can be improved through construction of the core collection, it is time-consuming and laborious to appraise phenotypic traits because most of them are largely subject to environmental and developmental influences. In the future, more attention should be paid to identification of genes and alleles linked with important phenotypic traits. The identification and evaluation of tea genetic resources may show a tendency from the phenotype to the genotype.

2.5.2 Isolation and Application of Functional Genes

Isolation of functional genes from the tea plant was initiated in 1992, and it has made significant progress, in particular large scale EST generation and analysis in recent years, which will provide useful DNA sequence information to efficiently isolate and identify the functional genes. Though dozens of genes have been isolated and cloned, their mechanisms of expression and regulation mostly remained unknown and unclear. Consequently, it is necessary to increase input and deepen functional genome research, in order to reveal specific genes and better understand the mechanism of genetic variation. It is very urgent to organize an international tea genome consortium to initiate the functional genome sequencing project as soon as possible (Chen *et al.*, 2009). We need identification of gene encoding enzymes for a quality-related component facilitated engineering metabolism pathway to either enhance or suppress expression of the target production. For example, identification and cloning of caffeine biosynthesis in tea and degradative genes in microorganisms open up the possibility of using genetic engineering to produce naturally decaffeinated tea through over-expressing caffeine degradative pathway genes or silencing the caffeine biosynthesis pathway (Yadav & Ahuja, 2007).

2.5.3 Marker-Assisted Selection

The construction of genome-wide molecular linkage maps will play a great role in carrying out genetic analysis of quantitative traits, identification of functional markers and cloning of new genes. The partial linkage maps of the tea plant have been developed (Ma JQ *et al.*, 2010a), but high density and precision of the linkage map is yet not to be achieved because of the small mapping population and deficient DNA markers. In particular, the absence of co-dominant DNA markers such as RFLP (restriction fragment length polymorphism) and SSR leads to difficulty in knowing the recessive loci related to important traits. Therefore, those constructed maps have a limited ability to locate QTLs linked to some important traits due to their low distribution of molecular markers. Although a practical genetic map is not available for the tea plant right now, the potential advantage of marker-assisted selection is clear and strong with the development and application of a large amount of co-dominant markers such as SSR and SNP (single nucleotide polymorphism). QTL mapping is generally based on linkage analysis in a given population, which needs to perform suitably designed crosses. This is a serious limitation on the use of DNA markers in some cases, because the desired crosses are difficult to achieve and require dozens of years in some plants like tea, and the mapping populations used are sometimes not sufficient, with only two alleles detected at a locus. In view of this, alternative methods have been developed and used to study the phenomenon of linkage and recombination. One such method is linkage disequilibrium (LD)—based association analysis that has received the increasing attention of plant geneticists during the last few years (Gupta *et al.*, 2005). The method of LD mapping is found in detail in Mackay and Powell's review of 2007 (Mackay & Powell, 2007). LD studies have now been conducted in more than a dozen plant systems, both at the individual gene level and at the level of the whole genome. One of the applications is to identify marker-trait association, which can help the development of a functional marker to use for marker-assisted selection in tea breeding.

2.5.4 Genetic Transformation

This is an important approach to get innovative breeding material through exogenous gene transformation in the tea plant. Though various transformation protocols were studied (Luo & Liang, 2000; Zhao *et al.*, 2001; Wu *et al.*, 2005) and transgenic tea plantlets have been developed (Mondal *et al.*, 2001), there have been technical bottlenecks which needed to be studied and resolved in genetic transformation of the tea plant. Future research should be focused on: (1) effective methods to induce transgenic callus into plantlets, (2) the construction of high efficiency vectors for target genes, and (3) the establishment of effective transformation methods.

2.6 Conclusions

China has an advantage in abundant genetic resources (wild tea species, landraces and mutants), which can provide diverse gene donors for tea breeding. Today, tea breeding is displaying hopeful trends, from phenotypic to genotypic selection. More and more genes (alleles) and DNA markers associated with target traits have been identified and developed. The new high-throughput genome sequencing technology (for example, 454 Life Science and/or Solexa) is available now. Application of even one sequencing run of this new technology would result in dramatically larger cDNA sequence resources (Chen *et al.*, 2009). Most recently, deep sequencing of tea transcriptome was conducted (Shi *et al.*, 2011).The whole genome sequencing of the tea plant is now ongoing. Transgenic techniques make it possible to genetically manipulate and control the tea plant. All of these technical advancements in tea breeding will be helpful in improving breeding efficiency and shorten the development time.

However, more and more challenges have to be faced to meet the requirements of the rapid development of the modern tea industry. It is yet necessary to answer questions as to how to more effectively make use of abundant genetic resources in breeding programs, how to further improve breeding efficiency and shorten breeding times, how to breed tea clones of extraordinary quality and high resistance, etc. We believe that, in the future, a breakthrough will be achieved in aspects such as (1) identification and application of new functional genes and alleles; (2) development of co-dominant DNA markers such as SSR and SNP; (3) construction of a high density molecular linkage map and detection of QTLs controlling important target traits; (4) development of functional markers and application to genotypic selection; (5) breeding of new clones with outstanding traits.

Acknowledgements

This work was supported, in part, by the Hi-Tech Research and Development Program of China (863 Plan; No. 2006AA10Z171), the Earmarked Fund for the China Agriculture Research System (CARS-023) to Liang Chen.

References

Cao PR, Chen TC, Wu DQ (2001) The influence of environmental factors on tea grafting. Journal of Guangdong Tea, (3): 21-23 (in Chinese).

Chang HT (1981) *Thea*—a section of beverage tea trees of the genus *Camellia*. Acta Scientiarum Naturalium Universitatis Sunyatseni, 20(1): 87-99(in Chinese).

Chang HT (1984) A revision of the tea resources plants. Acta Scientiarum Naturalium Universitatis Sunyatseni, 23(1): 1-12 (in Chinese).

Chang HT (1990) New species of Chinese Theaceae. Acta Scientiarum Naturalium Universitatis Sunyatseni, 29(2): 85-93 (in Chinese).

Chen HC, Xu N, Chen XF, Chen ZM, Yu FL (1996) The resistance mechanism of tea clones to pink tea rust mite. Acta Phytophylacica Sinica, 23(2): 137-142 (in Chinese).

Chen HC, Xu N, Chen ZM (2000) The relationship between content of free amino acid in tea shoot and resistance of tea tree to pink mite (*Acaphylla theae* Watt). Acta Phytophylacica Sinica, 27(4): 338-342 (in Chinese).

Chen J, Wang PS, Xia YM, Xu M, Pei SJ (2005) Genetic diversity and differentiation of *Camellia sinensis* L. (cultivated tea) and its wild relatives in

Yunnan province of China, revealed by morphology, biochemistry and allozyme studies. Genetic Resources and Crop Evolution, 52: 41-52.

Chen L, Yamaguchi S (2002) Genetic diversity and phylogeny of tea plant (*Camellia sinensis*) and its related species and varieties in the section *Thea* genus *Camellia* determined by randomly amplified polymorphic DNA analysis. Journal of Horticultural Science & Biotechnology, 77: 729-732.

Chen L, Yamaguchi S (2005) RAPD markers for discriminating tea germplasms at the inter-specific level in China. Plant Breeding, 124: 404-409.

Chen L, Zhou ZX (2005) Variations of main quality components of tea genetic resources preserved in China national germplasm tea repository. Plant Foods for Human Nutrition, 60: 31-35.

Chen L, Tong QQ, Gao QK, Shu JL, Yu FL (1997) Observations on pollen morphology of eight species and one variety in genus *Camellia*. Journal of Tea Science, 17(2): 183-187 (in Chinese).

Chen L, Yang YJ, Yu FL, Gao QK, Chen DM (1998) A study on genetic diversity of 15 tea cultivars (*Camellia sinensis* (L.) O. Kuntze) using RAPD markers. Journal of Tea Science, 18(1): 21-27 (in Chinese).

Chen L, Yu FL, Yang YJ, Chen DM, Xu CJ, Gao QK (1999) A study on genetic stability of excellent tea germplasms (*Camellia sinensis* (L.) O. Kuntze) using RAPD markers. Journal of Tea Science, 19(1): 13-16 (in Chinese).

Chen L, Yu FL, Tong QQ (2000) Discussions on phylogenetic classification and evolution of Sect. *Thea*. Journal of Tea Science, 20(2): 89-94 (in Chinese).

Chen L, Yang YJ, Yu FL (2004) Tea germplasm research in China: Recent progresses and prospects. Journal of Plant Genetic Resources, 5(4): 389-392 (in Chinese).

Chen L, Yu FL, Yang YJ (2005a) Descriptors and data standard for tea (*Camellia* spp.). Beijing: China Agriculture Press (in Chinese).

Chen L, Zhao LP, Gao QK (2005b) Generation and analysis of expressed sequence tags from the tender shoots cDNA library of tea plant (*Camellia sinensis*). Plant Science, 168: 359-363.

Chen L, Zhao LP, Gao QK (2005c) Sequencing of cDNA clones and analysis of the expressed sequence tags (ESTs) properties of young tea plant (*Camellia sinensis*) shoots. Journal of Agricultural Biotechnology, 13(1): 21-25 (in Chinese).

Chen L, Yao MZ, Zhao LP, Wang XC (2006) Recent research progresses on molecular biology of tea plant (*Camellia sinensis*). In: Teixeira do Silva JA (eds.) Floriculture, Ornamental and Plant Biotechnology: Advances and topical issues. London: Global Science Books, 4: 425-436.

Chen L, Zhou ZX, Yang YJ (2007a) Genetic improvement and breeding of tea plant (*Camellia sinensis*) in China: from individual selection to hybridization and molecular breeding. Euphytica, 154: 239-248.

Chen L, Yu FL, Yang YJ, Yao MZ, Wang XC, Zhao LP, Wang PS, Xu M, Qian YZ (2007b) NY/T 1312-2007, Technical Code for Evaluating Crop Germplasm-Tea Plant (*Camellia sinensis*). Beijing: China Standard Press (in Chinese).

Chen L, Yu FL, Yao MZ, Lv B, Yang K, Du YY (2008) Preparation of the UPOV guidelines for the conduct of tests for distinctness, uniformity and stability–Tea plant [*Camellia sinensis* (L.) O. Kuntze]. Agricultural Sciences in China,

7(2): 224-231.

Chen L, Zhao LP, Ma CL, Zhang YL, Liu Z, Qiao XY, Yao MZ, Wang XC (2009) Recent progress in the molecular biology of tea (*Camellia sinensis*) based on the expressed sequence tag strategy: A review. Journal of Horticultural Science and Biotechnology, 84(5): 476-485.

Chen L, Yao MZ, Wang XC, Ma CL, Jin JQ, Yang YJ, Jiang YW, Xiong XP, Wang PS (2011) NY/T 2031-2011, Evaluating Standards for Elite and Rare Germplasm Resources-Tea Plant (*Camellia sinensis*(L.) O. Kuntze). Beijing: China Standard Press (in Chinese).

Chen XQ (1980) Physiological and biochemical parameters to evaluate cold tolerance of tea plant. The Communication of Silkworm, Mulberry and Tea, (3): 11-14 (in Chinese).

Chen ZM (2000) The Grant Dictionary of Chinese Tea. Beijing: China Light Industry Press (in Chinese).

Cheng H, Zeng JM, Zhou J, Wang LY, Chang J, Ge Y, Yuan HB, Gu BJ, Zhang XF (2007) Rapid propagation of tea clonal seedlings in auto-controlled greenhouse. Journal of Tea Science, 27(3): 231-235 (in Chinese).

China Tea Varieties Compilation Committee (2001) China Tea Varieties. Shanghai: Shanghai Scientific and Technical Publishers, p.9 (in Chinese).

Dong L, He L (1991) The effects of mature degree of shoot on cutting propagation of tea plant. Tea Communication, 4: 28-31(in Chinese).

Du QZ, Li MJ, Liu WH, Wang HS (1990) Chemical and numerical taxonomies of plants *Thea* section plants. Journal of Tea Science, 10(2): 1-12 (in Chinese).

Fang WP, Zou ZW, Hou XL, Zang D, Duan YS, Yang YY, Li XH (2009) Cloning and sequence analysis of cold-induced H1-histone gene from *Camellia sinensis*. Acta Botanica Boreali-Occidentalia Sinica, 29(8): 1514-1519 (in Chinese) .

Feng HS (2010) Both production and income of Chinese tea industry increased in 2009. China Tea, 32(2): 1 (in Chinese).

Food and Agriculture Organization (FAO) (2011) http://faostat.fao.org/.

Gao QK, Hu C, Zhu LP (1997) Effects of tea varieties with different tea polyphenol contents on larval growth and development of *Ectropis obliqua* Prout. Journal of Wuyi Science, 13: 211-214 (in Chinese).

Guo CF, Tang YH, Sun Y, Chen CS, Chen RB, Zhang MQ (2008) ISSR analysis of genetic diversity of tea cultivars [*Camellia sinensis* (L.) O. Kuntze]. Chinese Journal of Tropical Crops, 29(2): 181-186 (in Chinese).

Guo JC, Yang RX, Ye NX, Chen ZH (2003) The breeding and application of 10 elite tea cultivars. Fujian Tea, (3): 18-20 (in Chinese).

Guo JC, Ye NX, He XY (1992) The relationship between genetic parameters and selection on yield of Oolong tea cultivars. Fujian Tea, (4): 15-18 (in Chinese).

Guo JC, Ye NX, He XY (2004) Genetic variation in the leaf-expansion period of the first hybrid generation of tea plants. Journal of Tea Science, 24(4): 255-259 (in Chinese).

Gupta PK, Rustgi S, Kulwal PL (2005) Linkage disequilibrium and association studies in higher plants: present status and future prospects. Plant Molecular Biology, 57: 461-485.

Hou YJ, He Q, Liang GL, Li PW, Peng P, Deng M (2006) ISSR analysis of the hybrid of the descendants of tea camellias. Journal of Southwest Agricultural University (Natural Science), 28(2): 267-270 (in Chinese).

Hou YJ, He Q, Li PW, Liang GL, Peng P, Deng M (2007) Genetic diversity of tea camellias germplasm by ISSR molecular marker. Southwest China Journal of Agricultural Sciences, 20(3): 462-465 (in Chinese).

Huang FP, Liang YR, Lu JL, Chen RB, Mamati GE, Sun QL (2004) Evaluation of genetic diversity in Oolong tea germplasms by AFLP fingerprinting. Journal of Tea Science, 24(3): 183-189 (in Chinese).

Huang FP, Liang YR, Lu JL, Chen RB (2006) Genetic mapping of first generation of backcross in tea by RAPD and ISSR markers. Journal of Tea Science, 26(3): 171-176 (in Chinese).

Huang HT, Liu ZS, Zhuang WF (1986) Studies on the physiology of cold resistance of tea plant-The enzymes and cell membrane permeability with respect to the cold resistance of tea plant. Journal of Tea Science, 6(1): 41-48 (in Chinese).

Huang JA (1990) Relationship between protective enzymes and the ability of cold resistance of tea plant. Journal of Tea Science, 10(1): 35-40 (in Chinese).

Huang JA, Li JX, Huang YH, Luo JW, Gong ZH, Liu ZH (2005) Construction of AFLP molecular markers linkage map in tea plant. Journal of Tea Science, 25(1): 7-15 (in Chinese).

Huang JA, Li JX, Huang YH, Luo JW, Gong ZH, Liu ZH (2006) Genetic diversity of tea (*Camellia sinensis* (L.) O. Kuntze) cultivars revealed by AFLP analysis. Acta Horticulturae Sinica, 33(2): 317-322 (in Chinese).

Huang YH, Zhang JW, Zhang YL, Yang Y, Wang YJ (1994) The correlation between resistance to *Myllocerinus aurolineatus* Voss and characteristic of tea cultivars. Tea Communication, (4): 5-6 (in Chinese).

Huang YH, Zhang JW, Zhang YL, Yang Y, Wang YJ (1998) Anatomical characteristics of leaf structure of tea plant resistant to leafhopper. Journal of Tea Science, 18(1): 35-38 (in Chinese).

International Union for the Protection of New Varieties of Plants (UPOV) (2008) TG/238/1: Guidelines for the Conduct of Tests for Distinctness, Uniformity and Stability of Tea (*Camellia sinensis* (L). O. Kuntze). Geneva.

Jin J, Luo YP, Ren MX, Kang ML (2003) Study on the photosynthetic characteristics of grafted tea plants. Journal of Tea, 29(2): 86-88 (in Chinese).

Jin JQ, Cui HR, Chen WY, Lu MZ, Yao YL, Xin Y, Gong XC (2006) Data mining for SSRs in ESTs and development of EST-SSR marker in tea plant (*Camellia sinensis*). Journal of Tea Science, 26(1): 17-23 (in Chinese).

Jin JQ, Cui HR, Gong XC, Chen WY, Xin Y (2007) Studies on tea plants (*Camellia sinensis*) germplasms using EST-SSR marker. Hereditas (Beijing), 29(1): 103-108 (in Chinese).

Li B, Chen XY, Chen GB, Wang JG (1986) The analysis of karyotype in tea plant. Journal of Tea Science, 6(2): 7-14 (in Chinese).

Li DZ, Pritchard HW (2009) The science and economics of *ex-situ* plant conservation. Trends in Plant Science, 14(11): 614-621.

Li GT (1983) The chromosome karyotype of tea plant and its implication on

classification of Sect *Thea*. Journal of Tea, 9 (4): 11-16 (in Chinese).
Li GT, Liang T (1990) Karyotype studies on six taxa of *Camellia* in China. Guihaia, 10(3): 189-197 (in Chinese).
Li SF (1996) Identification of aneuploid and euploid in tea germplasm. Journal of Tea Science, 16(1): 73-74 (in Chinese).
Li SF, Cheng H, Chen SR, Yu FL (1997) Climatic contributing factor, cell structure and biochemical analysis on the stage albescent phenomenon of Anjibaicha. In: Tea Research Institute Chinese Academy of Agricultural Sciences (eds.) Tea Science Research Proceedings (1991-1995), Shanghai: Shanghai Scientific and Technical Publishers, pp.38-44 (in Chinese).
Li XH, Shi ZP, Liu CL, Luo JW, Shen CW, Gong ZH (2001) Parentage identification of filial generation tea plants from 'Yunnan Daye' and 'Rucheng Baimao' with RAPD method. Journal of Tea Science, 21(2): 99-102 (in Chinese).
Li YH, Jiang CJ, Yang SL, Yu YB (2004) β-glucosidase cDNA cloning in the tea (*Camellia sinensis*) and its prokaryotic expression. Journal of Agricultural Biotechnology, 12(6): 625-629 (in Chinese).
Liang YR, Liu ZS (1988) Studies on chromosome number and karyotypes of five tea clones. Journal of Tea Science, 8(2): 37-41 (in Chinese).
Lin JK, Zheng JG, Chen RB, Chen CS (2005) Screening specific tea germplasm resources (*Camellia sinensis* (L.) O. Kuntze) with high EGCG content. Acta Agronomica Sinica, 31: 1511-1517 (in Chinese).
Liu BY, Zhou J, Xu M, Tang YC, Wang LY, Cheng H, Zhang XF, Wang PS (2008) Tissue culture of immature embryo and parentage identification of hybrids between *Camellia taliensis* (W.W.Smith) Melchior and *C. sinensis* 'Fuding Dabaicha'. Acta Horticulturae Sinica, 35(5): 735-740 (in Chinese).
Liu YQ, Ling ZF, Xie DX, Xu Z, Wu WW (1994) A study on the mechanism of resistance of tea varieties to *Polyphagotarsonemus latus*. Chinese Agricultural Science Bulletin, 27: 86-87 (in Chinese).
Liu Z, Wang XC, Zhao LP, Yao MZ, Wang PS, Xu M, Tang YC, Chen L (2008) Genetic diversity and relationship analysis of tea germplasms originated from southwestern China based on EST-SSR. Molecular Plant Breeding, 6(1): 100-110 (in Chinese).
Liu ZS, Zhou JG (1994) Progress in the field of tea breeding researches in the past 30 years in China. Journal of Tea Science, 14(2): 89-94 (in Chinese).
Liu ZS, Liang YR, Zhou JG, Zhao D, Lu JL (2005) A summation of 50-year research on the tea breeding and genetics. Journal of Tea, 31(1): 3-8 (in Chinese).
Lu DB, Luo YP, Tong, QQ (1992) The change of CAT activity during drought stress with relation to drought resistance of tea plant. Acta Agriculturae Universitatis Zhejiangensis, 18(S): 50-55 (in Chinese).
Lu DB, Tong QQ, Luo YP, Xu HR (1995) Comprehensive evaluation on the drought resistance of tea germplasm resources. Journal of Zhejiang Agricultural University, 21(5): 447-450 (in Chinese).
Lv WM, Lou YF (1991) Variation of resistance to pink mite among tea germplasm. China Tea, 13(5): 8-9 (in Chinese).
Lv WM, Lou YF, Hu HJ, Chen XR (1990) Variation of tea cultivars resistant to

Ectropis oblique Prout. China Tea, 14(2): 24-25 (in Chinese).
Lv WM, Luo HW, Lou YF, Hu HJ, Peng P, Zeng L, Li ZC, Chen XR (1991) Characterization for tea germplasm of tolerance to pests and diseases. In: Tea Research Institute Chinese Academy of Agricultural Sciences (eds.) Tea Science Research Proceedings, Shanghai: Shanghai Scientific and Technical Publishers, pp.70-76 (in Chinese).
Luo JW, Tang HP, Huang YH, Gong ZH, Xiao WJ (2001) Differences of activities of protective enzymes of tea plant varieties with different cold resistant abilities. Journal of Hunan Agricultural University (Natural Science), 27(2): 94-96 (in Chinese).
Luo JW, Shi ZP, Shen CW, Liu CL, Gong ZH, Huang YH (2002) Studies on genetic relationships of tea cultivars [*Camellia sinensis* (L.) O. Kuntze] by RAPD analysis. Journal of Tea Science, 22(2): 140-146 (in Chinese).
Luo Y, Liang Y (2000) Studies on the construction of Bt gene expression vector and its transformation in tea plant. Journal of Tea Science, 20(2): 141-147 (in Chinese).
Luo YP, Wu S, Wang LL, Qian LS (2000) The replanting effect of grafting on the old tea plant. Journal of Tea Science, 20(1): 36-39 (in Chinese).
Ma CL, Chen L (2007) Research progress on isolation and cloning of functional genes in tea plants. Frontiers of Agriculture in China, 1(4): 449-455.
Ma CL, Chen L (2009) Molecular cloning and expression of flavonol synthase gene from tea plant (*Camellia sinensis*). Genomics and Applied Biology, 28 (3): 433-438 (in Chinese).
Ma CL, Qiao XY, Chen L (2010) Cloning and expression analysis of leucoanthocyantin reducase gene of tea plant (*Camellia sinensis*). Journal of Tea Science, 30 (1): 27-36 (in Chinese).
Ma CL, Yao MZ, Wang XC, Jin JQ, Chen L (2011) Young shoot purple-related gene screening in tea plant (*Camellia sinensis*) using CDNA microarray. Journal of Tea Science, 31(1): 59-65 (in Chinese).
Ma JQ, Yao MZ, Chen L (2010a) Research progress in genetic map of tea plant (*Camellia sinensis*). Journal of Tea Science, 30(5): 329-335 (in Chinese).
Ma JQ, Zhou YH, Ma CL, Yao MZ, Jin JQ, Wang XC, Chen L (2010b) Identification and characterization of 74 novel polymorphic EST-SSR markers in the tea plant, *Camellia sinensis* (Theaceae). American Journal of Botany, 97(12): e153-e156.
Mackay I, Powell W (2007) Methods for linkage disequilibrium mapping in crops. Trends in Plant Science, 12: 57-63.
Mei JF, Wang XC, Yang YJ, Li XH (2007) Preliminary study on differential gene expression during cold acclimation in tea plant (*Camellia sinensis*). Journal of Tea Science, 27(4): 286-292 (in Chinese).
Ming TL (1992) A revision of *Camellia* sect. *Thea*. Acta Botanica Yunnanica, 14(2): 115-132 (in Chinese).
Ministry of Agriculture (MOA) (2008) Tea production and consumption in China. China tea, 30(6): 4-6 (in Chinese).
Mondal TK, Bhattacharya A, Ahuja PS, Chand PK (2001) Transgenic tea (*Camellia sinensis* (L.) O. Kuntze cv. *Kangra Jat*) plants obtained by

Agrobacterium-mediated transformation of somatic embryos. Plant Cell Reports, 20: 712-720.

Ni S, Yao MZ, Chen L, Zhao LP and Wang XC (2008) Germplasm and breeding research of tea plant based on DNA molecular marker approaches. Frontiers of Agriculture in China, 2(2): 200-207.

Nian DH (2002) Replanting of old tea bushes by grafting with elite tea clones. Seed, (1): 67-69 (in Chinese).

Peng P (1990) Identification of tea cultivars on resistance to *polyphagotarsonemus latus*. Plant Protection, 16(6): 12-13 (in Chinese).

Qiao TT, Ma CL, Zhou YH, Yao MZ, Liu R, Chen L (2010) EST-SSR Genetic diversity and population structure of tea landraces and developed cultivars (lines) in Zhejiang Province, China. Acta Agronomica Sinica, 36(5): 744-753 (in Chinese).

Qiao XY, Ma CL, Chen L (2009) Molecular cloning and real-time PCR analysis of flavone synthase II gene full-length cDNA from the tea plant. Journal of Tea Science, 29 (5): 347-354 (in Chinese).

Qin XJ, Lin CC, Chen XQ (2004) Studies on new techniques of cuttage reproduction and coming out from nursery rapidly of tea. Chinese Agricultural Science Bulletin, 20(6): 224-226 (in Chinese).

Ran LX, Yu XS, Zeng L, Xiao Q (2008) Study on the resistance of tea varieties to tea gall. Journal of Yunnan Agricultural University, 23(2): 250-256(in Chinese).

Rao HF, Cheng FY (1998) The propagation of tea clones by cutting with short branch under plastic film. Journal of Hubei Agricultural Science, (1): 51-52 (in Chinese).

Sealy JR (1958) A Revision of the Genus *Camellia*. The Royal Horticultural Society, London, 233p.

Shen CW, Huang YH, Huang JA, Luo JW, Liu CL, Liu DH (2007) RAPD analysis for genetic diversity of typical tea populations in Hunan province. Journal of Agricultural Biotechnology, 15(5): 855-860 (in Chinese).

Shen CW, Liu FZ, Luo JW (2001) Advances of methods and technologies on tea cutting propagation. Tea Communication, (1): 44-48 (in Chinese).

Shi CY, Yang H, Wei CL, Yu O, Zhang ZZ, Jiang CJ, Sun J, Li YY, Chen Q, Xia T, Wan XC (2011) Deep sequencing of the *Camellia sinensis* transcriptome revealed candidate genes for major metabolic pathways of tea-specific compounds. BMC Genomics, 12: 131.

Shu JL (1995) The theory and technology on characterization for anatomical structure of tea leaves. China Tea, 17(1): 2-4 (in Chinese).

Shu JL, Chen L (1996) Study on the evolution route of tea pollen morphology. Journal of Tea Science, 16(2): 115-118 (in Chinese).

Su YQ, Zhang JX (1997) A study on comparative anatomy and relationship with resistance of tea blades of 10 species. Journal of Northwest Forestry College, 13(4): 1-8 (in Chinese).

State Council Information Office of the People's Republic of China (2005) White Book of New Progress of Intellectual Property Right Protection in China (in Chinese).

Tan Z, Tong X, Fang C, Jiang CJ, Chen C (2009) Prokaryotic expression of α-tubulin gene of *Camellia sinensis* and preparation of α-tubulin polyclonal

antibody. Journal of Tea Science, 29 (5): 336-340 (in Chinese).
Tong QQ, Lu DB, Lou YP, Xu HR (1992) Enzyme parameters for screening drought resistance in tea germplasms. Acta Agriculturae Universitatis Zhejiangensis, 18(S): 108-111 (in Chinese).
Wang L, Yang SJ, Wang YS, Cheng H (1997) Indoor preservation of tea germplasm resources and studies on their hereditary stability. In: Tea Research Institute Chinese Academy of Agricultural Sciences (eds.) Tea Science Research Proceedings (1991-1995), Shanghai: Shanghai Scientific and Technical Publishers, pp.20-28 (in Chinese).
Wang L, Zhou MD, Zeng Q (1999) Study on storage characters of tea seed. Journal of Tea Science, 19(1): 25-28 (in Chinese).
Wang XC, Zhao LP, Yao MZ, Chen L, Yang YJ (2008) Preliminary study on gene expression differences between normal leaves and albino leaves of Anji Baicha (*Camellia sinensis* cv. *Baiye*1). Journal of Tea Science, 28(1): 50-55 (in Chinese).
Wang XC, Liu Z, Yao MZ, Ma CL, Chen L, Yang YJ (2009) Sampling strategy to establish a primary core collection of Chinese tea germplasms. Journal of Tea Science, 29(2): 159-167 (in Chinese).
Wang YS, Yang SJ, Cheng H (1990) Studies on tissue culture with immature embryo. Journal of Tea Science, 10: 11-18 (in Chinese).
Wang ZX, Jiang CJ, Li J, Li YY (2006a) Selection of an early-sprouting, cold-resistance and quality tea cultivar 'Chanong No.1'. Chinese Agricultural Science Bulletin, 22(4): 324-327 (in Chinese).
Wang ZX, Jiang CJ, Li YY (2006b) Characteristics of an improved tea cultivar 'Chanong No.8' selected by nitrogen ions implantation technique. Nonwood Forest Research, 24(1): 67-70 (in Chinese).
Weatherstone J (1992) Historical introduction. In: Willson KC, Clifford MN (eds.) Tea: Cultivation to Consumption. London: Chapman & Hall, pp.1-23.
Wei CL, Gao XF, Ye AH, Yang YQ, Jiang CJ (2007) Differential gene expression profiles analysis of tea plant induced by tea looper (*Ectropic oblique*) attack using DDRT-PCR. Journal of Tea Science, 27(2): 133-140 (in Chinese).
Wu JN (1987) Review on 'Cha Ching'. Beijing: Agriculture Press, pp.168-206 (in Chinese).
Wu S, Luo YP (2001) Effect of grafting on the major biochemical components in the two-year-old grafted tea plants. Journal of Tea, 27(2): 22-26 (in Chinese).
Wu S, Liang YR, Lu JL Li HY (2005) Combination of particle bombardment-mediated and *agrobacterium*-mediated transformation methods in tea plant. Journal of Tea Science, 25(4): 255-264 (in Chinese).
Xiao YS, Wang ZH (1991) Studies on correlation between tea quality and fine hairs. Journal of Anhui Agriculture College, 18(1): 39-44 (in Chinese).
Xu HR, Dong SS, Luo YP, Tong QQ (1997) Main physiological characteristics of tea germplasm. Journal of Tea Science, 17(S): 100-103 (in Chinese).
Xu Z, Li ZL, Hu X, Peng P, Hou YJ (2005) Comprehensive technology for cutting propagation of tea. Southwest Horticulture, 33: 4-6 (in Chinese).
Yadav SK, Ahuja PS (2007) Towards generating caffeine-free tea by metabolic engineering. Plant Foods for Human Nutrition, 62: 185-191.

Yamanishi T (1995) Special Issue on Tea. Food Reviews International, 11(3): 371-546.
Yang JB, Yang J, Li HT, Zhao Y, Yang SX (2009) Isolation and characterization of 15 microsatellite markers from wild tea plant (*Camellia taliensis*) using FIASCO method. Conservation Genetics, 10(5): 1621-1623.
Yang YH (1989) The application of TTC on the characterization of cold-resistance in tea plant. China Tea, 11(5): 15-16 (in Chinese).
Yang YH, Lin SQ (1987) The application of voltage resistance on the characterization of cold-resistance in tea plant. China Tea, 9(2): 37-38 (in Chinese).
Yang YH, Lin SQ (1992) The study on artificial mutation technique of tea plant. In: Tea Research Institute Chinese Academy of Agricultural Sciences (eds.) Tea Science Research Proceeding. Shanghai: Shanghai Scientific and Technical Publishers, pp.45-54 (in Chinese).
Yang YJ, Ying HJ (1990) Chemical evaluation on tea quality during early-stage of breeding program: relationship between the biochemical component content in shoot and the quality of black tea. Journal of Tea Science, 10(2): 59-64 (in Chinese).
Yang YJ, Ying HJ (1991) Chemical evaluation on tea quality during early-stage of breeding program: relationship between the biochemical component content in the shoots and quality of green tea. Journal of Tea Science, 11(2): 127-131 (in Chinese).
Yang YJ, Ying HJ (1993) Studies on the biological effect and economic benefit of tea cutting density in nursery. Journal of Tea Science, 13(1): 21-26 (in Chinese).
Yang YJ, Chen L, Yu FL (2003a) GB 11767-2003, Seedling of Tea Plant. Beijing: China Standard Press (in Chinese).
Yang YJ, Yang SJ, Wang YS, Zeng JM (2003b) Selection of early budding and high quality green-tea cultivar. Journal of Tea Science, 23(S): 9-15 (in Chinese).
Yang YJ, Yu FL, Chen L, Zeng JM, Yang SJ, Li SF, Shu JL, Shu AM, Zhang ZF, Wang YS, Wang HS, Wang PS, Xu M, Song WX, Guo JC, Yang RX, Zhang WJ, Chen ZH (2003c) Elite germplasm evaluation and genetic stability of tea plants. Journal of Tea Science, 23(S): 1-8 (in Chinese).
Yao MZ (2009) Studies on genetic diversity and structure of tea germplasm in China based on ISSR and EST-SSR markers. PhD Thesis, Zhejiang University, China (in Chinese).
Yao MZ, Huang HT, Yu JZ, Chen L (2005) Analysis of applicability of ISSR on molecular identification and relationship investigation of tea cultivars. Journal of Tea Science, 25(2): 153-157 (in Chinese).
Yao MZ, Chen L, Wang XC, Zhao LP, Yang YJ (2007) Genetic diversity and relationship of clonal tea cultivars in China revealed by ISSR markers. Acta Agronomica Sinica, 33(4): 598-604 (in Chinese).
Yao MZ, Chen L, Liang YR (2008a) Genetic diversity among tea cultivars from China, Japan and Kenya revealed by ISSR markers and its implication for parental selection in tea breeding programs. Plant Breeding, 127: 166-172.
Yao MZ, Guo HW, Wang XC, Xiao Q, Chen L (2008b) The variation of resistance to pink mite among tea germplasm and screening of high-resistant and excellent landraces from Wuyishan region in Fujian. Chinese Agricultural Science Bulletin, 24(9): 127-130 (in Chinese).

Yao MZ, Liu Z, Chen L, Wang XC, Ma CL, Liang YR (2009) Genetic diversity and structure of tea germplasm originated from region of north Yangtze River based on EST-SSR markers. Journal of Tea Science, 29(3): 243-250 (in Chinese).

Yao MZ, Qiao TT, Ma CL, Jin JQ, Chen L (2010) The association analysis of phenotypic traits with EST-SSR markers in tea plants. Journal of Tea Science, 30(1): 45-51 (in Chinese).

Ye NX, Chen XY, Chen GB, Wang JG (1990) Genetic analysis of net photosynthetic rate in tea plants. Journal of Tea Science, 10(2): 65-69 (in Chinese).

Ye NX, Yang RX, Guo JC, Yang JF (2004) Genetic resources diversity and cultivars enhancement of tea in Fujian. Journal of Fujian Agriculture and Forestry University (Natural Science), 33(2): 174-177 (in Chinese).

Yu FL (1986) Discussion on the originating place and the originating center of tea plants. Journal of Tea Science, 6(1): 1-8 (in Chinese).

Yu FL, Wang HS, Han ZF, Zhong WJ, Wang YX, Xu N, Wang PS, Zhang WW, Qin XJ, Hou YJ, Wei FH, Li Q, Chen SR (1991) The characterization on agronomic traits, quality and cold tolerance of tea germplasm. In: Tea Research Institute Chinese Academy of Agricultural Sciences (eds.) Tea Science Research Proceedings, Shanghai: Shanghai Scientific and Technical Publishers, pp.29-34 (in Chinese).

Yu FL, Yu YM, Li MJ, Shu JL, Liu WH, Lü WM, Wang HS, Han ZF, Zhong WJ (1992) Comprehensive evaluation and characterization of some well-performed tea germplasm. Journal of Tea Science, 12(2): 95-126 (in Chinese).

Yu JZ, Yang Y, Huang H, Chen L, Yao MZ (2009) The ISSR analysis of genetic diversity and relationship of half-sib tea cultivars in Fuding Dabaicha and Yunnan Dayecha. Genomics and Applied Biology, 28(2): 281-288 (in Chinese).

Yu M, Jiang CJ, Fang WP, Ye AH, Wang ZX, Li YY, Zhu L (2008a) Cloning and analysis of differential expression of a 14-3-3 protein gene from tea flower bud. Scientia Agricultura Sinica, 41(10): 2983-2991 (in Chinese).

Yu M, Jiang CJ, Ye AH, Wang ZX, Zhu L (2008b) Isolation and characterization of *CsPSP1*, a gene encoding pollen specific protein in tea plant (*Camellia sinensis*). Acta Laser Biology Sinica, 17(2): 206-212 (in Chinese).

Zeng L, Wang PS, Xu M (2001) Studies on the resistance of tea plant to leafhopper (*Empoasca vitis* Gothe). Journal of Tea Science, 21(2): 90-93 (in Chinese).

Zhan ZJ, Lin CQ, Chi YZ (1993) Study on the cytology and genetic background of 4 sterile tea clones. Journal of Tea Science, 13(2): 115-120 (in Chinese).

Zhang GH, Liang YR, Lu JL (2006a) *Agrobacterium rhizogenes*-mediated high frequency hairy root induction and genetic transformation in tea plant. Journal of Tea Science, 26(1): 1-10 (in Chinese).

Zhang GH, Liang YR, Lu JL, Dong JJ (2006b) Construction of tea caffeine synthase gene RNAi vector. Journal of Tea Science, 26(4): 243-248 (in Chinese).

Zhang JW, Wang YJ, Huang YH (1994a) The identification and screening of cultivars resistance to leafhopper. Tea Communication, (1): 2-5 (in Chinese).

Zhang JW, Wang YJ, Huang YH (1994b) Identification and screening of cultivars resistance to *Myllocerinus aurolineatus* Voss. Tea Communication, (3): 5-7 (in Chinese).

Zhang RG, Meng XS, Guo HY (2008) Studies on grafting with deep cutting stems of tea plant. Modern Science and Technology of Agriculture, (17): 18-19 (in Chinese).

Zhang YL, Zhang JW, Yang Y, Huang YH, Wang YJ (1994) The correlation between leafhopper resistance and tea characters. Tea Communication, (2): 4-5 (in Chinese).

Zhao D, Liu ZS, Lu JL, Qian LS, Tu YY, Xi B (2001) Study on *Agrobacterium tumefaciens*-mediated transformation of tea plant. Journal of Tea Science, 21(2): 108-111 (in Chinese).

Zhao LP, Chen L, Wang XC, Yao MZ (2006a) Quantitative detection of β-glucosidase and β-primeverosidase gene expressions in different leaves of tea plant (*Camellia sinensis*) by real-time PCR analysis. Journal of Tea Science, 26(1): 11-16 (in Chinese).

Zhao LP, Gao QK, Chen L, Wang XC, Yao MZ (2006b) Development and preliminary application of cDNA microarray of tea plant (*Camellia sinensis*). Journal of Tea Science, 26(3): 166-170 (in Chinese).

Zhao LP, Liu Z, Chen L, Yao MZ, Wang XC (2008a) Generation and characterization of 24 novel EST derived microsatellites from tea plant (*Camellia sinensis*) and cross-species amplification in its closely related species and varieties. Conservation Genetics, 9(5): 1327-1331.

Zhao LP, Ma CL, Chen L (2008b) Construction and expressed sequence tags analysis of young roots cDNA library of tea plant. Molecular Plant Breeding, 6(5): 893-898 (in Chinese).

Zhong LQ, Peng P, Hou YJ, Wei FH (1997) Resistance identification of tea varieties to grey-blight, *Pestalotia theae* Saw. Southwest China Journal of Agricultural Sciences, 10: 104-107 (in Chinese).

Zhou J, Cheng H, Wang LY (2005) The optimization research on tissue culture and rapid propagation of *Camellia sinensis*. Journal of Tea Science, 25(3): 172-176 (in Chinese).

Zhou JG, Liang YR, Xia GR (1993) Analysis for genetic parameters of chlorophyll contents in tea cultivars. Fujian Tea, (1): 10-13.

Zhu JQ (1992) Preliminary study on resistance of different tea varieties to small green leafhopper. Acta Phytophylacica Sinica, 19(1): 29-32 (in Chinese).

Zhu L, Deng WW, Ye AH, Yu M, Wang ZX, Jiang CJ (2008a) Cloning of two cDNAs encoding a family of ATP sulfurylase from *Camellia sinensis* related to selenium or sulfur metabolism and functional expression in *Escherichia coli*. Plant Physiology and Biochemistry, 46: 731-738.

Zhu L, Jiang CJ, Deng WW, Gao X, Wang RJ, Wan XC(2008b) Cloning and expression of selenocysteine methyltransferase cDNA from *Camellia sinensis*. Acta Physiology Plant, 30: 167-174 (in Chinese).

Zhu L, Jiang CJ, Ye AH, Li YY, Yu M, Deng WW, Fang WP (2008c) Construction of plant expressing vectors of ATP sulfurylase and selenocysteine methyltransferase genes from *Camellia sinensis*. Journal of Nanjing Agricultural University, 31 (2): 121-125 (in Chinese).

Zou ZW, Fang WP, Zhang D, Duan YS, Li XH (2008) Analysis of differential expression genes in cold-induced tea plant. Journal of Tea Science, 28(4): 249-254.

3

Breeding of the Tea Plant (*Camellia sinensis*) in India

Suresh Chandra Das, Sudripta Das*, Mridul Hazarika

Tocklai Experimental Station, Tea Research Association, Jorhat 785008, Assam, India
* E-mail: suresh_chdas@yahoo.co.in

Abstract: The indigenous Assam tea plant was discovered by Robert Bruce in 1823. However, commercial tea cultivation was initiated in India during 1834, using a China tea plant. The better cup quality produced by the Assam tea plant popularized it as an important planting material in the tea industries of the country. Seeds were the only source of propagation until the discovery of the vegetative method. As a result, seed *jat* populations of unknown parents created a wide range of genetic variation and resulting inconsistency in their performance. Considering the need for improved planting materials for the tea industry, Tocklai Experimental Station (TES) Tea Research Association, initiated a tea breeding programme in 1930, under which germplasms were collected based on trait specific phenotypic characteristics. Promising plants selected from heterogeneous *jat* populations as well as from wild tea patches were characterized and preserved in the gene bank of TES, along with some of the non-tea *Camellia* species for utilization in the breeding programmes. The Tea Research Foundation, United Planters' Association of Southern India (UPASI), initiated a similar program in 1963 and collected germplasms were preserved in their gene bank. The technique of vegetative propagation, standardized in 1955, provided scope for developing improved clonal cultivars as well as biclonal seed cultivars through hybridization. From the selected plants from old seed *jats* and progenies of biclonal hybrids, 153 locally adapted and 31 universal clones were developed for the tea industry. Under polyploid breeding triploids, tetraploids and aneuploids were produced through hybridization, out of which high yielding quality triploid plants were selected, cloned and made available for plantation. Water logging tolerant genotypes have been selected and used as parents in the breeding programme for the development of tolerant cultivars and progenies. EST's have been developed from *Camellia* species and a cDNA library was constructed. Marker development for draught resistance and blister blight disease is under progress using cDNA-AFLP and EST-SSR techniques. Transgenic technology has been developed and vector construction is completed to confer resistance against blister blight. The micropropagation technique has also been standardized for quick multiplication of these biotechnologically modified plantlets. Details of tea breeding in India are discussed in this chapter.

3.1 General Introduction to the Indian Tea Industry

The tea plant (*Camellia sinensis* (L.) O. Kuntze) of family Theaceae is the oldest crop plant used in the preparation of the non-alcoholic caffeine containing hot beverage which is popular all over the world. Most scientists believe the tea plant originated in southwest China. China was the first country to use tea as beverage (Barua, 1989). Tea as a beverage was, therefore, a monopoly of China until the establishment of the tea industry in Assam and North East India by the British from the middle of the 19th century. At present more than 50 countries grow tea as a commercial enterprise and a few more for their domestic consumption.

The tea plants of the Assam race were known to people in the hill tribes of the northeastern districts of Arunachal Pradesh, long before the idea of growing tea in this country was conceived in the minds of the British rulers. They used tender leaves from this broad leafed small tree species, for making some traditional drinks that have a refreshing effect on the human body and mind. These plants were growing in patches on the slopes of the hills in those regions. This fact was brought to the notice of the British Government when they initiated the process of tea cultivation in India. The inclusion of the indigenous tea plant in commercial cultivation revolutionized the production of high quality Assam tea which could acquire a unique position amongst consumers.

3.1.1 Historical Introduction to the Tea Industry

Discovery of Assam tea: Although there are number of claims as to who was the real discoverer of the indigenous Assam tea plants, the name of Robert Bruce was recorded first. He saw some patches of tea plants growing in the hills near the capital of the Ahome Dynasty, Rangpur (near Sibsagar town of today) in the year 1823 when he visited on an official tour (Ukers, 1935). Those plants were growing wild and were scattered. He advised one Singpho chief, who knew this plant and the art of preparing the traditional drink as a beverage, to raise some plants by collecting matured seeds to hand over to him on his next visit. Unfortunately, Robert Bruce died in the same year and could not collect the plants. His brother C. A. Bruce, who was a British army official, visited upper Assam on an official tour and met the Singpho chief and collected some plants and seeds.

C. A. Bruce grew the seeds in his kitchen garden. He developed a plantation area using the seedlings obtained and the plants collected from the Singpho chief near Sadiya. He took the initiative to confirm this plant as a tea plant and sent some leaves to the Government Botanical Garden in Kolkata. Dr. Willich, Superintendent of the Botanical Garden identified the plant as being a member of the genus *Camellia*. However, he did not consider this as being the same species of tea as that of the China tea plant (Barua, 1989). Bruce provided some plants for F. Jenkins, who was Commissioner of Assam at that time.

Initial efforts at tea growing in India: In connection with the government's decision regarding commercial tea plantation in India, while exploring suitable areas for tea growing, F. Jenkins showed the Assam wild tea growing areas to the Scientific Committee, with C. A. Bruce as guide. He sent complete specimen of the indigenous plant to the Botanical Garden, Calcutta, for final identification and it was confirmed by Willich that it was "not different from the tea plant of China" (Barua, 1989). After looking at the wild patches of the Assam tea plants in various locations of Assam, all members of the committee, except Dr. Griffith, felt it unnecessary to import tea seeds and seedlings from China any more. Dr. Griffith, of course, expressed his doubt as to whether the wild tea plant would produce tea as good as that prepared from the century old China tea plants. Therefore, the Assam tea plants were not included in the tea plantations initially.

Although there is controversy surrounding the true discoverer of the Assam tea plant, C. A. Bruce was honoured with a medal for his discovery of the Assam tea plant by the English Society of Arts through the Agricultural and Horticultural Society of Bengal (Barua, 1989). Another two gentlemen of the British Army, Major Jenkins and Captain Charlton, also claimed the credit and honour as the first discoverers. As a compromise, they were also given medals by the society. Robert Bruce, who was the real discoverer of Assam tea, however, did not receive any medal or honour (Ukars, 1935). Various sources in Assam and elsewhere claimed Maniram Dewan to be the discoverer of the Assam tea plant, who served under the British Government and worked in the Assam Company later. It could also be true that he brought the Assam plant to the notice of Robert Bruce during his visit to Rangpur in 1823 (Barua, 1989).

Commercial tea cultivation in India: Apart from the medicinal use of tea in China, its popularity as a beverage germinated the idea of utilizing the tea plant as an article of commerce, if its cultivation could be introduced in India and other territories being ruled by the British Empire. The Governor General of India, Lord William Bentinck, appointed a Tea Committee in 1834 to explore the possibility of growing tea in this country for commercial exploitation. The secretary of the

Committee, G. J. Gordon, was given the responsibility of procuring seeds, seedlings and workmen from China and he did the job accordingly. A scientific committee was also constituted to decide suitable locations for tea cultivation and the committee, after visiting the wild tea growing tracts, came to the opinion that the soils and climate of the northeastern region could be the most suitable location for tea plantation. Since C. A. Bruce led the team to various locations in the region, he showed the wild Assam tea plants growing luxuriantly and convinced the members about its authenticity as a tea plant and advocated its inclusion in the proposed cultivation. But due to confusions still remaining, the Scientific Committee decided to only use tea plants imported from China in the experimental gardens of the government.

The committee could not suggest a specific location for experimentation and plants produced in the Botanical Garden, Calcutta, from the imported China seeds were sent to various locations in India including Upper Assam, Dehra Dun, Nilgiri hills etc. (Barua, 1989). In Assam, plants did not perform well at their first site at Saikhowa but were successful when shifted to a new site near Chabua. Seedlings sent to the Himalayan region were first tried in Bhimtal and Almora and then experimental gardens were established in Kumayon, Garhwal and the Kangra district. Seedling samples sent to other places did not show encouraging performance.

The Superintendent of the Government Tea Plantation raised nurseries for indigenous plants and made every effort to explore the forests to discover new wild tea tracts. He also managed to collect leaves and manufacture tea from wild tea plants. The first tea sample sent to Calcutta in 1836 obtained positive comments and on January 10, 1889, eight chests of Assam tea were auctioned. This day was considered to be the most memorable, on which Assam tea plants received recognition and Assam tea established its worth as a beverage (Barua, 1989). Since then it has become popular as an important planting material all over the world and at present more tea is being prepared from the *assamica* type of plant.

3.1.2 Global Standing

Tea Estates and tea growing areas: In India the number of tea estates (TE) and tea growing areas, in terms of tea garden acreage is increasing. As per records of the 'Tea Statistics', the number of tea estates increased to 129,027 in 2004 from 112,010 in 2000 (Table 3.1). More new tea estates have been established during the last three years.

Table 3.1 The number of tea estates and tea growing areas (in ha) in the country

Zones	2000		2004		2006*	2008*
	No. TE	Area	No. TE	Area	Area	Area
North East India						
Assam	39,151	266,512	43,293	271,768	NA	NA
West Bengal	1,540	107,479	8,709	114,003	NA	NA
North India	4,511	16,915	8,627	20,419	NA	NA
South India						
Tamil Nadu	60,618	74,398	62,213	75,978	NA	NA
Kerala	6,153	36,940	6,153	37,107	NA	NA
Karnataka	37	2,122	32	2,128	NA	NA
Total All India	112,010	504,366	129,027	521,403	555,611	474,000

* Harvest acreage, data sources of 2006 and 2008: http://faostat.fao.org/faostat/

Tea production: The country's tea production in the years 2002, 2004, 2006 and 2008 is shown in Table 3.2. It is increasing progressively, from 838.47 to 892.97 kilotonnes in 2002 and 2004, and a further increase to 981.80 kilotonnes in 2006. The increase is due to the adoption of improved cultural practices, use of high yield quality planting materials and extension planting. However, there was slight decrease in production in the year 2008 (980.82 kilotonnes) due to prevailing unfavourable conditions. India could enhance the export considerably and the countrywide export of Indian tea from 2004 to 2007 is shown in Table 3.3 (ITC, 2009).

Table 3.2 Tea production in the country (in kilotonnes)

Zones	2002	2004	2006	2008
North India	631.75	662.19	753.24	733.92
South India	206.72	230.78	228.56	246.90
Total all India	838.47	892.97	981.80	980.82

Table 3.3 Countrywide export of total tea from India (excluding instant tea) (Qty in kilotonnes; Value in 1,000 US$)

Name of the countries	2004		2005		2006		2007	
	Qty	Value	Qty	Value	Qty	Value	Qty	Value
Europe	89.61	181,501	83.60	174,393	85.48	183,391	86.67	204,768
America	7.51	20,325	8.98	25,570	8.06	22,400	9.01	25,836
Africa	12.30	13,890	4.23	5,721	12.10	16,506	9.00	13,129
Asia	79.53	147,488	93.30	167,215	105.55	179,129	66.27	148,795
Oceania	4.95	18,144	4.93	19,507	4.47	18,595	4.89	23,105
Grand total	193.91	381,348	195.03	392,406	215.67	420,021	175.84	415,633

Age of industry in the country: The first privately owned tea company, the Assam Company, was formed in 1839 and the Government handed over its tea holding to the newly formed company in 1840 (Barua, 1989). Although the company faced many problems initially, it was successful within a short period of time due to hard work and skill. Being encouraged by the success gained and the confidence generated by the performance of the Assam Company, many others came forward to be involved in tea cultivation and invested more and more in the growth of the tea industry in India. In North East India, tea cultivation spread rapidly throughout the Brahmaputra valley. Within two and a half decades from the initiative taken by the British Government, the Indian tea industry was established on a strong footing. A tea plantation was initiated in the hill districts of Darjeeling in West Bengal by 1856. In Dehradun and Kangra, at the foot of the western Himalayas, tea plantations were started in 1840 with *sinensis* bushes. Although tea could be grown luxuriantly on experimental trials in the Nilgiris, an effort at commercial plantation was geared up from 1855 onwards. The age of the Indian tea industry can, therefore, be estimated to be about 170 years.

3.1.3 Type of Tea Produced

India is producing mainly black tea. Green tea is produced by only a few tea gardens on a demand basis. This country exports black tea and its internal consumers drink black tea as a common beverage. The black tea is produced in two forms; a major part is CTC (crush, tear and curl) tea which is around 85 per cent of the total production. India produces less than 15 per cent green tea of its total annual production. So far Oolong tea is not produced in this country.

The black tea is manufactured from the tender shoots of the growing buds and two adjacent leaves immediately after harvesting. The harvested green tea leaves are transported quickly to the factories since the changes in the biochemical components start instantly, immediately after detachment from the bushes. Indian black tea manufacturing involves the following stages.

Withering: Withering or partial drying of the tea leaves was carried out by spreading on meshes in thin layers under natural conditions in the past. However, in recent times withering troughs have been used in which the harvested tea leaves are exposed to a regulated flow of warm air. The bottom of the trough is fitted with a metallic mesh over which the leaves are spread in thin layers for withering. At this stage the water content of the shoots is reduced to 60 – 70 per cent depending on the type of manufacture. Studies of the moisture content of the withered leaf revealed that in case of orthodox manufacture it is around 60% on the plains, while it is even much below 50% at higher elevations. For the CTC manufacturing process, the desired moisture level in the withered leaf is around 70% (Gogoi, 1993). At the withering stage, some important biochemical reactions take place which influence the ultimate quality of the tea.

Rolling/Preconditioning: The withered leaves are subjected to rolling with pressure through a rolling machine. At this stage, the cells are disturbed along with the rolling of the withered leaves so that the chemical components of the leaf tissues come out as juice which facilitates its mixing up with the enzymes. For orthodox manufacture, a preconditioning roll is necessary (Gogoi, 1993) which is very gentle so that the leaf juice appears on the surface of the twisted particle. It is important because juice contributes to the development of the brilliant black tea color in the subsequent processing. For breaking down the leaf cells further, rolling is used. During this process, considerable heat is generated which is detrimental to the development of quality and it needs to be controlled.

CTC: Leaves, after preconditioning either in a rotorvane or rolling machine pass through three pairs of cylindrical rollers in the CTC machine. In each pair, two stainless steel rollers rotate in opposite directions at a speed differential of 1:10. The CTC machine completes the three actions of crushing, tearing and curling in the same machine in one go. Leaf appearance, make, grade percentage, fiber content, liquor and infusion depend on the cup obtained in the machine.

The process has been widely used by tea factories, especially in India and Africa. This style of manufacture has the advantage that the finished product (black tea) brews quickly, gives a dark infusion rapidly, is well suited for tea bags and gives more cups per kg. The CTC process gives better exposure of the substrates, mainly flavanols, to the oxidative enzymes present in the leaf and thereby accelerates the oxidation process during the fermentation stage of manufacturing.

Fermentation: This is a complex process of chemical reactions and it starts immediately after maceration of the withered shoots in rolling. Rolled or CTC leaves are either spread on a floor in a specially designed room or carried through a conveyor to a fermenting machine where oxidation and other associated reactions take place. Oxidation of polyphenols is the most important chemical reaction in the fermentation process. The rate of conversion of the polyphenols is a function of leaf type, temperature and oxygen concentration. The oxidation of the polyphenol by air is a slow process but the available enzyme polyphenol oxidase of the tea leaves accelerates the reaction rate. During fermentation, the color of the black tea develops. Orange-red theaflavins (TF) and the dark brown thearubigines (TR) are formed during fermentation and their optimum ratio needs to be maintained. Some flavor is also developed during fermentation. Carotenes and amino acids are believed to combine with the unoxidized polyphenols to develop the flavors.

Drying: The fermentation stage of black tea manufacturing is followed by 'drying' with the basic objective of terminating the oxidation reactions by deactivating enzymes application at high temperature. The second objective of this step is to bring down the moisture content of the processed leaf from about 70% to 2.5% – 3%. It is known that most of the enzymes in tea are deactivated at a temperature of 60 °C. Based on this fact, the temperature of the drying air in the top circuit of a tea dryer is adjusted to a maximum of 60 °C. The temperature of the hot air across the leaf bed in the dryer and the throughput time are important in tea drying (Gogoi, 1993). Both a lower temperature and excess drying are detrimental to the quality of the tea produced. The shelf life of the tea is dependent on the moisture content of the tea produced. A higher moisture content accelerates the deterioration process.

Sorting: Dryer Mouth Tea consists of a mixture of tea particles in different sizes along with some fibres and stalks. The amount of fibres and stalks found in the finished product basically depends on the plucking standard as well as on the climatic condition prevailing at the time of harvesting. At this stage, specially designed equipment is used to remove the fibres and stalks as well as machinery for separation of the tea into different particle sizes. In general, the sorting equipment consists of several sieves, of different mesh size, mounted one above the other. The mesh size of the topmost sieve is the largest and the smallest is at the bottom. The Dryer Mouth Tea is poured on the topmost sieve and the equipment shakes or vibrates the sieves at a measured frequency. Teas of different particle size become separated and are collected and stored.

3.1.4 Climate

The effect of climatic conditions on growth and productivity, as well as on the quality of the tea plant, is well known. Tea is grown between 11° N and 27°50′N latitudes and the climate varies widely between these two latitudes. In India, a temperature of 28 °C ± 2 °C is found to be ideal for higher production tea. The plant does not produce any visible growth when the minimum temperature goes below 13 °C for a period of at least 3 weeks. Well distributed uniform rainfall is proffered for productive growth of the tea plant under North East Indian conditions. Day length also plays an important role in the growth of this plant. The day length requirement is 11 h 15 min, below which the plant does not produce any flush of growth.

The tea bushes of the plantation areas, situated above 16° N latitude remain dormant during the winter season, when both day length and temperature go below the minimum requirement level (11 h 15 min and 13 °C respectively). Moreover, the rainfall also becomes light and is limited to a few showers. Due to the dormancy of the tea plants during the winter season, the tea producing season is shortened and is limited to between 6 and 7 months. The quality of tea produced is reduced during the middle of the cropping season when heavy rainfall prevails. The black tea produced from the second flush harvest during the spring season is the highest quality tea. The back-end crop harvested during autumn (at the beginning of the winter season) also produces comparatively better quality tea when compared to the rainy season's harvest.

The ambient temperature and rainfall data of some of the tea growing areas along with yield and quality are shown in Table 3.4.

Table 3.4 Climate of tea growing areas in different locations of North East and South India

Months	Tocklai[a] (Assam)			Thakurbari[b] (Assam)			Silcoorie[c] (Assam)			Gungaram[d] (West Bengal)			Ging[e] (West Bengal)			Dikom[f] (West Bengal)			Valparai[g] (Tamil Nadu)		
	Temp. (°C)		Rainfall (mm)	Temp. (°C)		Rainfall (mm)	Temp. (°C)		Rainfall (mm)	Temp. (°C)		Rainfall (mm)	Temp. (°C)		Rainfall (mm)	Temp. (°C)		Rainfall (mm)	Temp. (°C)		Rainfall (mm)
	Max.	Min.		Max.	Min.		Max.	Min.		Max.	Min.		Max.	Min.		Max.	Min.		Max.	Min.	
Jan.	22.3	9.6	20.9	23.9	8.6	15.3	25.6	10.8	13.7	22.5	8.8	13.0	16.4	8.2	17.5	22.9	9.4	34.5	27.7	8.9	8.4
Feb.	24.0	12.1	35.5	25.7	11.4	23.6	27.4	13.0	53.4	25.8	11.5	14.1	18.2	9.9	14.8	24.2	12.4	55.0	27.9	12.7	17
Mar.	27.3	15.8	74.7	28.9	15.1	51.4	30.5	16.9	141.8	30.8	16.0	28.9	21.9	13.0	28.8	26.3	15.8	129.8	27.5	14.3	221.2
Apr.	28.4	19.2	193.4	29.6	18.6	154.7	31.4	20.6	300.0	32.2	20.6	83.1	24.6	15.7	93.4	28.2	18.9	222.3	27.9	15.8	84.2
May	29.9	22.1	269.4	31.1	21.2	256.0	31.7	22.7	396.5	32.8	22.4	241.6	25.8	17.9	112.4	27.1	21.6	349.0	27.7	16.2	74
Jun.	31.6	24.4	313.6	32.0	23.6	427.8	32.0	24.5	542.5	32.4	24.0	591.8	26.1	19.7	383.3	31.6	24.2	395.3	24.4	17.6	569.6
Jul.	32.1	24.9	385.0	32.1	24.2	432.2	32.2	24.9	496.0	31.4	24.2	976.9	25.8	20.3	509.7	31.5	24.8	501.1	23.4	17.8	810.7
Aug.	32.1	24.9	332.4	32.6	24.2	321.6	32.6	24.9	412.0	32.5	24.6	626.2	26.2	19.9	458.5	31.8	24.8	400.3	24	18.1	711.8
Sep.	31.2	24.1	254.2	31.8	23.2	316.4	32.0	24.4	346.6	31.8	23.7	511.5	25.3	19.2	311.8	31.4	24.1	320.4	25.4	17	422.4
Oct.	29.4	21.1	119.9	30.6	19.8	138.9	31.5	22.5	179.4	31.7	20.5	140.7	24.0	16.4	84.4	29.8	20.9	125.7	25.8	17.8	327.4
Nov.	26.5	15.5	25.7	28.4	14.1	19.8	29.8	17.7	33.3	29.7	15.5	16.2	21.3	13.1	26.8	27.8	15.7	23.3	26.4	15.9	37
Dec.	23.4	10.7	12.3	25.3	9.7	13.8	27.1	12.7	14.6	26.6	11.0	12.9	18.3	9.9	14.8	25.3	10.1	4.1	27	13.8	68.6
Total			2,037			2,171.5			2,929.8			3,256.9			2,056.2			2,560.8			3,352.3

[a] Latitude: 26°47′ N, Longitude:94°12′ E; [b] Latitude: 26°48′35″ N, Longitude: 92°42′35″E; [c] Latitude: 24°50′ N, Longitude:92°48′ E; [d] Latitude: 26°38′ N, Longitude:88°48′ E; [e] Latitude: 27°17′ N, Longitude:88°19′ E; [f] Latitude: 27°30′ N, Longitude:95°08′ E; [g] Latitude:10°22′ N, Longitude: 76°58′ E

3.2 Tea Germplasm Collection, Conservation, Appraisal

To develop a tea gene bank i.e., a reservoir of diverse genetic materials collected from indigenous and exotic sources, a systematic and long term program was taken up by the Tocklai Experimental Station. Since then, a continuous effort at collecting germplasms with attributes of high yield, high quality, and resistance to biotic and abiotic stresses, has been made. So far more than two thousand one hundred valuable germplasms having a wide range of genetic and morphological diversity are being maintained along with some rare related *Camellias*. Many clonal cultivars and seed varieties were developed using these germplasms (Bezbaruah, 1974; Bezbaruah & Dutta, 1977; Bezbaruah & Singh, 1978; Singh, 1979, 1980a, 1980b).

In South India, the germplasm effort was initiated by Venkataramani (1963) and Sharma (1976), they collected more than 600 of diverse genetic and morphological characteristics in the first phase. UPASI (The United Planters' Association of Southern India) Tea Research Foundation collected more than 600 accessions germplasms in the second phase and maintained more than 1,250 accessions germplasms in their gene bank (Satyanarayana & Sharma, 1991).

3.2.1 Collection and Conservation

Germplasm collection is a routine task of Tocklai Experimental Station and promising plants are being collected regularly from various sources. The old tea areas planted with seed *jats* are the most valuable source of germplasms, which are now being uprooted gradually under the replantation program and replanted with clonal cultivars. It is, therefore, very important to select the plants bearing desired traits from those populations before erosion to be preserved in the gene bank as germplasm for future use in the breeding program of tea improvement.

With the development of civilization, the increase in population and the extension of agricultural and industrial activities, jungles and forest areas are being cleared gradually. Because of this, some wild patches with valuable tea germplasms have already been removed and the situation of the remaining tea tracts has become alarming. These seed grown populations of tea gardens and the wild patches of tea plants in the forest are recognized as a gold mine of tea genes and will be lost forever if uprooted (Bezbaruah, 1967a, 1968, 1971a, 1974; Singh & Bezbaruah, 1978; Singh, 1980a). Wight, therefore, opined that one plant out of 40,000 old seed grown bushes should be selected as a golden bush (Wight, 1958a, 1958b).

Apart from the diploid tea germplasms preserved in the gene bank, Tocklai has developed genetic stocks which are also maintained carefully. Some of the *Camellia* species, not used in making beverage tea (Table 3.5), are preserved and used for different purposes.

Table 3.5 Non tea germplasms preserved at Tocklai

Sl. No.	Name of the species/entities	No. of plants
1	*Camellia rosiflora*	3
2	*C. japonica*	47
3	*C. sasanqua*	28
4	*C. drupifera*	2
5	*Pyraneria baringtonifolia*	17
6	*Eurya japonica*	1
7	Cross *Camellia* tea	9
8	Pink flower tea	2
Total		109

Table 3.6 shows the tea germplasms collected till 1991 and maintained in the gene bank at different locations of Tocklai (Konwar, 1999). Collection of germplasm is a routine practice and those collected during 1992 – 2000 are shown in Table 3.7 (Konwar, 1999).

Table 3.6 Tea germplasms collected and maintained at Tocklai

Genotype	No. of genotypes/ clones	Characters	Location	Remarks
Rajgarh	1, 152 & 2 clones	Big leaf *assamica* type with considerable morphological variations in growth, leaf shape, size and odor	Sec 22, Borbheta, Clone 1/7/1 in expt plot & 1/61 in LTT, New Botanical Area (NBA)	Clone 1/7/1 was released as TV3
Kharikatia	2 clones	Light leaf *assamica* type, hardier, originally selected from Singloo jat	Uprooted, clones 2/23 & 2/1 in LTT, NBA	Clones under evaluations
Betjan	8,229 & 15 clones	Typical *assamica*, fast growing, high quality with large dependent leaves	Sec 27 & Expt 104 Borbheta. Released clones; Clone 3/248 in LTT; 3/28, 3/22, /161, 3/218, 3/242, 3/22, 3/161, 3/281, 3/242, 3/22, 3/161, W14, 3/28 & 3/218 in GP plot, NBA	Clones 20/ 23/1,, 20f7, 20/6,, 61/4, 3/77, 61/5 released as TV2, TV3, TV4, TV6, TV8, & TV12
Mesai Manipuri	4,225 & 3 clones	A dark leaf hardy *assamica* type with intermediate leaves	Sec 43, Borbheta Clone 4/6, 4/1 in LTT; 4/5, 4/6 in GP plot, NBA	Clone 4/6 released as TV 21
Upper Burma	30 & 1 clone	Vigorous *assamica* type with large bullated leaves	Plants as TS 132 in GP plot, clone 9/42 in LTT, NBA	—

(To be continued)

(Table 3.6)

Genotype	No. of genotypes/ clones	Characters	Location	Remarks
Burma	4, 885 & 3 clones	Closely resembles Manipuri jat, but with shorter and broader leaves, dark green in color	Clone 6/8, 6/38, 6/67 in LTT; 6/14/18 GP plot, NBA	
Naga	3 clones	Typical *assamica* type with narrow acuminate leaves	Clone 7/3 in GP plot; 7/61 & W73 in LTT, NBA	—
Lusai	3 clones	Extreme *assamica* type with big, glossy, dependent, light green leaves and prominent vein	Clone 8/5, 8/8, 8/3, 8/5 in GP plot, NBA	
Kachin	1 clone	A delicate *assamica* type	Uprooted, seeds of 13/3 as TS 106, Clone 106/1 at GP & LTT, NBA	
Sinensis hybrid	12 clones	Predominantly *sinensis* (seeds from China)	TV7 in clonal plot, 14/9, 14/1/1. 14/3/5, 14/5/1, 12/2/18, 14/3/13, 14/13/20, 14f7117 in LTT; 14/5/35, 14/6128 & 14/1 in GP plot, NBA	14/2/28 was released as TV7
Singlo	2 clones	Indigenous *assamica* type; delicate light green glossy leaves	Clone 15/5/1 and 15/6/2 in GP plot, NBA	—
Kalline	116 & 4 clones	A dark leaf hardy *assamica* variety	TS 16 in Stock Trial; clone 16/6/25 16/10/22 in LTT; 16/2/15 & 16/11/12 in GP Plot, NBA	
assamica-*sinensis* (Mrs A C Tunstall, 1918 collection)	10 clones	All hybrid types (most of them dried & uprooted)	Clone 19/29/13, 19/14/6, 19/31/14, in plot, 19/5/3, 19/15/13, 19/47/14, 19/48/16, 19/29/12 in LTT; 19/29/2, 19/22/4 in seed *baries*, NBA	19/23/13 released as TV1
Mizo	1 clone	Wild Lusai	NBA	—
Martinga	6 clones	Light leaf *assamica* with early flushing	NBA	—

(To be continued)

(Table 3.6)

Genotype	No. of genotypes/ clones	Characters	Location	Remarks
Samsing	6 clones	Dark leaf hardy *assamica*	NBA	—
Towkok	5 clones	Indigenous *assamica*	NBA	—
Cinnamara	1 clone	*assamica* hybrid with intermediate leaves	19/29/13 in clonal plot, NBA	Clone 19/23/13 released as TV1
Matelli	1 clone	*assamica* variety	NBA	—
Bazaloni	1 clone	Semi wild teas of *assamica*	Clone 24/9 in GP plot, NBA	—
assamica crosses	1 clone	*assamica* hybrid	6 plants of 1 clone at Tocklai	—
Tingamira 39, 779	1 clone	Light broad leaf typical *assamica*	Sce 1A, 5/1, 5A, 6, 11, 13, 42/1, 108, 112, Borbheta Clone 1/13 in LTT, NBA	—
Doomar Dullong	1 clone	Dark leaf, fast growing with prominent *lasiocaly* characters including anthocyanin pigmentation		
Goipani	1 clone	High pubescence, intermediate leaf type hybrids	Clone 112/8 in GP plot, NBA	—
Kumaon Hills 530	4 clones	Typical *sinensis* plants	Sec 3B, Borbheta, Clone 128/26/2, 128/27/16, 128/25/18, 128/26/4 in GP plot, NBA	—
Vah Tukvar	8 clones	Flavoury *sinensis* hybrids	Area IV A, Tocklai & GP plot, NBA	—
Lingia	595	An extremely variable *sinensis* hybrid	Sec 27, Borbheta mixed with Betjan	—
Tocklai	Biclonal stocks 37 (3326 plants)	Clone crosses	Experiments, NBA	—
	Advance clones 48	Hybridization followed by selection	Experiments, NBA	—
	Triploids 73 Polyploid 59 (tetraploids/ aneuploids)	Product of hybridization between a tetraploid stock and TV clones followed by selection Ploidy level confirmed by cytological study	Experiments, NBA	—

(To be continued)

(Table 3.6)

Genotype	No. of genotypes/ clones	Characters	Location	Remarks
	Tissue	10 plants rooted	Experiments, NBA	
	Culture 10 (clonal lines)	In pots were grown (extended area) in the field. Clonal multiplication of each plant and assessment		
	Somaclones	*In vitro* rooted plants grown in the fields	Tocklai campus	Extended variability
UPASI	Clones 11	UPASI selection in TEs, mostly *jats*	GP plot, NBA collected in 1999 – 2000	Released clones
Silcoorie	Clones 8	*assamica* hybrid	GP plot, NBA	District selection & germplasm
Langree	Clones 2	*assamica-sinensis* hybrid with drought tolerance	Evaluation at NBA	—
Gandrapara	Clones 4	*assamica-sinensis* hybrid plants	GP plot, NBA	—
Badamtam	Clones 16	*Sinensis-sinensis* hybrids	GP plot, NBA	—
Rupai Plot, NBA	Clone 1	*assamica-sinensis* hybrid with water logging tolerance	GP plot, NBA	—
Chandighat	Clone 1	Dangri Manipuri type	GP plots, NBA (9 plants)	
Arcuttipore	Clone 1	Garden own selection (Dangri Manipuri)	GP plots, NBA	
C. assamica ssp. *lasiocalyx*		A dark leaf *assamica* type with fast growth and early flushing. The plants differ markedly from other teas (1 triploid)	Clone 124/53/25, 124/41/42, 124/35/18 in LTT; 124/30/32, 124/26/4, 124/48/8, 124/24/1, 124/35/1 & 28/2 in GP plot, NBA	124/53/25 & 124/41/42 released as TV22 & TV23
C. irrawadiensis	Plants 6	A distinct species (from Upper Assam)	NBA	
C. kissi	Plants 8	A related species	NBA	from Imphal, 100-200 MSL and K & J hills, Meghalaya
C. barringtonifolia	Plants 13	A related species (from Burma)	NBA	
C. rosiglora	Plants 4	A related species	NBA	

(To be continued)

(Table 3.6)

Genotype	No. of genotypes/ clones	Characters	Location	Remarks
C. drupifira	Plants 2	A related species	NBA	
C. sp.	Plant 1		NBA	
C. japonica	Plants 47	Related species (from USA & Japan)	NBA	
C. sasanqua	Plants 29	Non tea flowering species	NBA	
Eurya spp.	Plants 2	Different genus (from Mikir hills)	NBA	

Table 3.7 Germplasms collected during 1992 – 2000

Year	Genotypes	No. of accessions	Conservation locations
1992 –1993	Old seed *jat* plants	68	Field germplasm plot
1993 – 1994	Old seed *jat* plants	35	-do-
	Old seed *jat*s	10	Crypreservation at NBPGR*
	Biclonal seed stocks	11	-do-
1994 – 1995	TRA/Garden clones	58	Field germplasm plot
1996 – 1997	Old seed *jat*s	26	Crypreservation at NBPGR
	Shan tea (Vietnam)	1	Field germplasm plot
	South Korean material	7	-do-
	Vietnamese genotypes	8	-do-
1997 – 1998	Old seed *jat* plants	8	-do-
1998 – 1999	UPASI clones	6	-do-
	TRA/garden clones	4	-do-
	Old seed *jat* plants	11	-do-
1999 – 2000	UPASI clones	5	-do-
	TRA/garden clones	2	-do-
	Old seed *jat* plants	28	-do-

*NBPGR: National Bureau of Plant Genetic Resources

***Selection*:** Since all tea plants of past and present plantations developed using plants derived from seeds are genetically dissimilar, such bush population shows great variation in its growth habits, branching pattern, size, shape texture and pose of the leaf as well as in the inherent quality and yield of the bush (Bezbaruah, 1975). There are records on the variations of productivity in the seed grown populations. It was shown that in the past, when tea plantations were initiated using seedlings raised from seeds having no standard, only 0.5 per cent of bushes produced more than 300% of the yield of an average bush in the commercial plantations of North East India, while 10% of bushes produced just 2% (Wight, 1939). Similar observations were also recorded in the tea plantations of Java and

Ceylon (now Sri Lanka). Another study of the Statistics Department, Tocklai Experimental Station, revealed up-to 500 per cent variation in the crop yield of individual bushes (unpublished data) between the lowest and highest yielding bushes. Approximately 67 per cent of the variations might be due to the environment and about 33 per cent due to genetic differences (Bezbaruah, 1975). A similar situation remains in respect of the quality of the tea produced.

In the situation mentioned above, it became important to evolve strategies for development of improved planting materials to meet the requirements of the tea industries as well as tea consumers. The selection procedure adopted earlier by people in tea growing countries was primarily the initial selection of bushes from the mature tea populations on the basis of visual estimation, i.e., by eye judgment of the bush size, frame and plucking point density (Visser & Kehl, 1958; Venkataramani, 1963; Barua, 1965; Bezbaruah, 1974). Strategies had been streamlined and North East Indian tea improvement was taken up based on mass selection, line breeding and clonal selection. Since the mass selection method proposed by Wight (1956) did not help in producing the desired results, a method of line breeding was adopted where each selection was maintained separately to establish a new seed *bari* from where further selections were made (Wight, 1961; Barua, 1963a, 1963b). The concept of developing biclonal varieties was evolved from this after standardization of the vegetative propagation technique for tea.

3.2.2 Appraisal and Utilization

Assam tea plant is a large leaf plant and was given the rank of variety of the *sinensis* species of genus *Camellia* and was included under var. *assamica* [*C. sinensis* (L.) O. Kuntze var. *assamica*] as described by Kitamura (1950). Wight (1962), however, disputed this classification and advocated a specific rank for the Assam tea plant (Barua, 1989). Wight strongly proposed the scientific name *C. assamica* (Masters) for the Assam plant since Masters (1844) first described this plant as a separate type. The important phenotypic characteristics of the *assamica* tea plants are as follows.

C. assamica (Masters) Chang is a small tree that grows up to 10 – 15 m in height. It has a distinct trunk, sometimes up to one third of its height, and has a robust branch system. In typical *assamica* type plants, the leaf blade is broadly elliptical, dependent, thin and glossy with more or less an acuminate apex and distinct marginal veins. The leaf blade is 8 – 20 cm long and 3.5 – 7.5 cm wide (Fig. 3.1), the base cuneate, the margin obscurely denticulate to bluntly wide-serrulate, glabrous or persistently hairy on the midrib below (Barua, 1989).

Flowers grow on the cataphyllary axils singly or in pairs (Fig. 3.2). Pedicels are with scars of 3 caducous bracteoles, smooth and green. Sepals 5 – 6, unequal, leathery and persistent. Petals 7 – 8, white, occasionally with pale yellow pigmentation at the base of petals and stamen numerous. Ovary 3 – 4, sometimes 5 locular, white and hairy. Style generally 3, sometimes up to 5, free up to one third of its length, stigma apical.

Fig. 3.1. *C. assamica.* (a) Free growing plant; (b) Leaves

Fig. 3.2. Flower of *C. assamica*

Among the anatomical features, sclereids of the *assamica* plant are short, with an acuminate upper end and thick secondary wall; the lumen is constricted at several points on the body. Sclereids are also characterized by a few spicules i.e., out-growths.

3.2.3 Assam Tea and Its Quality

Tea in Assam acquires special characteristics due to genetic and environmental interactions. Cultural practices and the manufacturing process also play an important role in the development of the unique characteristics of the end product. Roberts (1962) reported earlier that *C. assamica* was a light leaf variety and produces better quality tea than the dark leaf *sinensis* teas. The uniqueness of Assam tea can be attributed to the chemical constituents present in the leaves which produce the very important and specific Assam characteristics like very coppery infused leaf and the thick gutty liquor that creams down heavily on cooling with a full body and good aroma (Hazarika, 2008).

Tea in Assam is spread across two valleys, the Barak and Brahmaputra. The Brahmaputra valley has acquired a special position amongst plain teas. Tea produced in the Assam valley of North East India is normally referred to as Assam tea. Presently more than 366,753 ha are occupied by tea plantations on the North East Indian plains at a mean latitude of 26° N. In summer, the temperature is as high as 37 °C during June to August for 5 to 6 days. However, the average summer temperature generally remains at around 35 °C. Development of the qualifying Assam characteristics in the Brahmaputra valley is determined primarily by the polyphenols present, specifically catechins. Cream, which settles down on the cooling of a tea infusion, is a measure of Assam tea quality. The color of this cream varies from orange to brown. The creaming property is measured by the cream index which is a measure of the quantity of cream (Roberts, 1962). It has been established that the high caffeine and high polyphenol concentrations are responsible for the high creaming properties of Assam tea. This is a unique innate character of the leaf, which provide fullness to the Assam cup and tea produced from a good quality leaf of *C. assamica* is always superior. At present, many cultivars having an innate quality have been developed from Assam tea germplasms to produce tea of choice. TF and TR in Assam tea are invariably well balanced to produce tea with brightness, strength and fullness.

Extensive research activities in the recent past at Tocklai have helped in identifying some of the specific chemical constituents in Assam tea as well as their aroma, with the help of the taster's palate (Hazarika *et al.*, 2002). TF gives brightness to Assam tea. On the other hand, high molecular weight TR contributes to its color and briskness (Hazarika *et al.*, 1984). Moreover, the amount of high molecular weight TR could determine the quality of cream formed in a black tea brew which is a special characteristic of Assam tea. High catechins are associated with high phenylalanine ammonia lyase (PAL) activity (Jain *et al.*, 1979) and in Assam tea, high catechins, epigallocatechin gallate (EGCG) in particular, influence the level of TF and TR in the end product.

The expression of Assam tea characteristics is influenced to a great extent by the climatic conditions. Significant changes in the chemical constituents of *C. assamica* during the four flushes produced in different seasons of the year are recorded and the flush variation appears well marked. It has also been established that the chlorophylls and carotenoids also influence the quality of this tea considerably.

Assam tea with high polyphenols has also been utilized in the development of antioxidant tea pills. Assam tea is unique not only for the making of a hot cup of special tea but also in the development of many products which are likely to bring about a revolution in the utility of this plant type.

3.3 Tea Breeding and Selection Techniques

The scientific approach to the cultivation of tea was taken up after the establishment of the Tocklai Experimental Station, Indian Tea Research Association. To develop improved planting materials, a program of tea breeding was taken up during the period 1930 – 1939. Since there was no record of any such studies, the initial work was initiated with broad objectives, e.g.

(a) Production of pure lines by natural selfing.

(b) Breeding of clonal seed varieties superior to the commercial *jats* of tea.

(c) Selection of elite bushes and their vegetative multiplication as clones.

Hundreds of artificial pollinations were carried out under objectives (a) and (b), commercial *jat* populations were screened in search of plants having yield and quality attributes.

Based on the recommendation of an Enquiry Commission in 1935, planned breeding was initiated during 1936 (Singh, 1984). Due to various reasons, it was delayed and the actual work was started from 1946. In order to develop superior planting materials, priority was given to the following specific objectives: (1) selection of plants with the potential for producing a high yield and quality of tea as well as very high rooting ability, (2) screening of generative plants having most of the important traits for use in breeding programs for the development of improved seed varieties. Since then efforts have been put in place for the selection of promising plants, their evaluation and preservation as germplasm and several experiments have been carried out on hybridization, polyploidy and mutation breeding.

3.3.1 Breeding Techniques

Tea populations of North East India are highly heterogeneous due to indiscriminate introduction of diverse genetic materials from various sources, including wild sources, in the early years and free hybridization among them. Selection of promising natural variability from these heterogeneous populations, based on trait related morphological characteristics to utilize in the tea improvement program was advocated by the early tea scientists (Wight, 1956; Barua, 1963a, 1963b; Bezbaruah, 1967a, 1967b, 1968, 1969; Satyanarayana & Sharma, 1986). The standardization of the vegetative propagation technique with very high efficiency brought about the concept of cloning the selected genetic materials, having the potential to produce desired traits like high yield, high quality, and resistance to biotic and abiotic stresses, for the development of a homogenous population to use as a clonal cultivar. Controlled hybridization is another approach for breeding between selected parents to produce hybrid seed varieties of high yield and quality potential and the established cloning technology has strengthened the hands of tea breeders in the development of different breeding lines. Mutation, polyploidy, nursery grafting and tissue culture techniques are some of the nonconventional methods with prospects (Satyanarayana & Sharma, 1993).

There are some limitations to the success in developing hybrid progenies of desired characteristics through conventional/classical breeding technology. Since the characteristics of the plant species are governed by the genes of the chromosomes, in gamete formation through meiosis during the sexual reproduction phase, some genes remain linked which never separate. Similarly, some genes cannot be brought together in the offspring through crossing over. Therefore, it is not possible to eliminate some undesired characteristics or to combine some desired characteristics in the progeny.

The new technologies developed through biotechnological research are likely to help in overcoming these inconveniences and provide the means for developing plants of desired characteristics. Expression of the morphological characters is very much influenced by the environmental factors and, therefore, selection using morphological markers sometimes becomes misleading. Since the molecular characteristics do not change with changes in the environment, use of DNA markers can provide a better and reliable means for selection of promising germplasms provided trait related DNA markers are developed. Tissue culture technology is used in the development of 'somaclones', which generate heterogenic populations, from where potential plants of desired characteristics can also be selected to use for various purposes. The protoplast technology is likely to help in the development of 'cybrids' through somatic hybridization. The gene transfer technology, i.e., development of transgenic plants, has been adopted to insert genes responsible for the expression of some specific characteristics which are beyond the scope of conventional sexual breeding.

It is important to understand the genome of a plant to exploit the benefits of these modern technologies. DNA fingerprinting and the development of trait specific genetic markers need to be developed for meaningful selection of plants with desired characteristics. Success in these aspects may help in developing the modalities of marker-assisted selection (MAS) for quick identification of desired plants. The Tea Improvement Division, Tocklai Experimental Station, has put every effort into adopting all possible approaches and technologies, conventional as well as biotechnological, for tea improvement.

The old tea plantations, raised basically from seeds, are highly heterogeneous and these are the best source of genetic materials for the selection of elite clones. The original seed sources from which these populations were raised, no longer exist (Singh, 1984). Uprooting old plantings of tea, the routine practice of replacing old plants (60 years or more old), is causing the loss of many valuable germplasms every year which were the basic genetic source required for tea improvement through breeding. To preserve the important germplasms before uprooting the old seed populations of the tea estates under the replantation program so as to include newly developed improved planting materials, a scheme for selecting bushes having the promise of offering an important genetic resource was started during 1972. Since then this Tea Improvement Division has put every effort into the identification of promising plants from the seed grown populations of commercial tea estates under the different agro-climatic conditions of North East India.

A team of experts was constituted and the criteria for selection was prepared based on some morphological characteristics, having a positive correlation with desired characteristics, under a long term project 'Estate Selection Scheme' or 'District Selection Scheme'. Under this program, the seed derived populations with wide genetic variability were surveyed and the locally adapted elite plants were marked, to be developed into clonal cultivars and those having the most important but rare characteristics as valuable germplasms. For selection, the procedure described in the *Tea Encyclopedia* (Serial No. 163, Tocklai Experimental Station 1965) was followed. So far, 153 clones have been released in a separate series as 'TRA/Garden Series' (till 2005), which are location specific. The steps followed for selection of the clones, assessment and release for commercial exploitation are summarized in Fig. 3.3.

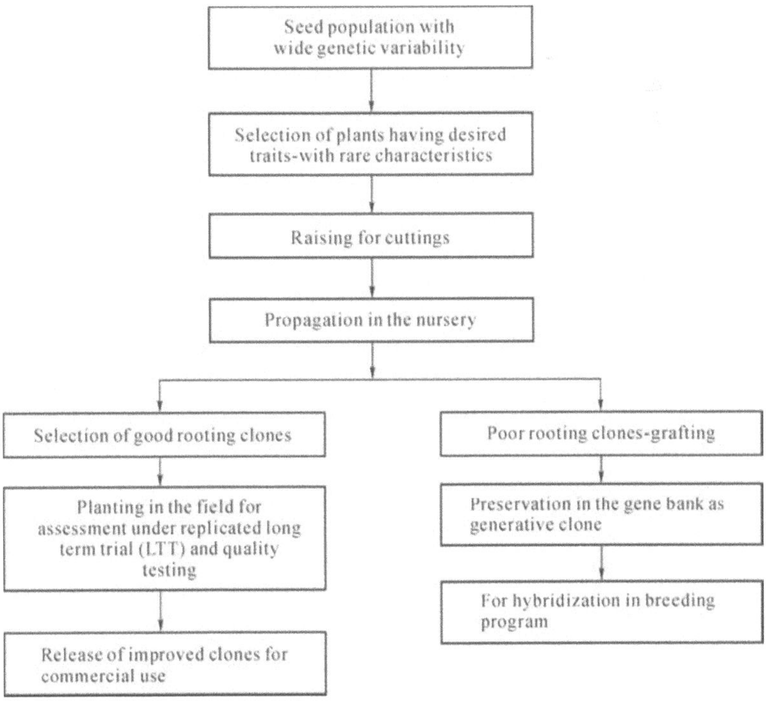

Fig. 3.3. Schematic diagram for clonal selection

3.3.1.1 Breeding for Quality Planting Materials

Till the discovery of the vegetative propagation method, seed *jat* populations of unknown parents were used in the tea plantations and there was no consistency in their performance since there was a wide range of genetic variation among the plants. The vegetative propagation method strengthened the breeders hand by providing a means of cloning to obtain homogeneous populations of selected parents, which are required for the development of the clonal seed cultivars in the seed *baries* under natural conditions. A tea breeding program was started at the Tocklai Experimental Station in 1939 using clonal parents.

So far, 14 clonal seed progenies, developed through crossing between two selected quality clonal parents, have been released as biclonal seed stock for commercial exploitation by the tea industries of North East India. Out of these seed cultivars, 4 were specifically for hilly areas including Darjeeling. The other 10 seed cultivars were suitable on the plains of Brahmaputra and the Barak valleys under different agro-climatic conditions. All the seed cultivars are extensively used by the tea industries in their plantations.

The early efforts at Tocklai were aimed at producing synthetic seed cultivars to replace the popular seed *jats*, grown in North East India. A group of selected clones were planted in isolated seed orchards to produce the seed under natural conditions. Such seed orchards were commonly known as ‘polyclonal seed *baries*’ and seed cultivars as ‘polyclonal seed stocks’ (Singh, 1984). After extensive testing, stock 203 was released to the industry in 1954 as ‘Gaurishankar’. However, polyclonal seed stocks were found to be of unpredictable performance and more unstable than the biclonal seed stocks. Still, development of such synthetic seed cultivars deserve attention in view of the wide genetic base and are expected to be more elastic in their adoptability than biclonal seed stocks.

The characteristics and details of the Tocklai released biclonal seed stocks are shown in Table 3.8. Among the recently released seed stocks, TS 462 produced up to 31 per cent higher yield than TS 449, under different agro-climatic regions and their overall quality was also found to be superior to the control (Singh, 1984).

The United Planters Association of Southern India (UPASI) has developed biclonal hybrid seed genotypes, out of which 6 were released as seed cultivars for commercial plantation. Among the released cultivars UPASI: BSS-1, UPASI: BSS-2, UPASI: BSS-3 and UPASI: BSS-4 were produced by combinations of a common parent TRI 2025 and UPASI-10, UPASI-2, UPASI-9 and UPASI-15 respectively. UPASI: BSS-5 was produced by crossing between parents CR-6017 and UPASI-8. UPASI-9 was used as parent for another breeding program where TRI-2026 was used as the second parent and the resulting progeny was released as TISS-1 for commercial cultivation.

Table 3.8 Parental combinations and characters of biclonal seed cultivars developed at Tocklai

Stock	Year	Parents	Leaf	Growth	Yield	Quality	Drought	Pests/Disease	Manufacture
TS 378	1968	14/5/35×14/6/28	Small	Uniform	Above average	Good	Good tolerance	—	Orthodox
TS 379	1989	14/5/35×14/12/16	Medium	Uniform	Above average	Very good	Good tolerance	—	Orthodox
TS 397	1976	19/29/13×19/35/2	Medium	Fairly uniform	Above average	Above average	Good tolerance	—	CTC, Orthodox
TS 449	1970	19/29/13×19/31/14	Medium	Fairly uniform	Above average	Above average	Good tolerance	—	CTC, Orthodox
TS 450	1970	20/23/1×270/2/13	Medium	Uniform, Vigorous	High	Above average	Tolerant	—	CTC, Orthodox
TS 462	1980	19/29/13×124/48/8	Medium	Uniform, Vigorous	Higher than TS 450	Above average	Fairly tolerant	—	CTC, Orthodox
TS 463	1984	19/29/13×107/14	Medium	Fairly uniform, Vigorous	Above average	Above average	Tolerant	—	CTC, Orthodox
TS 464	1984	19/29/13×19/29/2	Medium	Fairly uniform	High	Above average	Fairly tolerant	—	CTC, Orthodox
TS 491	1989	19/29/13×S3A/1	Medium, light coloured	Fairly uniform	Above average	Very good	Fairly tolerant	—	CTC, Orthodox
TS 506	1994	19/29/13×19/22/4	Medium	Fairly uniform	High	Above average	Fairly tolerant	Good tolerance to pests/diseases	CTC, Orthodox
TS 506	1994	19/29/13×19/22/4	Medium	Fairly uniform, Vigorous	High	Above average	Fairly tolerant	Tolerance to pests/diseases	CTC, Orthodox
TS 557	1996	TRA/AV2×TA17/1/54	Medium	Fairly uniform	Above average	Above average	Tolerant	—	Orthodox, suitable for high altitude areas like Darjeeling.
TS 569	1996	TRA/AV2×TRA/T78	Medium	Medium	Above average	Above average	Moder-ately tolerant	—	Orthodox, suitable for high altitude areas like Darjeeling.
TS 589	1996	468/3/13×HK22/14	Medium	Medium	High	Very good	Tolerant	—	Orthodox, CTC

3.3.1.2 Polyploid Breeding

Cultivated teas all over the world are diploid ($2n = 2x = 30$) in general. Wight (1962) identified a vigorous stock 28/2 out of germplasms Stock 28, collected in 1917 by Pasquier, and this was identified as triploid ($2n = 3x = 45$) by Bezbaruah (1971b, 1971c). The open pollinated progenies of this stock produced natural diploids, triploids, tetraploids and aneuploids (Annual Scientific Report, Tocklai Experimental Station, 1969 – 1970, 1976 – 1977, 1978 – 1979; Bezbaruah, 1968, 1971c; Singh, 1980a; Singh *et al.*, 1982). This genetic resource opened the avenue for polyploid breeding.

Open pollinated F_1 seeds of St. 28/2 were collected in 1958 from which 24 seedlings were obtained. In 1959, plants were renamed as St. 398 and transferred to the field and studied cytologically in 1964 – 1970. Out of these 24 stocks, 13 were confirmed as polyploid (Table 3.9).

Table 3.9 Ploidy level of the F_1 progenies of 28/2

Ploidy level	No. of plants
Triploid ($2n = 45$)	1
Tetraploid ($2n = 60$)	3
Aneuploid ($2n = 32, 33, 38, 42, 58, 59, 60\pm3$)	9

Open pollinated seeds were collected from the fertile diploid and tetraploid plants in 1964 – 1970 and grown in the nursery, from which 100 F_1 seedling progenies of St. 398 (F_2 of 28/2) were transferred to the germplasm block of the New Botanical Area at Tocklai during 1973 – 1974 and studied cytologically in 1976 – 1977 (Annual Scientific Report, Tocklai Experimental Station 1973 – 1974, 1974 – 1975). The ploidy level of these progenies is exhibited in Table 3.10.

Table 3.10 Ploidy level of the progenies of St. 398

Ploidy level	No. of plants
Diploid ($2n = 30$)	27
Triploid ($2n = 45$)	63
Tetraploid ($2n = 60$)	2
Pentaploid ($2n = 75$)	2
Aneuploid ($2n = 30 – 75$)	6
Total	100

The preliminary investigations showed poor quality in natural triploids and tetraploid teas. In order to bring both quality and vigour together in the progeny for the development of high yielding quality hybrid triploids, a controlled hybridization program was initiated during 1972 – 1975, between natural tetraploid and selected high quality diploids (Annual Scientific Report, Tocklai Experimental Station 1973 – 1974, 1974 – 1975). In this attempt, 116 polyploid stocks could be produced (Table 3.11) representing 103 euploids and 13 aneuploids (Singh *et al.*, 1982).

Table 3.11 State of polyploidy breeding at Tocklai

Types of polyploids	No. of Stocks	No. of Living plants
I. Euploid		
Triploid	96	277
Tetraploid	5	5
Pentaploid	2	2
II. Aneuploid ($2n\pm1$)	13	30
Total	116	304

The initial assessment on triploids and tetraploids, carried out by Singh (1980b) and Singh *et al.* (1982), revealed their superiority regarding higher rooting ability, bigger leaf size and higher leaf dry weight over diploids. The pentaploids and most of the aneuploids were poorer in rooting ability with smaller leaves of lighter weight than the diploids, triploids or tetraploids.

Sarmah and Bezbaruah (1984) produced twenty triploid lines of tea from four cross combinations carried out by artificial pollination, using two vigorous tetraploids (St. 398/2 and St. 398/4) as female and quality diploid clones (TV1, TV3 and TV7) as male parents. Their comprehensive studies established that the tetraploids have good rooting ability and vigorous growth and produced a higher yield than the normal seed cultivars. These stocks varied widely in their chemical constituents due to parental influence. The quality of made tea from triploids was found to be comparable with that of TV1, TV3 and TV7. Considering yield and quality performances, promising stocks were selected and brought under systematic study to screen out highly productive planting materials.

Estimation of some of the biochemical components involved in the development of cup quality showed variations within and between the stocks. The caffeine, tannin, theaflavin (TF) and thearubigin (TR) contents in parents and their F_1 hybrids, determined by Sarmah and Bezbaruah (1984) are shown in Table 3.12. They advocated that the marked variations in the chemical constituents among the hybrids might be due to their parental influence.

Tetraploid stocks were further crossed with 4 diploid clones of proven merit and their reciprocals to produce superior triploid clones. A total of 101 breeding stocks were produced through controlled breeding between tetraploid (398/24, 398/2 and 398/4) and diploid (TV1, TV19, TV20, 14/12/16, 124/48/8, S3A/3) out of which 72 triploids, 4 tetraploids and 9 diploids were identified (Annual Scientific Report, Tocklai Experimental Station, 1993 – 1994) following a new technique for quick identification of the ploidy level through the counts of chloroplast in the guard cells of leaf stomata (Ahmed & Singh, 1993). From the crosses between tetraploid and diploid, made in 1992 – 1993, 108 seeds were produced. The seedling plants were examined cytologically for their ploidy level. No seeds were produced in crosses between diploid and triploid. A vigorous triploid plant was identified from the earlier crosses based on morphological, anatomical and cytological observations. The plant is under study for commercial exploitation. Biochemical analysis of triploid progenies showed a wide range of variations in the chemical composition of TF, TR, total color and brightness (Annual Scientific Report, Tocklai Experimental Station, 1997 – 1998).

Table 3.12 Caffeine, tannin, theaflavin (TF) and thearubigin (TR) contents in parents and their F_1 hybrids (%)

Parents/ F_1 hybrids (Stocks)	Caffeine	Tannin	Theaflavin (TF)	Thearubigin (TR)
498	3.74	12.66	0.97	12.24
499	5.75	10.62	1.15	14.60
500/1	4.21	11.62	0.83	12.84
500/2	5.61	12.28	1.11	13.81
500/3	3.95	10.62	1.21	12.44
500/4	3.87	11.78	1.21	12.82
500/5	4.22	10.95	0.88	11.83
500/6	3.86	10.62	1.21	12.20
500/7	5.75	11.12	1.25	11.75
501/1	4.04	12.45	0.89	10.00
501/2	3.78	13.59	0.82	10.93
501/3	5.75	11.12	0.73	10.69
501/4	4.07	12.28	0.73	11.96
501/5	3.43	12.77	1.03	11.61
501/6	3.28	13.40	0.73	11.00
501/7	5.76	12.94	0.81	12.00
501/8	3.49	13.76	1.09	10.82
501/9	3.44	12.11	0.78	10.96
501/10	5.63	13.27	0.90	11.00
501/11	3.82	12.28	0.73	10.45

All triploids exhibited higher values for various leaf characteristics than their diploid parents. Chlorophyll-a and chlorophyll-b contents were higher in the triploid progenies than in their diploid parents. Similar values were recorded for phenolics as well as carotenoids. Heterosis for the leaf area and carotenoid content was expressed by some F_1 progenies and some possessed lower phenolics than the mid-parental value, as indicated by their negative heterosis.

During 1990 – 1996, 104 hybrid seedlings were developed from crossings between diploids (TV1, TV17, TV19, TV20, S3A/l, 14/12/16 and 124/48/8) and tetraploids (398/2, 398/4 and 398/24), out of which 81 were identified as triploid.

3.3.1.3 Mutation Breeding

Reports on mutation breeding in tea are very limited and no natural mutant has been identified so far. Tocklai made a series of systematic attempts to induce mutation by treating the seeds and cuttings with chemical mutagens as well as mutagenic radiations since 1968 – 1969 and could attain little success (Annual Scientific Report, Tocklai Experimental Station, 1968 – 1969, 1969 – 1970, 1974 – 1975). To induce polyploidy, seeds and cuttings were exposed to X-rays and gamma radiations of different energy. Chemical mutagens such as colchicine and ethyl methane sulphonate (EMS) were also used for this purpose. However, these efforts did not produce convincing positive results.

To induce desirable mutation through gamma irradiation for the development of tea mutants with better cup quality and resistance to biotic and abiotic factors, further attempts were made to standardize the technique of irradiation (Annual Scientific Report, Tocklai Experimental Station, 1979 – 1980; Singh, 1980b; Singh & Sharma, 1982). This study showed that an exposure of 2 Kr is the upper limit for the survival of the stem cuttings and mutation induction. Significant clonal difference in response to the doses of gamma radiation was recorded. Clones TV1 and TV18 showed the highest tolerance to gamma irradiation beyond 2 Kr while it was lowest in TV23 among the TV clones studied.

The results of the colchiploidy experiments are so far not very encouraging since autotetraploidy could not be induced. Colchicine treatment on axillary buds (Annual Scientific Report, Tocklai Experimental Station, 1993 – 1994) mostly caused death of the primordial cells. However, a few aneuploids could be obtained. In some cases, morphological abnormalities were also observed.

3.3.1.4 Molecular Characterization

***Selection and characterization of waterlogging resistant cultivars of tea*:** Around 11,458 ha of waterlogged plantation area from 26 tea estates spread over Assam and West Bengal have been surveyed and 113 germplasms of old seed *jats* were selected for evaluation and screening. The passport data on habit, stem type, branch angle, leaf shape, leaf angle, leaf curvature angle, leaf venation and leaf bulletins of 50 germplasms have been completed. The rate of photosynthesis, transpiration loss, water use efficiency, stomatal conductance, internal CO_2, leaf water potential and leaf temperature were evaluated. Proline, epicuticular waxes, root starch reserve, chlorophyll-a and chlorophyll-b content have been estimated.

***Generation and analysis of ESTs from camellia Species*:** Two standard cDNA libraries were made, each with 'two and a bud' (library I : TBSDTV13) and a 3rd and 4th leaf (library II : RBSDTV13) and a 3rd library constructed from the root system of a 3-year old seedling (library III: ADSDTV13). Subtractive library construction was completed for traits such as drought resistance, blister blight resistance and *Helopeltis* resistance. So far, 10,874 sequences have been obtained out of which 4,741 sequences from the standard library have been submitted to NCBI (National Center for Biotechnology Information). Functional classification of the ESTs is shown in Fig. 3.4.

***Development of an integrated genetic linkage map & marker-assisted selection in tea*:** Screening of tea germplasm for identification of appropriate clones of contrasting genotypes for drought and blister blight disease and the development of F_1 populations for future mapping of traits have been completed and the progenies are in the nursery. Data generated for the framework map construction will be completed with markers from TE-AFLP and SSRs using TS-463 progenies. Considerable progress has been made in the development of markers both for drought and blister blight using cDNA-AFLP and EST-SSR techniques. Some of the SSR marker bands analyzed in tea genotypes are presented in Fig. 3.5.

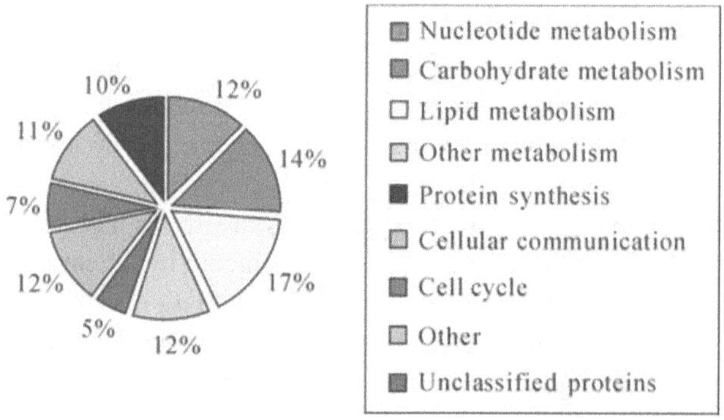

Fig. 3.4. ESTs from *Camellia* species

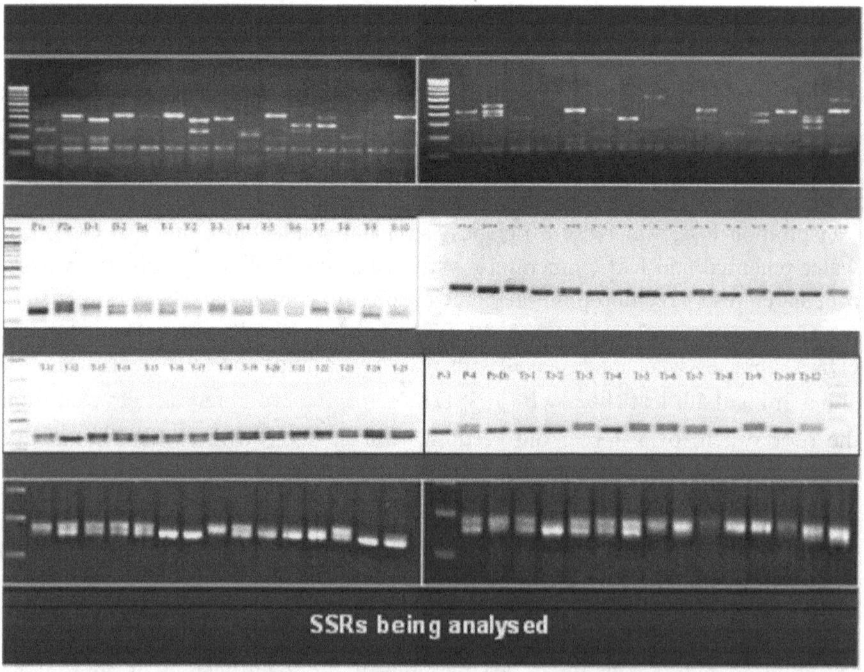

Fig. 3.5. Simple sequence repeat (SSR) marker bands of tea

Development of transgenic tea (C. sinensis) to confer resistance against blister blight: Optimization of an efficient *in vitro* and routine *Agrobacterium* mediated transformation system in tea was achieved. *Chitinase I* and *β-1, 3-glucanase* have been cloned from potato and *Streptomyces* and vector construction completed (Fig. 3.6). Regular transformation is ongoing. Analysis and bioassay will come later.

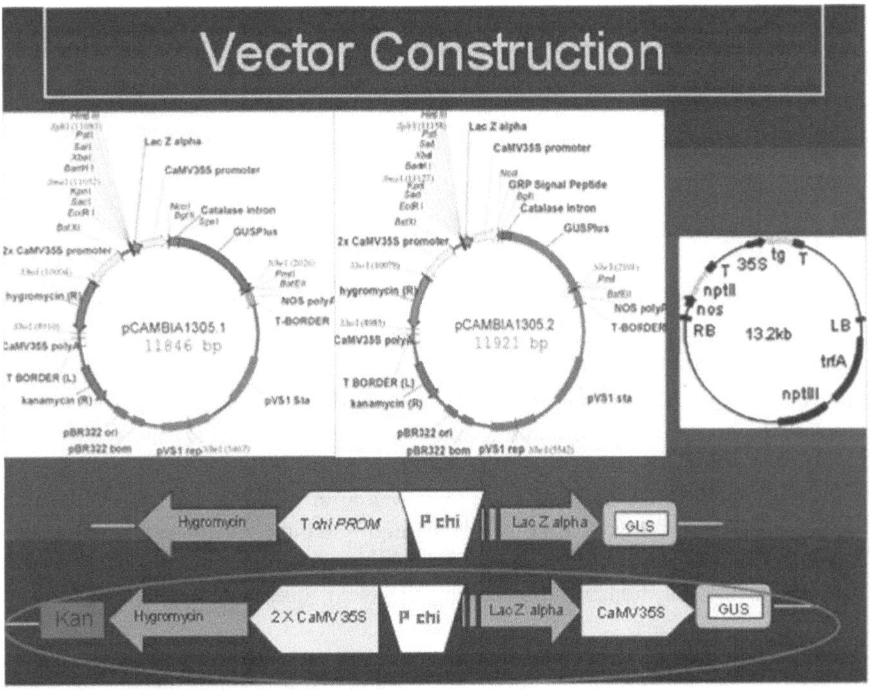

Fig. 3.6. Vector construction for *chitinase I* and *β-1, 3-glucanase*

Production of terpenoids in normal and transformed cell, organ cultures and whole plants in tea: Establishment of normal cell cultures has been accomplished but is a continuous process and more lines are being established. Some of the lines have been screened for production of monoterpenes and analysis of carotenoid carried out. Cloning of the genes has been achieved; vector construction for β-primeverosidase and linalool synthase genes were completed for transformation. Transformed organ cultures (Fig. 3.7) have been established which need to be screened for production of terpenoids.

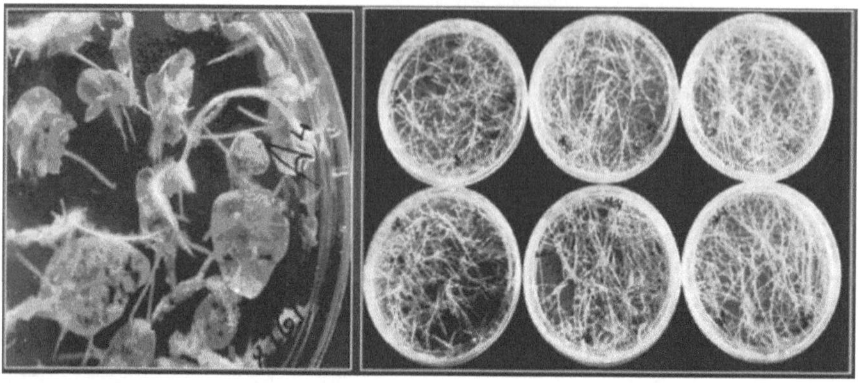

Fig. 3.7. Transformed organ cultures—root cultures from B-157 leaves

3.3.2 Selection Techniques Used for Yield, Resistance and Quality

High yielding quality planting materials, having resistance/tolerance to biotic (pests and diseases) and abiotic (drought, water-logging, frost etc.) stresses, are used in the tea plantations. Planting materials are vegetative clones, hybrid seed stocks and old seed *jats*. Originally, seedlings raised from seed *jats* of unknown parents (natural variability) were the only source of planting materials. After the discovery of indigenous Assam tea plants, high yielding quality seed stocks were developed through cross breeding and the hybrid seeds (created variability) are released to the tea industry after testing the yield, quality and other parameters under field conditions.

The technique of vegetative propagation (VP) was developed to multiply tea plants with uniformity in morphological, anatomical, cytological and all other characteristics. This technique of propagation is utilized extensively in tea culture all over the world as one of the most efficient VP techniques. The promising plants from old seed *jats* and progenies of breeding populations are selected taking into consideration some ideal morphological and numerical parameters that have a positive correlation with the desired traits (high yield and quality potential, resistance/tolerance to biotic and abiotic factors etc.). Then the selected plant is cloned through vegetative propagation.

The basic difference between populations, raised from seed and by cloning, is that every plant of the seed population of a particular race is different (heterogenous) but in the case of a clonal population of a particular clone all the plants are uniform in all respects (homogenous).

3.3.2.1 Mature Bush Yield

The yield of mature tea bushes depends on a number of factors. This is an important issue and no measures can be worked out without a proper understanding of the productivity related parameters. To know the degree of correlation of the yield to the parameters, e.g. a visual estimate of plucking points, the actual number of plucking points, the fresh weight of pruning, the number of pruning sticks, the weight of tipping and number of tipped primaries, extensive studies were made (Annual Scientific Report, Tocklai Experimental Station, 1966 – 1967). Rapid selection, on the basis of visual assessment, for high-yielding bushes in a mature field has limitations. Therefore, screening at different stages needs to be carried out before final selection of the bushes. The correlation studies showed positive indications that plucking point density is related to the shoot yielding capacity of a bush (Barua, 1989). Pruning weight also provides reliable support to the characteristics of the yield potential of the bush. For the selection of a high yielding mature bush the following criteria are generally considered.

Size of surface area: The area of the plucking surface of a plucked bush has a positive correlation with the yield of the bush. A bush with a higher area of plucking table can produce more yield. Visser in 1961 advocated the possibility of utilizing the surface area of a bush as an indication of yield potential. There are reports on a positive correlation obtained between the pruned or plucked surface area and the yield of a mature bush (Cohen Stuart, 1929, 1930; Visser, 1969; Satyanarayana & Sarma, 1982). In selection, therefore, the size of the surface area of the selected bush needs to be judged, taking into consideration the closest spacing of planting recommended. Nowadays, bushes with a medium surface area of the plucking tables have been considered for selection.

Leafiness: The total leaf area of the mature leaves in the maintenance canopy is also considered as a criterion for the assessment of yield capacity. Since the harvested 'two and a bud' shoots, which is the crop, are developed depending on the food produced by the maintenance foliage, reduction of the area of these leaves must have some negative effect on the yield of shoots harvested.

Leaf size: The size of the leaf of a growing 'two and a bud' shoot has been used as another criterion in the selection of a higher yield of tea. Earlier workers, Mamedov (1961) from the Union of Soviet Soualist Republics (USSR) and Toyao (1966) from Japan, observed a positive correlation between leaf size and yield i.e., a higher yield in shoots with bigger leaves. A positive correlation was also obtained between the length of the growing shoots and the yield (Amma, 1975). However, these correlations were confined to the small leaf *sinensis* and its hybrid population. Bezbaruah (1968) and Visser (1969) could not establish such a correlation in a large leaf tea plant. Hence, leaf size can be a criterion in the selection of small-leaf tea.

Plucking point density: Density of plucking points: The number of harvestable shoots per unit area, has a strong positive correlation with the yield. It is an important criterion for quick selection of high yielding bushes by eye assessment. In most of the tea bushes the maximum concentration of plucking points occurs at the central zone which becomes thinner towards the peripheral zone (Barua & Dutta, 1971). In some bushes, however, the density of plucking points is more or less uniform all over the plucking table and such bushes can be selected confidently through eye assessment.

Pruning weight: Apart from the criteria discussed above, some other parameters like pruning weight (weight of pruning litters), thickness and distribution of pruned sticks, number of bud breaks per stick, evenness of flashing, also contribute towards a high density of plucking points.

Selection of seedling for yield: Height, diameter of the stem and leaf area of the seedling have been advocated to be reliable indices of bush vigour. Green (1971) found a positive correlation between height, girth, root weight and branch angle of nursery seedlings and size and yield of the same plant at maturity (Barua, 1989). However, there was controversy surrounding Green's report since many workers from other parts of the tea growing countries, including India, did not find seedling vigour to be a reliable characteristic for predicting yield at maturity (Grice, 1969). The cause of these controversial results might be due to the differential age group of the nursery seedling, taken for study by workers of various countries. Based on his observations of diminishing seedling weight as the age increases, Barua (1961) opined that two year or older seedlings are likely to provide better information regarding their yield at maturity than seedlings of lesser age (Barua, 1989).

It appears that no reliable criteria for selection of promising seedlings for yield at the nursery stage have been developed. It is also not certain that a high yielding bush will produce a progeny which will also give a higher yield.

3.3.2.2 Selection Criteria for Quality

Virtually, no simple criterion is available for identification of a tea bush having superior quality potential. Some morphological characters, claimed to have positive correlation with the quality of made tea, are being used in the selection of quality tea bushes.

Pubescence: Pubescence is the hair on the under surface of young tea leaves. Pubescence varies widely between genotypes. In some plants hairs develop only on the midrib while in highly pubescent plants they develop abundantly on the undersurface and form a dense covering on the entire lamina. There are reports of where positive correlation was obtained with quality (Wight 1958b; Wight & Barua, 1954; Wight & Gilchrist, 1961; Venkataramani, 1963; Wight *et al.*, 1963; Venkataramani & Padmanabhan, 1964). Wight and Gilchrist refused any casual connection of pubescence with quality but it produces tips in orthodox manufacturing. In the CTC method of manufacturing, however, no correlation could be obtained between pubescence and quality.

Colour of leaf: Greenness of leaf is also considered as a phenotypic criterion for selection of a quality tea bush although leaf color alone cannot be taken as an indicator of quality. In *assamica* type, tea the light-green leaf produces better quality tea in CTC manufacturing than the dark green leaf type (Venkataramani & Padmanabhan, 1964). Harler (1964) also stated that bushes with yellow-green shoots provide quality. Wight *et al.* (1963) ascertained the degree of optimum greenness for high quality potential.

***Chloroform test*:** This is a chemical test and can be performed in the field. As claimed by the Japanese scientist, the test indicates the quality of black tea (Toyao *et al.*, 1971). This test is adopted in many countries for screening quality bushes. Sanderson (1963) developed this test to assess the fermentation ability of tea bushes. In this test, the young tea leaves are exposed to chloroform vapor and as a result the green color of the leaf changes to brown. The rapidity of changing color indicates the fermentation rate and the intensity of browning measures the extent of fermentation. The fermentation efficiency of the tea bush plays an important role in determining the quality of made tea.

3.3.2.3 Selection Criteria for Abiotic Stress

Plants of the popular seed population, producing healthy growth without any expression of adverse symptoms indicating suffering, are selected from the areas known as stress prone. Plants are assessed based on the physiological parameters related to stress tolerance and promising bushes are raised for propagation. The finally selected plants are multiplied by vegetative propagation for further systematic evaluation as per physiological criteria and assessment through exposure to an artificially created stress situation. The resistance of the clones is confirmed in the multi-locational trial carried out in stress prone areas. Among the physiological parameters, leaf/shoot water potential, cuticular wax, water use efficiency, rate of photosynthesis, rate respiration, rate of transpiration, stress hormone abscisic acid, proline amino acid etc. established a positive correlation with resistance/tolerance to abiotic stress. Efforts on the development of molecular markers for the selection of stress resistant genotypes are being carried out and significant progress has been made.

3.3.2.4 Selection Criteria for Biotic Stress

Different pests/insects and diseases appear in tea plantations at different seasons of the year. In seed grown heterogenous plant populations of old seed *jats* as well as biclonal seed cultivars; it is observed that some bushes are not infested by pests. Similarly, some buses are not infected by some diseases. Such bushes are likely to have a genetic mechanism for building up the defensive potential. In the peak infestation/infection season such plants are selected and tested through systematic laboratory and field experimentation by rearing the pests and infecting them with the pathogens. Physiological and molecular markers are being developed for quick assessment of the resistance potential of the genotypes to biotic stresses.

3.3.3 Local Adaptability Tests and Registration Systems

All required parameters of the selected cultivars are studied in the clonal proving centers of Tocklai and other centers in North East India. In the next step, performance and adaptability of the materials are tested through multi-locational trials under diverse agro-climatic conditions.

After systematic assessment of the seed and clonal cultivars, the finally selected materials are released provisionally for commercial exploitation. The assessment results are scrutinized by the institutional expert forum and approved for release. Approved cultivars are forwarded to the Tea Board of India for registration and inclusion in the list of approved tea cultivars. The unique cultivars are registered under the National Bureau of Plant Genetic Resources (NBPGR) of India.

3.3.4 Value Added Products and Cultivars Used

Considering the convincing interspecific cross compatibility between cultivated and non-cultivated camellias, Wight and Barua (1957) crossed Wilson's *Camellia* with *C. irrawadiensis*. Their systematic analysis on the morphological characteristics, growth behaviour and cup quality of the F_1 progenies revealed a morphological resemblance to cultivated tea but inferior in quality. They, however, recorded extreme vigor in the progenies as compared to the parents, which is indicative of their possible utilization in further breeding (Bezbaruah, 1971b). In crosses between *C. sinensis* and *C. japonica* regular meiosis and high fertility was observed (Bezbaruah & Gogoi, 1972). The morphological similarity of these hybrids indicated the possibility of the existence of species hybrids in cultivated tea (Singh, 1984).

In recent years (Annual Scientific Report, Tocklai Experimental Station, 2007 – 2008) cultivated teas were crossed with *C. irrawadiensis* where negligible caffeine content was recorded, and progenies were produced to develop cultivars for the preparation of tea with low caffeine. However, hybridization results revealed that *C. irrawadiensis* is not successful as a female parent. Progenies are grown in the nursery for systematic studies and further evaluation.

Development of value added tea has received special importance and efforts have been made to produce tea with flavor and medicinal properties with natural resources and Tocklai has taken the lead. Tea Research Association (TRA) has been devoted to developing diversified products, specifically for the young generations, and numbers of formulations of different products have already been developed which will be made available soon for the consumers. Some such products that achieved excellence according to tasters and evaluation authorities, are 'Tea-Cola', soft drinks from black and green tea, 'Tea Toffi',

'Tea Tablets' for hot and cold cups etc. Formalities required for commercialization have already been completed.

3.4 Propagation and Extension System of New Cultivars

The tea plant can use both sexual and asexual methods to propagate the off-spring, such as seeds, cuttings, grafting and tissue culture. To encourage the industry to replace their unproductive plantations with newly developed high quality cultivars under replantation programs, the Government of India provides a subsidy to the tea industry for the uprooted plants as compensation, through the Tea Board of India.

3.4.1 Propagation Techniques

Like any other seed bearing plants, the Indian tea industry was initiated using seedlings raised from tea seeds which were initially brought from China. Seeds were the only source of propagation till the development of vegetative reproduction techniques.

3.4.1.1 Propagation through Seeds/Seedlings

The new seed cultivars are basically biclonal seed cultivars developed through crossings between two clonal parents. At present the old seed *jat* cultivars, developed through crossings between unknown parents, are rarely used in new plantations. After release of the biclonal seed cultivars, seeds are initially distributed among the tea estates of different agro-climatic regions to generate information on their performance. Once the performance is proved satisfactory, the generative clones are provided to the seed producers for the establishment of commercial orchards so that the tea farmers can obtain their required seed easily. So far Tocklai has developed 14 biclonal seed cultivars (stocks) and made them available for the tea industry of North East India. Four out of these were for the high altitude areas of the hill districts.

3.4.1.2 Cutting

Seeds were the only source of propagation in tea till the discovery of the vegetative propagation method using single node cuttings. A tea farmer in Fujian

Province, China, successfully developed the first cuttings for the vegetative propagation method in the 1780s. Shortly afterwards, a famous Oolong tea clone 'Tieguanyin' was successfully bred. Late in 1857, a famous green and black tea clone 'Fuding Dabaicha' was bred successfully. Currently, they are both the predominant tea cultivars in Chinese tea gardens (China Tea Varieties Compilation Committee, 2001).

Tunstall (1931a; 1931b) reported the best rooting success of the cuttings from the current year's growth, with a leaf on the node and some amount of internodes below. The method was perfected further (Wight, 1955) and by other tea scientists of different tea growing countries. Because of its efficiency, this method is used throughout the world for commercial propagation. Development of this propagation technique was considered as one of the major breakthroughs responsible for the development of the clonal varieties. Considering the high degree of success and efficiency, this method is basically adopted all over the world for clonal multiplication of the tea plant.

The single node cuttings, i.e., the leaf with the attached node and some portion of the internode above and below, are used for propagation. Firstly, healthy bushes of the clones (free from pests and diseases) proposed for multiplication are to be selected. The mother bushes are clean pruned in the cold weather. After spring propagation during April to June, bushes are deep skiffed retaining 10 – 12 cm of new wood to take cuttings for autumn propagation. To obtain good quality cuttings, the thin *'banjhi'* shoots, weak branches, dead and diseased branches are also removed by knife cleaning. The developed new shoots are allowed to grow freely till throwing lateral branches. The semi-hard cuttings taken from the primaries generally have the best success. Propagation can be done in nursery beds or in soil filled polythene sleeves.

Sandy loam soils rich in organic matter are most suitable. Beds are prepared in an east-west direction at least 4 – 6 weeks before planting of cuttings in North East India. For sleeves propagation, polythene tubes are filled with specially prepared soils 4 – 6 weeks prior to planting of cuttings.

In a cutting nursery, shade is indispensable. The overhead shades are generally made either slanting, opening towards the north and sloping towards the south, or are flat depending upon the shade materials used. Different materials, e.g. bamboo laths, thatch, green colored nylon/polythene nets, are commonly used as shade materials. The height of the overhead shade is made in such a way that the workmen can move freely.

For best success, cuttings should be obtained from the hard/semi-hard, green, middle parts of freely growing primaries. A cutting should contain one healthy mother leaf and swollen/dormant axillary buds. Hard brown cuttings with overgrown axillary buds, however, do not show good success (Fig. 3.8).

Fig. 3.8. Stages of vegetative propagation using single node cuttings

Cuttings should be prepared under shade. At the top, the stem is cut above the bud parallel to the leaf blade and the basal cut is made almost parallel to the top cut at least 2.5 cm below the node. Before planting, the prepared cuttings can be treated with 0.1% zinc sulphate solution which helps achieve better success. After preparation, cuttings are planted immediately. Plantation of the cuttings is preferred during the cooler part of the day and exposure to direct sunlight at the time of planting needs to be avoided. Cuttings are planted on the prepared beds keeping the nodes with axillary buds about 4 – 5 mm above the ground and the mother leaf upright. To minimize possible sun scorch damage, cuttings are planted in such a direction that the tip of the mother leaf is pointing north or west. Planted cuttings should be firmly fixed on the bed.

The soil moisture of the propagation beds must be maintained at field capacity. Light watering after planting using a clean hand sprayer helps to improve success. Over watering causes excess callusing which inhibits root induction. The beds should also not be allowed to go dry. Although the newly planted cuttings do not require additional fertilizer, $N:P_2O_5:K_2O$ (2:1:2) mixer with dry soil (1:9) can be applied at 5 g per sleeve around the collar of the plants after they produce 4 – 5 foliage leaves at monthly intervals till transfer for plantation in the field. Pests and diseases are to be controlled as and when required.

3.4.1.3 Grafting

Grafting is a method of propagation through which the vegetative parts (organs) of a plant can be mechanically fixed to another plant. The grafted part is called a 'scion' and the basal part on which it is grafted is termed a 'root stock' or 'stock'. In grafted plants the 'scion' provides the canopy of the bush, maintaining all phenotypic and genotypic characteristics, while the 'stock' nourishes the scion derived canopy by absorbing and supplying water and nutrients from the soil with its strong root system. After successful union it behaves like a single plant.

Grafting was first tried in Java in the late 1920s as a method of propagation of the tea plants and then in North East India from the middle 1930s. In the initial years, only the patch budding technique was used: the other most effective techniques were developed later. Presently, cleft grafting is frequently used in tea for various purposes and rind grafting occasionally. Bud grafting is rarely used.

In tea culture, grafting techniques can be adopted for various purposes, e.g. (a) rapid multiplication of newly released clones for quick establishment of nucleus plots of mother bushes, (b) converting an old and unproductive seed *bari* into a new one, (c) early seed production on generative clones under experimental breeding schemes, (d) developing composite plants by grafting the cuttings of poor rooting quality clones (scions) on vigorously growing clones with a strong root system (stock), (e) for top replacement of undesired mature bushes with 'scions' of quality clones and, (f) for quick establishment of 'tissue culture' derived small plantlets in the field.

Grafting is a widely adopted method in converting old tea seed *baries* to a new one using a different parental combination within a very short period of time. It is also used frequently by the tea industry in the quick establishment of the nucleus plots of mother bushes of new clones. The strong root system of the stocks generally boosts growth and yield performance of the scions without affecting the quality and other parameters. No doubt, the tea industry is more or less experienced at present with grafting techniques.

Very often, questions of graft failure have been raised in spite of performing the grafting operation with all the required care. Grafting success sometimes becomes very poor and it might be due to the "compatibility barrier" between the stock and scion.

For best success in grafting, the scion and stock should generally have a close botanical relationship. Matching the magnitude of vigor between the clones proposed to be used as root stock and scion can be a satisfactory index of grafting success. However, compatibility can be tested by grafting a few scions of the selected clones on two/three stock plants before going for large scale operations.

Grafting can be done throughout the year. However, the best time for grafting has proved to be during November to February under North-East Indian environmental conditions.

Preparation of stock: In this method, a horizontal clean cut is given to the stump/branches of the bush on which a cleft is made and the scion, 1 – 2 leaf portion of a growing primary, has the basal tapering end inserted into the cleft. After successful union through callusing, the axillary buds of the scion grow into shoots and produce a composite plant.

Healthy vigorous bushes are selected and the thin weak branches removed leaving 5 – 6 stout and healthy branches for grafting. Selected branches are cut horizontally at a suitable height based on the purpose of grafting and made smooth using a sharp 'pruning knife'. Prepared stock plants need to be covered immediately with overhead shade (1.75 – 2.0 m height) and side walls with a door in the northern side.

Preparation of scions: Healthy plants free from disease and pest infestations are selected for taking scions and raised to produce primaries. Scions from the healthy primaries of pencil thickness are used for grafting. The semi-hard part of the primaries is suitable for scions and one to two leaf cuttings with healthy axillary buds are taken. Slashing cuts from both sides are made to the internode, 0.5 – 1.0 cm below the lower leaf, to form a tapering end.

Grafting operation: A cleft at the center of the clean cut stock branch is made using a sharp pruning knife. For widening the cleft, a bamboo wedge can be inserted at the center. Two prepared scions are inserted in the two ends of the cleft made on the stock branch in such a way that the cambium layers of both scion and stock come in close contact. After removing the bamboo wedge carefully the entire cleft is tied with jute thread and the exposed grafting surface of the stock is covered with moist coconut coir. Finally, the entire graft is covered with a polythene bag carefully to keep the air saturated with moisture surrounding the grafted scions (Fig. 3.9).

(a) (b)

Fig. 3.9. Grafted tea plant. (a) Scions covered with polythene bags; (b) Successful grafting union

Precautions during the operation: All steps of the grafting operation need special care to achieve the best success. The cleft should not be made too deep. The tapering cut of the scion should be smooth and made in a single go with the help of a sharp grafting knife. Tying of the cleft should not be very tight and water used in moistening the coconut coir must be clean. The coir used for covering the exposed grafting surface should not contain excess water.

Post grafting care: No sun flecks should be allowed to reach the grafted scions. The coconut coir is to be moistened time to time if it dries. If more water droplets accumulate inside the covered polythene bags they should be removed. The polythene bags are to be removed gradually as soon as the developing shoots from the grafted scion touch the top. In this process of removing the bags, developing shoots should not be allowed to wilt and should be replaced if symptoms appear. The soil moisture of the grafting area needs to be maintained at field capacity (20% moisture). Pests and diseases should be controlled immediately using specific chemicals. However, the frequent use of chemicals on the growing scions is detrimental. When the new growth from the axillary buds of the scions produces 2 – 3 flushes of growth, the shade should be gradually thinned out and finally removed. The graft union juncture can be vulnerable to wind damage in the first year in the case of seed *bari* grafting and should be supported by tying with firm bamboo sticks. Any branch that develops from the stock, below the grafting point, should be removed.

3.4.1.4 Tissue Culture

Micropropagation is an alternative method to vegetative propagation, commonly used in tea. Limited success can be achieved in developing a protocol for commercial exploitation of the technique in tea, anywhere in the world. Although the evidence of using this technique for clonal propagation in crop plants as well as in perennial plants is scanty, in plant species where vegetative propagation is difficult or not possible at all, the *in vitro* (tissue culture) method of mass propagation is tried. Since the conventional method of vegetative propagation in tea is extremely simple, easy and efficient for producing a high degree of homogeneity, the *in vitro* technique has very limited scope in tea propagation commercially. However, this technique will be required for multiplication of plantlets developed with biotechnological tools.

3.4.1.4.1 Micropropagation

Multiple shoot production: The technique of tea micropropagation has been developed for clonal multiplication of tea cultivars as well as plantlets developed through tissue culture (Das & Barman, 1988b; Borchetia *et al.*, 2009). For surface sterilization of the experimental materials isolated from the field-grown tea bushes, Das and Barman (1988a) used 0.01% Tween-20 and 0.05% mercuric chloride. They advocated collection of materials from the fields in the sunny days to avoid a heavy contamination problem. One minute treatment with a 1% solution of streptomycin sulphate for materials collected during rainy days was suggested. Borchetia *et al.* (2009) collected experimental materials in the morning hours of the sunny days from the pretreated tea bushes with Bavistin at 1000×10^{-6} concentration. Collected materials were washed thoroughly with a commercial detergent/wetting agent 'Nocidet' (NOCI Ltd. Bombay). The washed materials were surface sterilized with Bavistin 0.3% – 0.5%), mercuric chloride (0.1%) and alcohol (70%) before placing on the pre-sterilized culture medium under sterilized conditions. This method further improved the success of the materials on the culture medium.

Proliferation of shoots per cultured explants (terminal bud and axillary bud) varied between clones and explants. It also varied between the materials used from field grown bushes and taken from the shoots growing on the culture medium. Borchetia *et al.* (2009) obtained 6 – 15 shoots per explant (Fig. 3.10) used from field grown clonal bushes while Das and Barman (1988b) could induce proliferation of 20 – 40 shoots per explant (nodal axillary buds) taken from the shoots developing on the tissue culture medium. The lower rate of shoot proliferation in the explants from the field grown bushes might be due to the inhibitory effect of the decontaminating agents used to eliminate microbial contamination. In the case of the explants taken from tissue culture derived shoots, the decontamination step was not required. The shoot proliferation rate was found to be much higher in axillary bud explants than in terminal bud explants.

(a) (b)

Fig. 3.10. Proliferated shoots from axillary bud explant at different stages of proliferation

For multiple shoot proliferation, different basal media formulations, such as MS medium (Murashige & Skoog, 1962), White's medium (White, 1963), WPM (Lloyd & McCown, 1980), SH (Schenk & Hildebrandt, 1972) etc. were used, supplemented with various concentrations and combinations of growth regulators like BAP (6-benzylamino purine), Kinetin [N-(2-furfurylamino)-1H-purine-6-amine), GA_3 (giberellic acid), IBA (1H-indole-3-butyric acid), IAA (1H-indole-3-acetic acid), NAA (1-naphthalenacetic acid), Zeatin, Ads (adenine sulphate) and TDZ (thidiazuron) depending on the type of explant materials used. As an antioxidant of polyphenol exuded from cut surfaces of tea shoots, ascorbic acid was found to be effective, at a minimum concentration of 25 mg/L. BAP (6 mg/L) in combination with GA_3 (1 mg/L), in inducing higher shoot proliferation. The multiple shoot proliferation rate was significantly higher in *lasiocalyx* clones than in the *assamica* and *sinensis* clones. The combined effect of cytokinin and auxin on shoot multiplication was different to that of cytokinin alone. For maximum shoot proliferation, the hormone requirement was found to be different in different clones. Therefore, a single medium formulation cannot be used as a universal medium for micropropagation of any clone; hence modification needs to be made accordingly. The success of Tocklai in *in vitro* mass multiplication of shoots using axillary buds of nodal explants from *in vitro* regenerated shoots as well as in the field grown plants has been reported since 1990 (Annual Scientific Report, Tocklai Experimental Station 1989 – 1990, 1990 – 1991, 1991 – 1992, 1992 – 1993, 1993 – 1994, 1994 – 1995, 1995 – 1996).

***Rooting of regenerated shoots*:** Although *in vitro* tea shoots have been produced all over the world, success in rooting is limited. Das *et al.* (1990) reported rooting of *in vitro* regenerated tea shoots using three different techniques. Direct rooting excised shoots were pre-treated with 50 mg/L IBA for 30 – 45 min and then planted in a pot containing soil and peat-moss mixture (1:1). To maintain high moisture (100%), the pots were covered with polybags and incubated at low temperature (22 °C) and low light intensity. For liquid shake culture, the shoots were inoculated in a liquid MS medium supplemented with 2 mg/L pyridoxine HCl, 2 mg/L nicotinic acid, 1 mg/L thiamin HCl and 0.5 mg/L IBA and incubated by rotary shaking (70 – 80 rpm) under culture room conditions. In another technique, shoots were cultured on a filter paper bridge in the same medium. About 75% of shoots rooted directly in soil of a peat-moss mixture (Table 3.13). *In vitro* rooting in the liquid shake culture was poorer (25%) than on the paper bridge (43%). Although the initial nursery survival was high, plantlets failed to establish themselves in the field. Jain *et al.* (1991) modified the rooting technique with little improvement. Konwar *et al.* (1999) further standardized the rooting technique and could improve it by up to 93% (Fig. 3.11).

Table 3.13 Success in rooting of the *in vitro* regenerated shoots and survival in nursery

Rooting media	Rooting (%)	Survival in nursery (%)
Soil : Peat-moss (1:1)	75	80
With MS on wet paper bridge	43	55
With MS in liquid shake culture	25	23

Fig. 3.11. *In vitro* rooted plantlets

Hardening and transfer of plants to soil: *In vitro* rooted plantlets were transferred to earthen/plastic pots containing sand, soil, sawdust and cow dung in different combinations (Table 3.14). The greatest success of the rooted plantlets was recorded in the pot mixture of soil/sawdust (1:1). After establishment in the pots, the plantlets were acclimatized in makeshift facilities made of bamboo laths/thatch/poly-nets etc. (Fig. 3.12). The hardened and acclimatized plants were then kept in the nursery under ambient conditions and allowed to grow up to plantable height. The 40 – 45 cm tall plants were planted in the field for assessment (Fig. 3.13).

Table 3.14 Suitability of pot mixtures in the establishment of *in vitro* rooted plantlets

Pot mixtures (1:1:1) / (1:1)	No. of pots	Plantlets		
		Transferred No.	Established No.	Establishment (%)
Sand : Soil : Sawdust	20	36	28	77.8
Sand : Soil : Compost	20	36	18	50.0
Sand : Soil : Cow dung	20	36	11	30.5
Soil : Compost : Cow dung	20	36	8	22.2
Soil : Cow dung	20	36	16	44.4
Soil : Sawdust	20	36	31	86.1
Soil : Mukta	20	36	3	8.3
Soil : Sand	20	36	23	63.9

Fig. 3.12. Hardened micropropagated plants

Fig. 3.13. *In vitro* derived micro-shoot grafted on seedling

***Micro-shoot grafting method of tissue culture derived plantlets*:** Das *et al.* (2005) standardized the technique of cleft grafting, widely used in tea culture, for hardening and establishment of the micropropagated as well as the tissue culture derived plantlets and its success was very encouraging. The grafted micro-shoots and the cleft of the stock completed callusing at the grafting area within two to three weeks and the scion was fully established on the stock after two months (Fig. 3.14). Grafting success was uniform in all the three seed stock populations tested (Table 3.15) and the overall success was 65%. The grafted plants were established in the field with 100% success (Fig. 3.15).

Table 3.15 Micro-shoot grafting of tissue culture derived clones

Stock	No. of shoots grafted	No. of shoots established	Success (%)
TS 520	85	56	66
TS 506	51	33	65
TS 463	57	43	64
Total	203	132	65

Fig. 3.14. Micro-shoot established on the stock

Fig. 3.15. *In vitro* derived microshoot grafted plants established in the field

In view of the hardening problem and difficulties in field establishment of the tissue culture derived plantlets, this technique has provided an alternative and easy way of transferring the plants to the field.

3.4.1.4.2 Genetic Fidelity of the Micropropagated Plants

It is very important to confirm genetic stability in the population developed through micropropagation, otherwise the entire effort of mass multiplication of the clonal materials through the tissue culture technique is defeated. Examples of somaclonal variations at phenotypic, cytological, biochemical and molecular levels among the micropropagated plants have been cited in many plant species. Analysis of such variations in various cultural processes helps detection at an early stage which can help in avoiding such variations through suitable modifications in the protocols. Therefore, it is important to establish a particular micropropagation system for the production of genetically identical and stable plants before it is released for commercial exploitation.

The cytological studies of the *in vitro* derived micropropagated plants revealed $2n = 2x = 30$ chromosomes which were similar to the mother plants of all the experimental clones (Fig. 3.16). However, the karyotypic analysis is yet to be done.

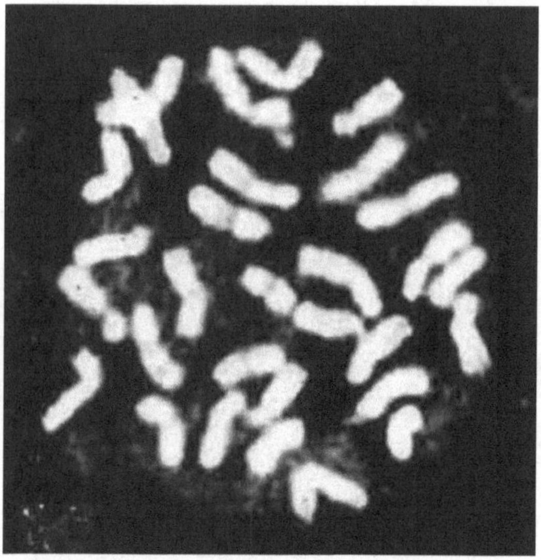

Fig. 3.16. Chromosomes of the micropropagated plants

Borchetia *et al.* (2009) studied the genetic fidelity in clones TV25 (*lasiocalyx* type), TV21 (*assamica* type) and T78 (*sinensis* type) taking 7 micropropagated plantlets from each clone along with the donor plants. Three marker systems viz. RAPD, ISSR and SSR were used considering 4, 6 and 7 primers respectively. Good amplification of all the RAPD primers, 3 ISSR and 5 SSR primers was obtained. The amplified primers generated clear reproducible bands.

Out of the 3 marker system studied, SSR primers showed complete genetic fidelity among the proliferated micro-shoots. There was no phenotypic abnormality among the micropropagated population which suggests that the genetic fidelity of the regenerated plantlets is maintained with the appropriate use of medium components. These results have convinced us that the genetic status of the plants derived from axillary buds as well as through the adventitious mode of propagation, was genetically true to the genotype. However, there are still limitations in the molecular techniques applied.

3.4.2 Extension Organization and Structure

The seed and clonal cultivars developed by TRA, after systematic studies and experimentation at the nursery level as well as in the field under long term trials (LTT), are approved for release by the Agricultural Committee, which is constituted by the TRA Management with a panel of experts in various fields. The cultivars are released by TRA for commercial exploitation in the tea industry. The released cultivars are then registered with the Tea Board of India and included in their approved list of planting materials.

The newly released clonal cultivars are distributed among the tea planters/ farmers in the form of cuttings, scions as well as sleeve-grown plants. Initially, a limited number of materials have been provided so that nucleus plots can be established in the individual tea gardens/estates which can be the source of genuine planting materials for large scale plantation in the gardens. However, some private clonal nurseries are also appearing and producing clonal plants on a large scale from which tea growers, the small stake holders in particular, also collect clonal plants to use in their plantations.

In case of the seed cultivars, after completion of all the formalities, the cuttings/scions of respective parents of the seed cultivars are provided to the registered seed growers, both private and in tea gardens. Seeds of required cultivars are collected by the planters/farmers from the certified seed growers only for commercial utilization. The authenticity of the seed cultivars is verified periodically, particularly at the time of renewal of the certificates through physical verification. The seed growers develop seed orchards (*bari*) for different stocks considering the demand of planters.

3.4.3 Strategy for Promotion of New Cultivars

The new cultivars developed through breeding and selection contain some specialities and promise an improved tea industry. Therefore, all cultivars in the plantations must suit diverse agro-climatic conditions. In the present scenario, unused vacant land is scarce for the extension of new plantations and the industry is also reluctant to replace the old plantations, even though the yield is not economic, because of the long gestation period required by the new plantations to attain the productive stage. A package of replantation, at the rate of 1.5 – 2.0 per cent, without any significant loss, had already been recommended for the tea industry.

Inclusion of high yielding quality cultivars is very important for the growth of the industry and the country's economy. To encourage the industry to replace their unproductive plantations with newly developed high quality cultivars under a replantation program, the Government of India provides a subsidy to the tea industry for the uprooted plants as compensation, through the Tea Board of India. The small tea growers have recently occupied a significant position in tea production and in the growth of the economy of the country by using new cultivars with the support of government and funding agencies and technical support from tea research.

To promote new cultivars, the speciality, usefulness and potential of the planting materials are explained to the planting community through various awareness programs. Routine refresher courses being conducted by the Tea Research Institutes help the tea growers to learn the potential of the new planting materials and encourage the planters to include them in their plantations.

3.4.4 Cultivar Specific Issues

A wide range of variations is generally observed among tea genotypes in rooting efficiency, field establishment, flushing behavior, leaf periods, etc. All these issues are considered at the time of developing a cultivar. Rooting ability of a newly selected genotype is assessed first at the nursery level itself and poor rooters are not considered as suitable for commercial cultivation. However, considering other important characteristics, if most of the trait related parameters are expressed, such genotypes are crossed with strong rooting clones to develop hybrid progenies with acceptable rooting ability so that the successful can be utilized as cultivars. Efficient grafting techniques have been developed to prepare composite plants with a high yield and quality potential and the results of recent studies have shown that the vigour and yield of the quality clone TV1 have been increased significantly when grafted onto strong rooting high yielding clone TV18 (Das *et al.*, 2005). This method of propagation is applicable for all high quality clonal cultivates in field establishment and boosts yield potential.

In general, the leaf period, i.e., rate of leaf unfolding, is shorter in *sinensis* cultivars as compared to the *assamica* and *lasiocalyx* varieties (Das, 1984). Hence, the *sinensis* cultivars produce harvestable 'two and a bud' shoots, required for black tea manufacturing, faster than the other varieties and the plucking round (harvesting) is also shorter accordingly. The tea estates produce quality black tea from clonal cultivars. This is an important issue that needs to be taken care of.

In general, Indian tea is harvested by manual plucking which is found to be efficient in maintaining the standard of harvesting shoots. In the peak season, all tea estates face the problem of completing routine harvesting due to the heavy shoot development of the *assamica* and *lasiocalyx* cultivars and the paucity of additional labor. Use of mechanical plucking aids substantially improves the pluckers' productivity. The wheel mounted plucking machines, operated by 4 persons, can harvest 400 kg of green leaf per day at an optimum plucking speed of 2 km per hour. It increases the pluckers' efficiency by 300%. Use of plucking shears is known to enhance labor productivity by up to 50%.

3.5 Research and Development

There are several research institutions, such as Tocklai Experimental Station, UPASI Tea Research Foundation, the Institute of Himalayan Bioresource Technology, and some universities to support specific tea growing areas of India.

3.5.1 R&D Institutions

Tocklai Experimental Station, Tea Research Association, Jorhat, Assam was started in the year 1911 to provide scientific and technological support to the tea industry. Since then this station has been growing and helping the tea industry in a big way. This institute is responsible for technological support to over 366,000 ha of tea growing areas of North East India in particular. This institute is funded by the tea industries through membership contributions and by the Ministry of Commerce, Government of India, through the Tea Board of India. For studies in specific basic research areas, government funding agencies, viz. the Department of Biotechnology (DBT), Department of Science and Technology (DST), Council of Scientific and Industrial Research (CSIR), Tea Board of India and also some other private agencies provide financial support in the form of research projects. Some funds are also generated through royalties from seed orchards producing seed varieties developed by this station, from analytical services, sales of planting materials and extension services.

For South Indian tea plantations there exists the Tea Research Institute, UPASI, Tea Research Foundation, Valparai, Coimbatore, Tamil Nadu, South India, managed with a fund from membership contributions and the Government of India. This also generates some funding through royalties and extension services. Some R & D work on tea is also being done in the Institute of Himalayan Bioresource Technology, Plumper, Himachal Pradesh, for the tea plantations of the Kangra valley. There are other universities and institutions which have undertaken certain specific research on tea, funded by different funding agencies.

The Tea Board of India has their own R & D center at Kurseong, to undertake research in some specific areas of Darjeeling plantations.

3.5.2 Success Stories of Research in Breeding and Selection

Tocklai Experimental Station, Tea Research Association, has made a tremendous contribution to the tea industry since its inception, particularly in the area of the development of improved planting materials, which are extensively used by the tea industry with success. The success of the planting materials is highly reflected in the manifold increase in tea production today, i.e., 944.7 kilotonnes (2007) compared to the production of earlier days, i.e., 354.4 kilotonnes (1961) from seed populations of unknown sources.

Tocklai has collected valuable tea germplasms from wild sources in remote forests and from the old seed populations under cultivation in the tea estates, by selection. More than 2,200 accessions having important traits have been preserved in the gene bank of the station at Tocklai.

Tocklai has developed 14 biclonal seed stocks for the tea industry through breeding, 4 of them specifically for the high altitude areas. All the seed cultivars are accepted by the tea industry and extensively used in the plantations. Many clonal cultivars have been developed, selecting the parents from seed *jat* populations of old plantations of the commercial tea gardens and progeny populations developed through hybridization in the breeding program. So far 31 TV (Tocklai Vegetative) series clones and 153 location specific Garden Series clones have been released to the tea industry for commercial exploitation. In the present scenario, these newly developed cultivars are successively replacing the old and unproductive plantations of the tea gardens. So far, 60% of the total tea growing areas of North East India are covered by planting materials released by Tocklai.

The clonal cultivars are released under 3 categories viz. Standard (both quality of made tea and productivity is above average); Quality (made tea quality very high and productivity above average); Yield (productivity very high with above average quality) with indications of resistance/tolerance to biotic and abiotic stresses, to facilitate the estates in selecting cultivars of their choice.

The UPASI Tea Research Institute has collected and maintained 1,250 accessions of tea germplasm from various sources (Satyanarayana & Sharma, 1991). Out of these genetic resources, UPASI has developed 31 clones (UPASI 1-28, TRF 1-3) having high yield and quality potential and released as cultivars for commercial cultivation by the tea industry. Six biclonal seed stocks (UPASI BSS 1-5, TTSS-1) were also developed using selected parents from the collected genetic resources and released for the tea industry as a potential seed source (Satyanarayana & Sharma, 1991). The germplasm site is recognized as the 41st National Active Germplasm Site (NAGS) by the National Bureau of Plant Genetic Resources.

3.6 Conclusions

Consumers taste perceptions are ever changing and the needs are multidimensional. It is the planting material which is most important in developing specific product categories to satisfy consumers' needs. Quality in recent times has emerged as an important dimension. Quality has many attributes and to have precursors for all the quality attributes is a stupendous task. Breeders in India have great challenges ahead in planting material which will satisfy a cross section of consumers with diverse taste perceptions. The impact of biotic and abiotic stresses has been ever increasing in view of the global impact on climate. Planting materials need to combat stress. Momentum on breeding for stress tolerant planting material needs to be increased. The cost of input has gone up and planting materials with a low input response continues to be a demand for the tea industry in India.

Considering the requirements and preferences as traits, the selection criteria discussed in chapter 3.3, were developed after a series of systematic correlation studies and using tea bushes selected from heterogeneous populations. After thorough study of the selected genotypes at various steps of experimentation under nursery and field conditions, they were released as cultivars for commercial exploitation. Although this method of selection is found to be effective, the disadvantage of the long time involved has become a subject of concern for breeders. A reduction in the selection time has been attempted by the breeders in North East India for some time. Another disadvantage of this conventional selection technique is limited choice from a large selected population and warrants more accurate and focused selection techniques.

In tea, the classical breeding technologies applied in the development of hybrid progenies sometimes fail to bring some important trait related characteristics to the progenies from the parents, due to the crossing barrier. Similarly, elimination of some unwanted characteristics also stands as a hurdle in the presently utilized hybridization methods for the same reason. In tea, the characteristics for producing high yield are negatively correlated to high quality characteristics. Tea plants with genetic potential for the expression of some desired characteristics are sometimes impossible to locate in the existing source of germplasm or they are not even available in nature. Such characteristics are often seen expressed in many other plant species which are incompatible for crossing with tea, under conventional breeding methods. The recent advances in biotechnological research have opened avenues to overcome such problems. Molecular marker technologies have been developed to pinpoint selection of potential genotypes where the inherent characteristics will be expressed. For efficient selection and reduction of the assessment period, efforts have been made to develop trait specific markers so that scheme like 'marker-assisted selection' (MAS) can be adopted. Genetic modification with the help of biotechnological tools is another approach being initiated to overcome the hurdle of crossing the barrier in the development of genotypes with the combination of desired traits.

Our future aim in the tea improvement program through conventional breeding and with non-conventional molecular methods is to develop elite cultivars which can be grown suitably in the tea growing areas under diverse agro-climatic regions. Tea produced by this country must fulfill the requirements of consumers with diverse taste perceptions. It is also important that we do not lose focus of the fact that quality cannot be compromised beyond a certain point.

Acknowledgements

The data for weather conditions, germplasm and cultivars of Southern India were kindly provided by Dr. P. Mohan Kumar, the Director of UPASI Tea Research Foundation.

References

Annual Scientific Report, Tocklai Experimental Station 1966-1967, 1968-1969, 1969-1970, 1973-1974, 1974-1975, 1976-1977, 1978-1979, 1979-1980, 1989-1990, 1990-1991, 1991-1992, 1992-1993, 1993-1994, 1994-1995, 1995-1996, 1997-1998, 2007-2008.

Ahmed N, Singh ID (1993) A technique for rapid identification of ploidy levels in tea. Two and A Bud, 40(2): 31-33.

Amma S (1975) Selection of high yielding clones of tea. Japan Agricultural Research Quarterly, 8: 214-218.

Barua DN (1961) The significance of seed-size in cultivated tea (*Camellia sinensis* L.). Empire Journal of Experimental Agriculture, 29: 143-52.

Barua DN (1963a) Use and abuse of clones. Two and A Bud, 10(1): 3-6.

Barua DN (1963b) Characteristics of the Tocklai released clones. Two and A Bud, 10(1): 26-28.

Barua DN (1989) Science and practice in tea culture. Tea Research Association, Calcutta.

Barua DN, Dutta KN (1971) Distribution of shoots on the plucking surface of a tea bush and its relation to spacing: Part I. Two and A Bud, 18: 8-11.

Barua PK (1965) Classification of the tea plant species hybrids. Two and A Bud, 12: 13-27.

Bezbaruah HP (1967a) Cytogenetics and breeding of tea. Proceedings of 23rd Tocklai Conference.

Bezbaruah HP (1967b) Some thoughts on tea breeding at Tocklai. Two and A Bud, 14(1): 1-5.

Bezbaruah HP (1968) Genetic improvement of tea in N. E. India—its problems and possibilities. Indian Journal of Genetics, 28A: 126-134.

Bezbaruah HP (1969) Economics of a selection scheme. Proceedings of 24th

Tocklai Conference.

Bezbaruah HP (1971a) Preserve the valuable tea plants before they become extinct. Two and A Bud, 18(2): 20-21.

Bezbaruah HP (1971b) Cytological studies on *Thea* and related *Camellia*. PhD Thesis, Gauhati University, India.

Bezbaruah HP (1971c) Cytological investigations in the family Theaceae I. Chromosome number in some *Camellia* species and allied genera. Caryologia, 24(4): 421-426.

Bezbaruah HP (1974) Tea breeding—a review. Indian Journal of Genetics, 34A(S): 90-100.

Bezbaruah HP (1975) Tea breeding—a review. Two and A Bud, 22(2): 123-131.

Bezbaruah HP, Gogoi SC (1972) An interspecific hybrid between tea (*Camellia sinensis* L.) and *C. japonica*. Proceedings of Indian Academy of Science, 76: 219-220.

Bezbaruah HP, Dutta AC (1977) Tea germplasm collection at Tocklai Experimental Station. Two and A Bud, 24: 22-30.

Bezbaruah HP, Singh ID (1978) Current status of tea germplasm in India. Proceedings of National Symposium: Plant and Animal Genetic Resources. 28-30 December, Indian Agricultural Research Institute, New Delhi.

Borchetia S, Das SC, Handique PJ, Das S (2009) High multiplication frequency and genetic stability in the three varieties of tea plants (*Camellia* spp.) through RAPD and microsatellite markers. Scientia Horticulturae, 120: 544-550.

China Tea Varieties Compilation Committee (2001) China Tea Varieties. Shanghai: Shanghai Scientific and Technical Publishers, p.9 (in Chinese).

Cohen Stuart CP (1929) Researches on the leaf-yielding capacity of tea plants. Arch Theecult Ned Ind, 3: 245-321.

Cohen Stuart CP (1930) Researches on leaf-yielding capacity of tea plants. Arch Theecult Ned Ind, 3: 276-288.

Das SC (1984) Rate of leaf unfolding in plucked and unplucked clonal tea bushes of North East India. Two and A Bud, 31: 7-11.

Das SC, Barman TS (1988a) Current state and future potential of tissue culture in tea. Proceedings of 30th Tocklai Conference, pp.90-94.

Das SC, Barman TS (1988b) Tea shoots regeneration from embryo callus. In: Subba Rao NS, Balagopalan C, Ramakrishna SV (eds.) New Trends in Biotechnology, 3-4 June, 1988, Trivandrum, New Delhi: Oxford & IBH Co. Pvt. Ltd.

Das SC, Barman TS, Singh R (1990) Tissue culture of tea: Plant regeneration and establishment in the nursery. Proceedings of International Conference on Tea Research: Global Perspective. 11-12 January, 1990, Tea Research Association, Calcutta, pp.19-22.

Das SC, Bordoloi SC, Bordoloi RK, Sarma AK, Dutta RK, Dutta PK, Ubhadia IB, Goswami BK, Barua DC (2005) Cleft grafting as tool for tea improvement. In: Proceedings of the 34th Tocklai Conference on "Strategies" for Quality 28-30 November, 2005, Tocklai Experimental Station, Tea Research Association, Jorhat, pp.222-232.

Food and Agriculture Organization (FAO) (2006 – 2008) Http://faostat.fao.org/.
Gogoi MN (1993) Black tea manufacture—The Principle and the Process Engineering Aspects. In: Lecture Course on Manufacturing of Tea, August, 1993. Tocklai Experimental Station, Tea Research Association, Jorhat, pp.40-44.
Green MJ (1971) An evaluation of some criteria used in selecting large-yielding tea clones. The Journal of Agricultural Science, 76: 143-56.
Grice WJ (1969) A second progress report on the yield of tea clones receiving two levels of nitrogen at maturity. Two and A Bud, 31: 46-48.
Harler CR (1964) The Culture and Marketing of Tea. (3rd Edition). London: Oxford University Press.
Hazarika M (2008) Uniqueness of Assam Tea. In: Jain NK, Rahman F, Peter Baker (eds.) Economic Crisis in Tea Industry. LLC, USA: Studium Press, pp.265-273.
Hazarika M, Chakravorty SK, Mahanta PK (1984) Studies on thearubigin pigments in black tea manufacturing systems. Journal of the Science of Food and Agriculture, 35: 1208-1218.
Hazarika M, Goswami MR, Tamuly P, Sabhapandit S, Barua S, Gogoi MN (2002) Quality management in tea: Biochemists review. Two and A Bud, 49: 3-8.
International Tea Committee (ITC) (2009) Annual Bulletin of Statistics, London.
Jain JC, Ullah MR, Devchoudhury MN (1979) L-phenylalanine ammonia lyase activity. Two and A Bud, 26(2): 67.
Jain SM, Das SC, Barman TS (1991) Induction ofroots from regenerated shoots of tea (Camelliasinensis L.). In: Mascherpa JM, Moncousin CH(eds.) InternationalSymposium on Plant Biotechnology and its Contribution to Plant Development,Multiplication and Improvement.Geneva, Switzerland. ActaHorticulturae, 289: 339-340.
Kitamura S (1950) Acta Phytotax and Geobot. Kyoto, 14: 56.
Konwar BK (1999) Biodiversity of tea in North East India and its conservation at Tocklai. Two and A Bud, 46(2): 7-12.
Konwar BK, Bordoloi BJ, Dutta RK, Das SC (1999) Rooting of *in vitro* shoots and field establishment of tissue culture-derived tea plants. Two and A Bud, 6(2): 26-32.
Lloyd GB, McCown BH (1980) Commercially-feasible micropropagation of mountain laurel, *Kalmia latifolia*, by use of shoot-tip culture. Proceedings of International Plant Propagator's Society, 30: 421-427.
Mamedov MA (1961) Tea selection in Azerbaidzan. Agrobiologia, 1: 62-67.
Masters JW (1844) The Assam tea plant compared with the tea plant of China. Journal Agriculture and Horticulture Society of India, 3: 61.
Murashige T, Skoog F (1962) A revised medium for rapid growth and bioassays with tobacco cultures. Physiologia Plantarum, 15: 473-497.
Roberts EAH (1962) Assessment of quality in teas by chemical analysis. Two and A Bud, 9(3): 3.
Sanderson GW (1963) On the nature of the enzyme catechol oxidase in tea. Tea Quarterly, 36: 103-11.

Sarmah PC, Bezbaruah HP (1984) Triploid breeding in tea. Two and A Bud, 31(2): 55-59.

Satyanarayana N, Sarma VS (1982) Biometric basis for yield prediction-tea clonal selection. In: Proceedings of PLACROSYM-IV. Indian Society of Plantation Crops, pp.237-243.

Satyanarayana N, Sharma VS (1986) Tea (*Camellia* L. spp.) germplasm in South India. In: Srivastava HC (eds.) Plantation crops. Vol. II. New Delhi: Oxford and IBH Publishing Co. Pvt. Ltd., pp.173-179.

Satyanarayana N, Sharma VS (1991) Tea plant improvement in South India. United Planters' Association of Southern India, Scientific Department Bulletin, 44: 69-70.

Satyanarayana N, Sharma VS (1993) An overview of tea plant improvement in South India. In: Tea Culture, Processing and Marketing, pp.36-44.

Schenk RU, Hildebrandt AC (1972) Medium and techniques for induction and growth of monocotyledonous and dicotyledonous plant cell cultures. Canadian Journal of Botany, 50: 199-204.

Sharma VS (1976) V.P. tea nursery. Planters' Chronicle, 71: 211-213.

Singh ID (1979) Indian tea germplasm and its contribution to the world tea industry. Two and A Bud, 26: 23-26.

Singh ID (1980a) Tea germplasm in India. Plant Genetic Resources News. FAO/IBPGR, Rome, 43: 12-16.

Singh ID (1980b) Non-conventional approaches to the breeding of tea in North-East India. Two and A Bud, 27: 3-6.

Singh ID (1984) Advances in tea breeding in North East India. In: Iyer RD (eds.) Proceedings of PLACROSYM-V. 15-18 December, 1982, Indian Society of Plantation Crops, CPCRI, Kasargod, pp.88-106.

Singh ID, Bezbaruah HP (1978) Collection of tea genetic resources-problems and procedures. Proceedings of National Symposium on Plant and Animal Genetic Resources, 28-30 December, Indian Agricultural Research Institute, New Delhi.

Singh ID, Sharma PC (1982) Studies on radiation breeding in the tea plant. In: Vishveshwaras (eds.) PLACROSYM-IV. Proceedings of 4th Annual Symposium on Plantation Crops: Genetics, Plant Breeding and Horticulture. 3-5 December, 1981, Mysore, pp.295-301.

Singh ID, Sharma PC, Bezbaruah HP (1982) Breeding tea polyploids. In: Vishveshwara S (eds.) PLACROSYM-IV. Proceedings of 4th Annual Symposium on Plantation Crops: Genetics, Plant Breeding and Horticulture. 3-5 December, 1981, Mysore, pp.74-78.

Toyao T (1966) Studies on Koro tea I. The inheritance of Koro type characters and the estimation of the degree of selfing of the tea plant. Study of Tea, 32: 18-22.

Toyao T, Katsuo K, Kayumi S, Matsushita S, Amma, S (1971). Improvements in the methods of early selection of black tea quality. Bulletin of Tea Research Station, Ministry of Agriculture and Forestry, Japan, (7): 1-55.

Tunstall AC (1931a) A note on the propagation of tea by green shoots cuttings.

Quarterly Journal of Indian Tea Association, pp.49-51.
Tunstall AC (1931b) Observations on the yield of individual bushes. Report of Indian Tea Association for 1930, pp.101-112.
Ukers WH (1935) All about Tea. Vol. I. New York: Tea and Coffee Trade Journal Co.
Venkataramani KS (1963) The principles of tea clonal selection and propagation and some practical considerations in clonal planting. United Planters' Association of South Indian, Scientific Department Bulletin, 22: 2-14.
Venkataramani KS, Padmanabhan TS (1964) A preliminary assessment of the relationship between certain leaf characteristics and cup quality. Annual Report United Planters Association of Southern India, Scientific Department, Tea Section, 1963-64: 50-63.
Visser T (1969) Tea. In: Ferwewerda FP, Wit F (eds.) Outlines of Perennial Crop Breeding in the Tropics. Wageningen: H. Veenman & Zonen, pp.459-493.
Visser T, Kehl FH (1958) Selection and vegetative propagation of tea. Tea Quarterly, 29: 76-86.
White PR (1963) The cultivation of animal and plant cells (2nd Edition). New York: Ronald Press, p.228.
Wight W (1939) Annual report of Indian Tea Association. Scientific Department, Tocklai.
Wight W (1955) Tea breeding in India. Two and A Bud, 2(3): 9-11.
Wight W (1956) Commercial selection and breeding of tea in India. World Crop, 8: 263-268.
Wight W (1958a) Selection policy in tea estates of North East India. Report of Tocklai Experimental Station, 1957.
Wight W (1958b) The agrotype concept in tea taxonomy. Nature, 181: 893-895.
Wight W (1961) Combiners for tea breeding. Two and A Bud, 8(3): 19-21.
Wight W (1962) Tea classification revised. Current Science, 31: 298-299.
Wight W, Barua PK (1954) Morphological basis of quality in tea. Nature, 173: 630-631.
Wight W, Barua PK (1957) What is tea? Nature, 179: 506-507.
Wight W, Gilchrist RCJH (1961) The concept of kind of tea. Nature, 161: 14-16.
Wight W, Gilchrist RCJH, Wight J (1963) Note on color and quality of a tea leaf. Empire Journal of Experimental Agriculture, 31(122): 124-126.

4

Tea Plant (*Camellia sinensis*) Breeding in Sri Lanka

M. Trixie K. Gunasekare

Plant Breeding Division, Tea Research Institute of Sri Lanka, Talawakelle, Sri Lanka
E-mail: kgunasekare@yahoo.co.uk

Abstract: Tea is one of the main foreign exchange earners of Sri Lanka. Sri Lanka ranks as one of the largest exporters of black tea in the world. Sri Lanka is well renowned for its high quality orthodox type (95% of the total production) black tea in the international market. The name "Ceylon tea" or "Sri Lankan tea" has been regarded as a sign of high quality throughout the world for a long time and even today, as reflected by the prices. In order to maintain the position of "Ceylon tea" in the world and to serve the local tea industry, it is imperative to focus on efficient tea crop improvement strategies to develop grower acceptable tea cultivars. Commencement of the tea crop improvement program at the Tea Research Institute of Sri Lanka dates back to the 19th century. Since then many significant achievements have been made in developing new tea cultivars, with salient milestones over the years. Growers have benefited immensely by using the improved tea cultivars developed by the Institute, though certain improvements have yet to be accomplished using the modifications made in the current breeding program.This chapter highlights significant achievements in the areas of germplasm collection, characterization and evaluation, and their use in the tea breeding program, with prominence given to cost effective complementary strategies adopted in germplasm conservation and the holistic approach adopted in germplasm characterization. Tea breeding strategies, priorities and the significant contribution made to the industry by developing improved tea cultivars over the years are discussed, emphasizing the diverse needs of the growers in multi-regional tea growing areas in the country. The chapter also focuses on the application of biotechnological tools for breeding and crop improvement, highlighting the recent advances made in the development of new technologies and their practical applications in facilitating conventional tea crop improvement programs. Overall, the chapter reviews the achievements, challenges and perspectives of tea breeding in Sri Lanka and overviews the future trends, aims and goals of breeding improved cultivars acceptable to the growers. There is a focus on producing improved tea cultivars that thrive better under changing environments to meet the ever-changing demands of the end-users. Future perspectives on the application of new technologies to address key challenges faced by the industry locally and globally, by focusing on the integration of advanced biotechnological tools and decentralized participatory approaches to the current tea breeding program are also discussed.

4.1 General Introduction to the Sri Lanka Tea Industry

The tea plant (*Camellia sinensis* (L.) O. Kuntze), is a woody-perennial plant of which the tender shoots are used to make the end product. Tea is a popular healthy beverage worldwide and ranks next to water. The Asian countries, mainly China, India and Sri Lanka generate more than half of the world tea production. It is an important revenue source for tea producing countries both in terms of earning foreign exchange and generating employment in the community.

4.1.1 Tea Growing Areas in Sri Lanka

Sri Lanka is an island located in the Indian Ocean off the southeast tip of India and separated from the Indian peninsular by the Palk Strait. It is located between latitudes 5°55′ and 9°51′ N and longitudes 79°41′ and 81°53′ E and covers a total extent of 65,610 km^2.

In Sri Lanka, tea growing areas are widely distributed over a range of agro-climatic regions. Tea grows from almost sea level to about 1,800 m above mean sea level (amsl). Depending on the elevation, tea growing areas are broadly divided into three regions; low country (below 600 m), mid country (600 to 1,200 m) and up country (above 1,200 m). Tea growing areas are mainly concentrated in the central highland and southern inland areas of the island. The terrain and topography of tea fields in the up country (Fig. 4.1) and the low country (Fig. 4.2) are quite different.

Fig. 4.1. A tea plantation in the up country region

Fig. 4.2. A tea plantation in the low country region

The total extent of land under tea cultivation at present is approximately 221,969 ha. The total area covered by tea in the major tea growing regions in Sri Lanka and their total productions are given in Table 4.1.

Table 4.1 Extent planted with tea in major tea growing regions and their production levels

Region	Area (ha)*	Production (kilotonnes made tea/year)**
Up country	41,137	84.2
Mid country	71,018	49.0
Low country	109,814	185.5
Total	221,969	318.7

Data based on *Census of Agriculture, 2002; ** Statistical Information on Plantation Crops, 2008

4.1.2 Historical Introduction to Sri Lanka Tea

Tea was first introduced to Sri Lanka (then Ceylon) from Kolkata Botanical Garden in India by Dr. Robert Bruce. The seeds brought were planted in the Royal Botanical Garden in Peradeniya in 1824. We also tend to believe to a certain extend that first tea plant to Sri Lanka was from China. Because, most of the old seedling teas that still exist are mainly the var. *sinensis* type (small, dark green leaf type). However, the first commercial plantation of tea was established by a Scottish planter, James Taylor, in 1867 at Loolecondera Estate, Hewaheta, which is located in the mid country region. At the time, coffee was the mainstay of the country's economy and was facing a crisis owing to coffee rust disease caused by the fungus *Hemileia vastatrix.* As a result, most of the coffee plantations were replaced with tea within a short period and tea has been grown with success since then.

By 1875, an extent of nearly 4,000 ha was under tea cultivation which showed a steady increase over the years. Since the inception of the tea industry until the 1950s, tea plants were raised from seed, mainly the var. *sinensis* type. The advent of commercially viable vegetative propagation techniques coupled with the introduction of the Tea Replanting Subsidy Scheme by the government in 1958 encouraged planting of clonal tea cultivars. Low production seedling fields have been replaced with high yielding clonal cultivars quite rapidly. At present, over 45% of the total tea acreage consists of seedlings of mixed types and 55% are clonal cultivars (Wijeratne, 2004). Many of the seedling teas are more than 80 to100 years old and some tea fields are no longer considered agriculturally productive. Those fields are being earmarked for replanting with new tea cultivars aiming at high productivity.

4.1.3 Climate, Soil Conditions and Productivity Levels of Tea Growing Area

Sri Lanka, being situated near the equator, does not have marked seasonal differences. Because of its location, tea grown in Sri Lanka can be harvested all year round. However, the geography of the island perpetuates its climate which, in turn, has a direct bearing on the tea productivity levels in the different tea growing regions of the country.

Topography plays a major role in determining the rainfall distribution over the island. The major rainfall zones in Sri Lanka are the wet (W), intermediate (I) and dry (D) zones. However, tea can only be grown in the wet zone and certain regions of the intermediate zone. Those two zones can be further divided into several agro-ecological regions based on their elevation, Low Country (L), Mid Country (M) and Up Country (U). Taking these two parameters (rainfall and the elevation) into consideration, the tea growing agro-ecological regions have been demarcated. Accordingly, there are 14 agro-ecological regions where tea can be grown successfully (Fig. 4.3) (Tea Research Institute of Sri Lanka, 2008). The soils of the agro-ecological regions suitable for tea cultivation in Sri Lanka fall into 3 major soil groups which are further sub-divided into soil series. Climatic and soil factors of the agro-ecological regions and principal planting districts falling under each agro-ecological region are given in Table 4.2. The influence of climatic conditions is partly responsible for the diversity of the characteristics present in tea produced from different tea growing regions in Sri Lanka.

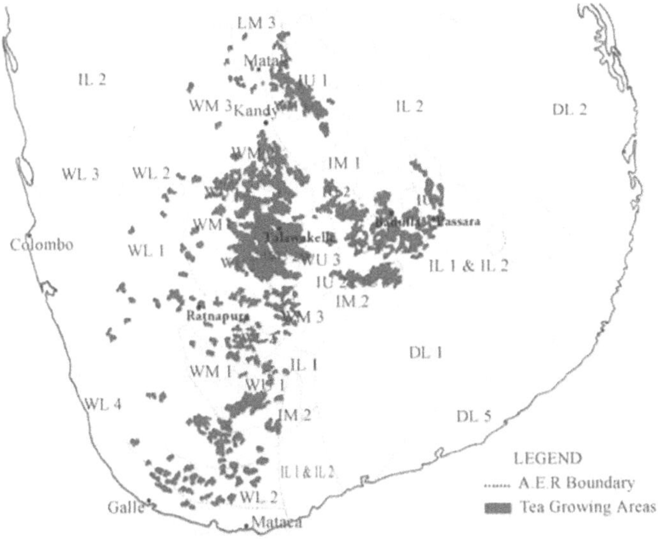

Fig. 4.3. Tea growing agro-ecological regions in Sri Lanka

Table 4.2 Climatic and soil factors of the tea growing agro-ecological regions (AER) and major planting districts in Sri Lanka

[1]AER*	[2]Annual rainfall (mm)	[3]Major soil group	[4]Soil series**	Planting Districts
WU 1	> 3,100	RYP; RBL	Maskeliya, Kandy	Hatton
WU 2	> 2,200	RYP; RBL	Kandy, Mattakelle, Maskeliya	Pussellawa, Pundaluoya, Dimbula, Dickoya
WU 3	> 1,800	RYP	Nuwara Eliya	Nuwara Eliya
WM 1	> 2,900	RYP	Weddagala, Malaboda	Morawak Korale, Ratnapura
WM 2	> 1,800	RBL	Kandy	Dolosbage
WM 3	> 1,400	RYP; RBL	Kandy, Malaboda	Kandy, Balangoda, Rakwana
WL 1	> 2,800	RYP	Pallegoda, Boralu, Dodangoda, Homagam, Malaboda	Ratnapura, Kalutara, Kelani Vally, Galle
WL 2	> 2,200	RYP	Galigamuwa, Malaboda, Dodangolla, Boralu	Kegalle, Ratnapura, Galle
IU 1	> 2,400	RYP; RBL	Ukuwella, Hunnasgiriya	Knuckels, Rangala, Kellebokka
IU 2	> 2,100	RYP	Ragala	Udapussellawa
IU 3	> 1,600	RYP	Badulla, Bandarawella	Badulla
IM 1	> 1,300	RYP	Badulla	Badulla
IM 2	> 1,600	RBL	Mahawalatenne	Madulsima, Passara
IM 3	> 1,100	RBL; IBL	Kundasale, Rikilagaskada	Kundasale, Hanguranketha

* Rainfall zones: W—Wet zone; I—Intermediate zone; Elevational categories: U—Up country; M—Mid country; L—Low country

** RYP—Red yellow podzolic; RBL—Reddish brown latosolic; IBL—Immature brown latosolic.

References: [1] Panabokke & Kannangara, 1975; [2] Punyawardena *et al*., 2003; [3] De Alwis & Panabokke, 1972; [4] Dissanayaka *et al*., 1999

The country's average yield is around 1,356 kg of made tea/(ha·annum) (Anon, 2008). Agro-ecology productivity levels vary considerably and accurate statistics are not available for productivity levels of the different agro-ecological regions.

4.1.4 Sri Lankan Tea Industry in the International Market

Tea is one of the main foreign exchange earners of Sri Lanka. Sri Lanka ranks as one of the largest exporters of black tea in the world and about 95% of the tea produced is exported. Sri Lanka, as the third largest tea producing country in the world, has a production share of 8.4% and is one of the world's leading tea exporters with a share of around 18% of the exports (ITC, 2009). The total tea production was 318.7 kilotonnes of made tea in 2008 (ITC, 2009). Foreign exchanged earned by exporting tea was over 1.2 billion US dollars in 2008 (ITC, 2009). Tea contributes about 14% of the total foreign exchange earnings, accounting for about 1.2% of the gross domestic product (GDP) annually (Central Bank, 2007). Statistics on tea acreage, production, exports and foreign exchange earnings over the past few decades demonstrate the growth of the tea industry in Sri Lanka (Table 4.3). Furthermore, the tea industry provides approximately 1 million employment opportunities in the country. Thus, the competitiveness of the Sri Lankan tea industry in the global tea market is crucial for the economy of the country

Sri Lanka is well renowned for its high quality orthodox type black tea in the international market. Major buyers of Sri Lankan tea are Russia, other former Commonwealth of Independence States (C.I.S.) countries and Middle East countries. Tea produced in Sri Lanka is mainly sold through Colombo tea auctions. The Colombo tea auction prices are the highest in the world. Average auction prices in the 3 years (2006, 2007 & 2008) being 1.89, 2.51 and 2.83 US dollars per kg, respectively (ITC, 2009). The name "Ceylon tea" or "Sri Lankan tea" has been regarded as a sign of quality throughout the world for a long time, as reflected by the prices.

Of the total exports about 60% are exported in bulk, while 40% are in value added forms such as tea packets, tea bags and instant tea. Most of the Sri Lankan tea exporters now focus on adding more value to the exports rather than exporting raw bulk tea.

Table 4.3 Growth of the tea industry in Sri Lanka over the last few decades

Year	1970	1975	1980	1985	1990	1995	2000	2005	2008	2009
Acreage (ha)	241,799	241,877	244,715	231,650	221,758	—	188,971	188,480	188,323	188,175
Production (kilotonnes)	212.2	213.7	191.4	215.3	234.1	246.4	306.8	317.2	318.7	289.8
Exports (kilotonnes)	208.3	212.4	184.5	197.6	215.3	235.1	280.1	298.8	317.3	279.8
Foreign exchange earnings (million US$)	187.5	273.0	371.9	438.7	492.4	462.6	662.3	769.4	1270.1	1145.1

Source: Annual Bulletin of Statistics, ITC

4.1.5 Types of Tea Produced

Tea produced in Sri Lanka is mainly of the orthodox type, accounting for about 95% of the total production and the balance is CTC type. The black tea produced is categorized into high grown, medium grown and low grown teas. Tea grown at each elevation has its own distinctive characteristics in appearance, flavor, aroma and color. High grown teas from Sri Lanka are reputed for their taste and aroma giving golden liquor and an intense powerful flavor. The medium grown teas are rich in flavor and give a good color. The teas produced in low grown areas are mainly of leafy grades and have good color and strength.

4.1.5.1 Regional Teas

In Sri Lanka, significant climatic variation can be seen within a short distance. The prevailing climate and soil factors in different tea growing regions, together with the predominant cultivars in that area and the processing methods used, influence the type of tea produced in different areas which is known as "Regional Tea".

***Dimbula Tea*:** Tea produced in the agro-ecological region WU 2 with an elevation of 800 to 1,400 m above mean sea level and having red yellow podzolic and reddish brown latosolic soils and a predominant cultivar combination of *C. sinensis* var. *sinensis* (L.) O. Kuntze hybrids and *C. sinensis* var. *assamica* (Masters) Kitamura hybrids (TRI 2025, TRI 2023, N 2, K 145, DT 1, DN) has the finest characteristics and is rich in color. The orthodox/rotavane type tea produced in this region has small particles and hence is suitable for selling as bulk tea as well as in tea bags. Major destinations for this tea are the UK, Europe and Japan.

***Nuwara Eliya Tea*:** This tea has an exquisite flavor and aroma with a comparatively light liquor (brew). It is produced from cultivars of var. *sinensis* hybrids (PK 2, K 145) grown in agro-ecological region WU 3 with its red-yellow podzolic soils. This tea is produced using orthodox rotavane and has small particles. The main markets for this tea are Japan and Germany.

***Uva Tea*:** This tea is produced in the agro-ecological region IM 2 from old seedling tea of mixed *jats*. The majority cultivars are var. *assamica* hybrid (TRI 2025, TRI 2024, NAY 3, TRI 2023) and var. *sinensis* hybrids (CY 9, DN). This tea is grown on red yellow podzolic soils and has quite a remarkable flavor. The main markets for this tea are Japan and Germany.

***Udapussellawa Tea*:** This tea is produced in the agro-ecological region IU 2, with an elevation ranging from 800 to 1,400 m, from old seedling tea and a combination of var. *assamica* and var. *sinensis* hybrids (TRI 2025, CY 9, KEN 16/3, TRI 2023 and DN). Tea produced from this area has a medium body and a rosy taste and the main destination is Europe.

***Ruhuna Tea*:** This tea is produced through orthodox processing methods and is a large leafy grade and is therefore only suitable for selling as bulk tea. This tea is produced mainly from cultivars of var. *assamica* hybrids (TRI 2026, TRI 2025,

TRI 2022, H 1/58, S 106) in the agro-ecological region WL 1 with an elevation ranging from 30 to 60 m and red-yellow podzolic soils. This tea has strong liquor/brew characteristics. It is much sought after particularly by countries in the Middle East and by Russia.

***Kandy Tea*:** This tea is produced in the agro-ecological regions WM 2 and WM3 (elevation 600 to 1,200 m) with red-yellow podzolic and reddish brown latosolic soil. It is produced from a combination of cultivars of var. *assamica* hybrid and var. *sinensis* hybrid characteristics (TRI 2025, TRI 2023, TRI 2026, DG 39, DG 7). This tea has a good color brew and medium flavor. Australia and New Zealand are the main consumer countries for this tea.

4.1.5.2 Seasonal Teas

Apart from the regional teas, there are two types of "seasonal teas" produced in the country in the regions of "Uva" and "Dimbula". Tea produced during the dry season in these two areas is world renowned for its exquisite flavor and aroma. Climatic factors such as hot days, cold nights and desiccating winds cause a stress on the tea bushes which in turn increases the volatile compounds in the tea leaves and which contributes to the special characteristics in those teas. There is a high demand for these specialty teas in Japan, Germany and UK.

4.1.5.3 Specialty Teas

Very small quantities of a special kind of tea known as "silver tips" (Fig. 4.4) are produced mainly in the low grown areas. This tea produces a very pale, straw-colored liquor. This unique type of tea fetches premium prices in the Middle East market. This particular type of tea is produced mainly from the cultivar TRI 2043 (Fig. 4.5), which is a *lasiocalyx* hybrid with dense pubescence in the unfurled bud and anthocyanin pigmentation in immature leaves.

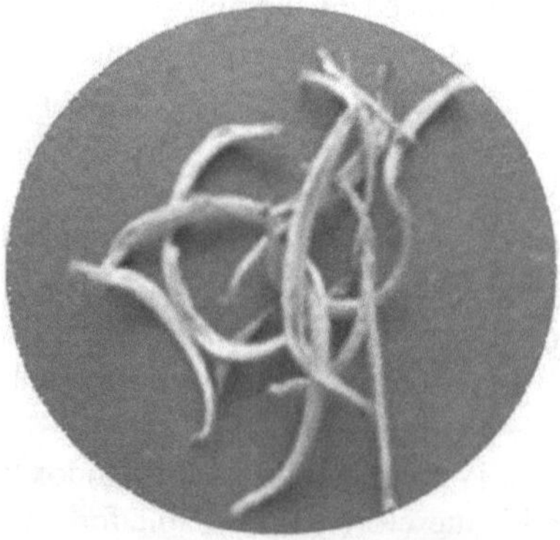

Fig. 4.4. Silver tips made from cultivar TRI 2043

Fig. 4.5. A bush of cultivar TRI 2043

Because of the diversity of tea produced in Sri Lanka, there is an advantage in catering to different markets around the world. Hence the heterogeneity of Sri Lankan teas, which are promoted through their inherent qualities, has contributed immensely to maintaining the position of Ceylon tea in the world.

Sri Lanka at present is the largest exporter of bulk tea. However, the size of value added exports, such as in the form of tea packets, tea bags and instant tea (Anon, 2008), is rising steadily (Fig. 4.6). The amount of green tea produced in the country is considered negligible as compared with the total production of black tea.

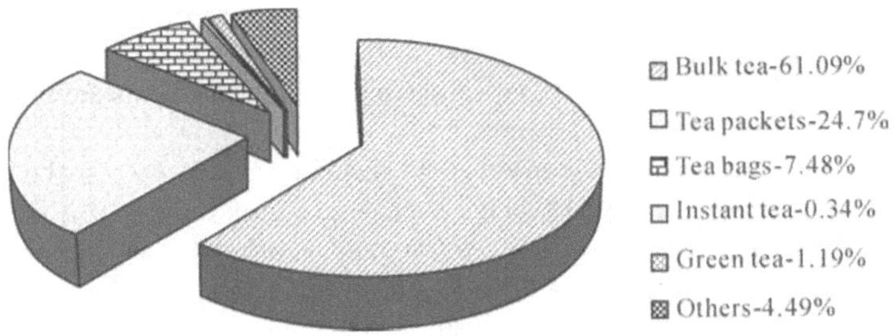

Fig. 4.6. Percentage of different types of tea, including value added tea exported from Sri

4.2 Tea Germplasm Collection, Conservation and Appraisal

A germplasm is an array of genetic resources that serves as the plant breeders' raw material in their endeavors to identify or develop new cultivars suitable for end-users current and future needs. Thus the genetic variation acquired in the germplasm collection is a vital aspect of any breeding and crop improvement program. It is therefore important to capture and conserve all possible variations present in the species for utilization in the breeding program to reach the desired objectives.

4.2.1 Origin of Planting Materials Used in the Country

Tea is a crop which was introduced to Sri Lanka. Hence, there are no wild relatives or landraces of tea present in the country. Imported seeds were the only source of planting material available during the early years of tea cultivation in Sri Lanka (Tubbs, 1939). Seeds were imported mainly from India and China and hence seedling populations were of mixed types containing *jats* of different sizes. Those seedlings were therefore highly heterogeneous.

Although clonal selection started in Sri Lanka as early as 1937, until the mid-fifties the preferred planting material on commercial tea plantations was the seed. It was only in 1955 that use of clonal high yielding cultivars was recommended for planting on estates (Richards, 1967).

4.2.2 Collection and Conservation Efforts

Conservation aspects of tea germplasm in an *ex-situ* gene bank as a living collection was initiated in 1986. Currently, about 500 accessions are being maintained in a field gene bank at the Tea Research Institute of Sri Lanka (TRISL). Accessions are maintained in gene banks at the main station at Talawakelle and regional stations at Ratnapura, Passara and Galle. Some accessions are maintained as duplicates in more than one location.

The tea genetic resources conserved in the field gene bank can be broadly categorized into two groups, beverage type and non-beverage type (Fig. 4.7). Beverage types comprise all tea accessions and can be further sub-divided as "Introductions", "Estate Selections" and "Improved Cultivars" primarily based on available pedigree data and breeding history (Ranathunga & Gunasekare, 2008b). "Estate Selections" are the ones that have been selected from old seedling populations on tea estates from various agro-ecological regions in the country. Through the long term estate cultivar selection scheme, adopted by the institute since 1905, a total of 688 estate selections have originated from old seedling tea populations (Gunasekare & Kumara, 2005). Of those, 45% of the accessions have already been secured by conserving them in field gene banks at TRISL at 4 different locations.

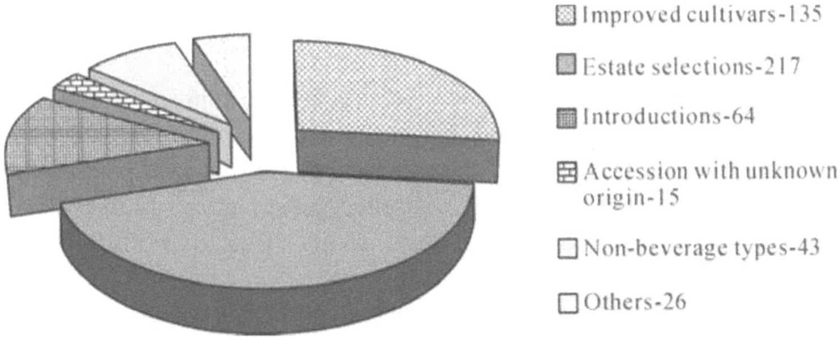

Fig. 4.7. Germplasm collection conserved at field gene bank at TRISL

The improved cultivars included in the collection were derived from selections made from open pollinated or controlled pollinated seed progenies. Non-beverage type and related species conserved in the gene bank include *C. sasanqua* Thunb., *C. japonica* L., *C. rosaeflora* Hooker and *C. lutenscens* Dyer, which show promise in interspecific hybridisation programs. The germplasm collection is, however, under-represented by exotic types and allied species of tea. Nevertheless, acquisition of diverse and elite germplasm from exotic sources is critical in this regard. Therefore, the available germplasm is the only source for the crop improvement program in the near future.

Accessions in the gene bank include both adapted and unadapted germplasm, as well as advanced breeding lines. The systematic breeding activities undertaken in keeping with varied objectives have resulted in improved cultivars. Obsolete varieties were also conserved in the gene bank. In addition, some of the new elite germplasm generated through breeding, but which did not enter the final stage of evaluation for release as potential cultivars, has also been conserved as pre-breeding material for future use.

4.2.2.1 Special Type of Germplasm

Though many cultivars developed and selected at the TRISL are diploids ($2n = 2x = 30$), there are some natural triploid cultivars which were selected from seedling tea populations existing on estates (Anandappa, 1973). They are HS 10A and GF 5 and are available in the germplasm collection at the main gene bank (Gunasekare & Ranathunga, 2003). Moreover, TRI 3069, which is a colchicine induced tetraploid cultivar recommended for commercial planting in the low country, is also one of the special kind of germplasms available in the collection. A non-fermenting tea cultivar, TRI 9, which is found to be deficient in enzymatic copper (Ramaswamy, 1960) is a unique type of local germplasm available in Sri Lanka. TRI 2043, which is an exclusive germplasm having properties for producing "silver tips" is also another special type of germplasm found only in Sri Lanka.

4.2.2.2 Sustainable Germplasm Conservation Efforts

Sri Lankan tea has very limited genetic diversity, as it is an introduced crop (Singh & Gunasekare, 2000). It was shown that the genetic base of cultivated tea is very narrow due to the recurrent use of the same parents in the tea breeding program over the years (Singh *et al.*, 2003). The limited genetic diversity of tea renders it more vulnerable to stress conditions such as the effect of pests, diseases and drought.

Fortunately, some old seedling tea fields are still retained on some tea estates in the country. Those seedlings possess countless genetic variation. In the face of rapid genetic erosion due to uprooting of old seedling tea for replanting programs, it has become important to collect and preserve them before loosing them for ever (Singh & Gunasekare, 2000). Thus, adoption of appropriate and effective conservation strategies has become one of the important aspects of the breeding program.

Painstaking evaluation and characterization of germplasm to maximize the genetic diversity may be counterproductive if the rate of erosion is so high that the diversity is lost while planning is in progress. Under such circumstances, undertaking a collecting expedition at a more accelerated pace is of prime importance. It has often become necessary to collect germplasm before it can be ascertained that it contains genes of genotypes not already present in the collection. Hence, strategic and complementary approaches have been adopted in conserving the germplasm.

A new strategy has been introduced to conserve old seedling teas that are still in existence on many of the tea estates in Sri Lanka (Gunasekare *et al.*, 2003), before they are uprooted and replanted with more uniform improved cultivars. This approach is similar to *in-situ* or on-farm conservation in which the growing community plays a pivotal role. This also provides the opportunity for selecting potential tea cultivars from existing old seed tea populations which are well adapted for specific locations (Gunasekare, 2007b). As it entails a large number of samples, initial planning should follow the method described by Fu (2000) and Park *et al.* (2002), which involves molecular marker analysis using pooled samples from bushes in each *in-situ* site to identify the site/sites having individuals with high genetic diversity (Gunasekare, 2007a).

4.2.2.3 Biotechnological Approaches in Germplasm Conservation

Tea is propagated vegetatively and therefore conserved as a living collection as whole plants in a field gene bank. However, there are several constraints related to managerial aspects of field gene banks, such as exposure to natural disasters, attack by pests and pathogens and high maintenance costs. Alternatively, *in vitro* preservation of germplasm can be used as a safe and complementary option to field gene banks. Among the cold storage methods, cryopreservation has proven to be a potential method for long term preservation as it requires minimum space, labor and maintenance.

For seeds derived from controlled hybridization programs that have proven valuable, some kind of storage methods are required until they are included in the cultivar evaluation program. Research carried out at TRISL enabled the establishment of a protocol for encapsulation of zygotic embryonic axes using alginic acid (Seran *et al.*, 2005) and efficient plant recovery was obtained after 3 months of storage at low temperature (Seran *et al.*, 2006a). It was shown that there is a potential for using such material for *in vitro* preservation while maintaining the genetic integrity. Research activities are also focused on using somatic embryos for encapsulation to use in cold storage.

4.2.3 Characterization and Evaluation of Germplasm

The analysis and characterization of genetic diversity is fundamental to any germplasm collection for its rational utilization in breeding programs and other related disciplines. Hence, the accurate identification of material held in the gene bank is arguably the most essential part of the germplasm characterization process. With concerted efforts made on germplasm conservation using complementary approaches discussed above, characterization and evaluation of the germplasm are receiving priority in the current tea breeding program.

4.2.3.1 Morphological Characterization

In the past, morphological characteristics have been used mainly for the purpose of cultivar identification (Richards & Sebastiampillai, 1964; Wickramaratne, 1981a). Recently, a more systematic approach was adopted using tea descriptors developed by the International Plant Genetic Resource Institute (IPGRI, 1997) to arrive at an internationally accepted format for germplasm characterization, evaluation and documentation. Initial studies were focused on identifying a minimum list of descriptors or highly discriminating descriptors relevant for rational characterization of local germplasm (Gunasekare *et al.*, 2001; Gunasekare & Pieris, 2006; Piyasundara *et al.*, 2006). Of the 35 vegetative descriptors proposed by the IPGRI, 6 descriptors (viz. type of serration of leaf margin, waviness of the leaf margin, pigmentations in the young leaf, pigmentations in the leaf petiole, size of the leaf and leaf angle) were found adequate in defining the phenotypic variation of the local collection and can be used in distinguishing accessions into phenotypically diverse groups. Slight modifications were also made in some vegetative descriptor scales to better suit the descriptors in characterizing the collection (Piyasundara, 2008). So far, over 200 accessions have been characterized phenotypically and discriminated into distinct groups using standardized descriptors (Piyasundara *et al.*, 2008; Piyasundara & Gunasekare, 2008).

As reproductive traits are known to be less affected by the environment than the vegetative traits, some investigations were conducted to analyze the genetic diversity present in the collection using floral characteristics (Sundaravathany *et al*., 2005), especially the pistil traits (Ariyarathna & Gunasekare, 2008). It was revealed that pistil traits such as the number of arms in the style, length of style arms and length of the style column, represent wide variations in the germplasm accessions. The remarkably high variation in pistil morphology captured in the study also confirmed that the local germplasm collection was a result of extensive inter-hybridization among the 3 major tea taxa (Fig. 4.8).

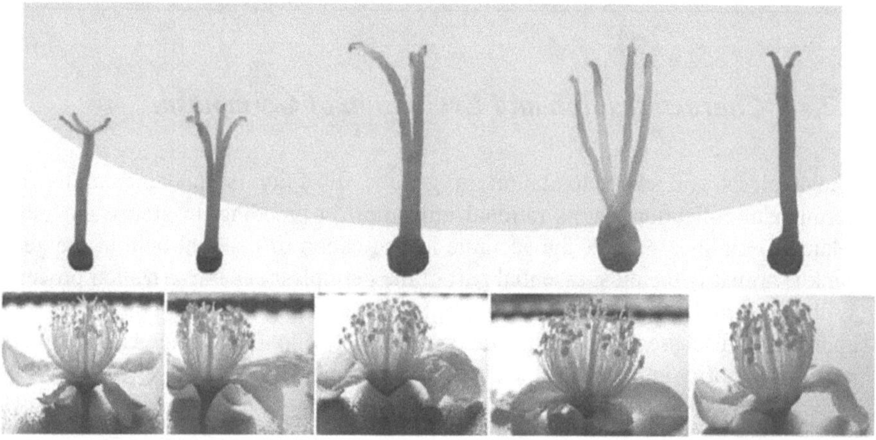

Fig. 4.8. Distinct pistil morphs observed in accessions in germplasm collection at TRISL

4.2.3.2 Isozyme Polymorphism and DNA-Based Characterization

Phenotypic variation is positively associated with genetic diversity, yet it also depends on environmental factors and the interaction between the genotypes and the environment. Hence, estimating genetic diversity using phenotypic markers has several limitations in a perennial crop, like tea, which is also grown across diverse environments. As such, attention was focused on using protein-based techniques. Polymorphism in isozyme variations was limited to very few isozymes with inadequate polymorphism (Liyanage *et al*., 1999). Due to paucity of isozyme loci, efforts have been shifted towards DNA-based marker systems. Work along this line was initiated primarily to measure the genetic diversity of 85 tea accessions, to identify parents for future breeding programs using RAPD markers (Mewan *et al*., 2005; Goonatilake *et al*., 2006). The recommended cultivars and advanced breeding lines exhibited a high degree of molecular similarity and the similarity was ascribed to their common pedigree. A few accessions such as TRI 777, TRI 2016, China and Yabukita, proved genetically diverse from the rest of the accessions studied. Results also revealed that estate cultivars, which were derived from seedling tea populations in the 1960s, were more diverse than the cultivars developed during later breeding programs in the 1980s from open pollinated and controlled hybridized progenies.

4.2.3.3 Genealogical Approaches in genetic Diversity Analysis

Genealogical approaches for measuring genetic diversity combined with statistical methods have not been carried out so far for woody perennials such as tea, locally or internationally. The first report on co-efficient of parentage (COP) analysis in tea, based on pedigree information to measure the genetic diversity among commercial cultivars and their parental lines, revealed that ASM 4/10 and CY 9 were the main ancestral lines that contributed to the cultivated gene pool (Ariyarathna & Gunasekare, 2006). Two major COP groups identified provide the basis for categorizing cultivars into genealogically distinct groups and therefore promote effective utilization of germplasm in future breeding programs. The report further highlighted that there is a substantial amount of genetic potential to be exploited from some of the existing commercial cultivars and unadapted germplasm.

4.2.3.4 Need for Holistic Approach in Germplasm Characterization

Over the years the methods for detecting genetic diversity have expanded from an analysis of discrete morphological variants, to biochemical approaches and the co-efficient of pedigree analysis, to methods based on DNA markers. This kind of approach, where genetic diversity assessment is supplemented with a combination of criteria, provides a sounder base for efficient conservation, and maintenance and utilization of existing genetic diversity in the collection, than an approach restricted to a single criterion.

4.2.3.5 Evaluation of Germplasm for Agronomically Important Traits

Though concerted efforts have been made to characterize and study the genetic diversity of the collection, systematic evaluation of the germplasm for desirable agronomic traits, mainly the unadapted and exotic types, remains an unaccomplished task in the past breeding program (Gunasekare, 2007b). The quest to increase the yield in cultivars developed during the 1960s has overlooked the other valuable traits present in the germplasm. In cases where there is evidence that the pool of available variation for biotic and abiotic stress-resistant genes within the recurrently used adapted lines is limited, search for novel sources of resistance is required. In such a situation, a wide range of accessions, whether they are adapted or unadapted, must be screened to find adequate levels of resistance to major pests and diseases (Gunasekare, 2007b). Furthermore, with restrictions on acquiring exotic genetic resources, evaluation of the available germplasm to maximize its use in the breeding program has been receiving a large amount of attention in the present tea breeding program.

Germplasm is being extensively evaluated for various important traits by a multi-disciplinary team. Evaluation for pest, shot-hole borer (SHB), has already commenced (Walgama *et al.*, 2008) and results revealed that an adequate amount of resistance to SHB can be tapped from unadapted germplasms compared to the adapted breeding lines. It is planned to extend the evaluation for other biotic stresses to find accessions possessing multiple pest and disease resistance for inclusion in the breeding program.

4.2.3.6 Documenting Information on Germplasm Descriptions

When allocating priorities for germplasm activities, not enough time or resources have often been assigned to documenting descriptions and their evaluated characteristics. This will inevitably lead to an ineffective utilization of genetic resources, though information was generated regarding them. Collating and documenting passport and collection descriptors of accessions originating from estate selections made since the 1930s have been accomplished. Accessions were stratified by agro-ecological region and presented according to the estate from which they were selected in order to be able to access the information in a usable form (Gunasekare & Kumara, 2005). A map indicating spatial distribution and origin of the "Estate Selections" were prepared. According to the information, a total of 688 tea accessions originated from old seedling populations that existed on various estates covering 10 agro-ecological regions. A comprehensive document was prepared incorporating information on all evaluated traits and this information was also used to formulate a database using Microsoft Access for easy retrieval of information in a more user-friendly manner.

4.2.3.7 Facilitating Germplasm Utilization and Managerial Aspects

Although the core collection concept (Frankel, 1984) aims primarily at facilitating germplasm managerial aspects, it also greatly facilitates the detailed characterization and evaluation and the search for desirable new characteristics in the collection. Assembling a core collection of germplasm at TRISL was initiated with the main aim being to create more descriptive information regarding the accession. Construction of a diversity tree of germplasm was completed using information already collected regarding taxonomy, pedigree, passport and collection descriptors (Ranathunga & Gunasekare, 2008b). The total collection was rationally stratified into smaller groups (sub-sets) using a stepwise procedure to develop a hierarchy and a "diversity tree" following a method proposed by Boukema *et al.* (1997). The results generated were of considerable importance to narrow down the breeding stock and to move towards assembling a core collection of the germplasm.

In some crop species, assembling a core collection was exclusively based on the molecular data. However, our approach is to complement the molecular marker data with characterization based on morphological, biochemical description and pedigree information, providing accurate and detailed information when making decisions on forming the core collection (Gunasekare, 2007a). This would also help to assemble a more meaningful core collection with enhanced and known breeding values to facilitate developing new cultivars which cater to the needs of the end-users.

4.2.4 Germplasm Utilization

The role of germplasm in crop improvement, though well recognized yet lacking sufficient information on the performance of the germplasm collection, has led to its limited use. There is a growing apprehension that planting large areas with few popular cultivars may lead to genetic uniformity and consequently to genetic vulnerability of the crop to biotic and abiotic stresses. Since 1960, the year when the controlled hybridization program commenced at the Institute, only 22 parents were used recurrently until late 1990 (Fig. 4.9). With the growing concern of a narrow genetic base in the cultivated gene pool, the attention was focused on using diverse parents in the breeding program rather than using limited progenitors recurrently over several years in the program. Rational selection of progenitors was made possible due to availability of information generated over recent years with regard to characterization and evaluation. The main goal of the recently performed crossing program was to use genetically diverse parents while eliminating the parents used over many years recurrently in the breeding program. As a result, such utilization of germplasm in breeding has been enhanced considerably and 46 new progenitors have been utilized in the breeding program at TRISL since 2004 (Fig. 4.9). Many new progenies have been generated from those crosses and have been entered into the current cultivar evaluation program.

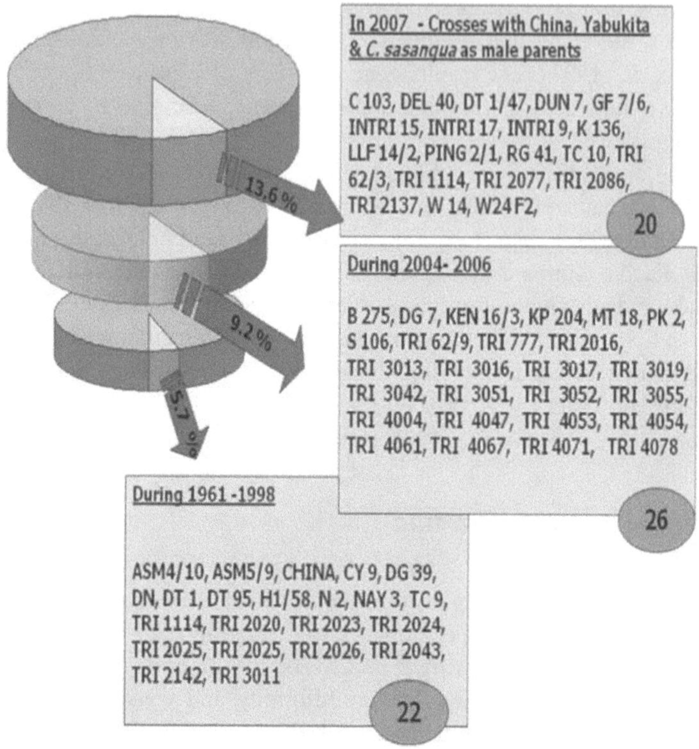

Fig. 4.9. Utilization of germplasm (as a percentage of total collection) in breeding of new cultivars in different era

4.2.5 Germplasm Enhancement and Pre-Breeding

In the past, breeders needed only 1 or 2 elite traits to be incorporated into their new cultivars. Nowadays, materials with more desirable traits have become important due to changing climate, agronomic and cultural practices. Narrowing the genetic base of the cultivars at commercial plantations is an unavoidable consequence of successful plant breeding. In keeping with the above concerns, genetically more diverse parents, such as 'Yabukita' and extreme 'var. *sinensis*' types, have been included in the breeding program.

Allied species of tea can be used as donors of gene or gene combinations required for further improving existing well adapted genotypes lacking in one or more desirable traits. Furthermore, introgressing new traits to the cultivated gene pool needs pre-breeding which would also contribute to germplasm enhancement and broadening of the genetic base. This is currently receiving attention. The limited exotic germplasm has been evaluated and *C. sasanqua* Thunb. was identified as one of the potential non-tea species for introgression breeding to incorporate improved resistance to blister blight disease as well as a low caffeine trait. A few inter-specific hybrids have already been recovered using *in vitro* techniques and they are being multiplied *in vitro* to raise clonal lines (Gunasekare, 2007c).

Though utilization of allied species and the bringing of an unadapted genetic background to an adapted genetic background has certain limitations, it is hoped that these issues could be addressed in future when the molecular markers linked to those traits are available for marker-assisted introgression breeding.

In general, germplasm activities at TRISL in the recent past have shifted from collection and conservation efforts towards proper characterization, evaluation and rational utilization of the germplasm in the current tea breeding program.

4.2.6 Other Issues Related to the Germplasm

The emergence of new Intellectual Property Rights (IPR) regimes, in relation to tea genetic resources necessitates working out modalities for benefit sharing between countries concerned. As the genetic resources of a particular crop are not distributed uniformly all over the world, no country can claim self-sufficiency in tea genetic resources. Even with the most efficient management programs, no single country can assemble resources which would satisfy all present and future needs. Therefore, it is necessary to foster collaboration at national and international levels for the acquisition and conservation of the tea germplasm, the sharing of information and technologies and a mechanism to access the safe movement of genetic resources and the sharing of benefits arising from their utilization, facilitated by the appropriate regulatory mechanism. This would certainly help to foster crop improvement programs for tea, not only at a national level but also internationally.

4.3 Tea Breeding and Selection

A classical tea crop improvement program consists of two main aspects: i.e., creation of genetic variability through controlled hybridization between selected progenitors and the pursuit of a selection criteria and evaluation of the new progeny generated, to identify promising individuals with desirable traits aimed at improving the productivity of the crop. Tea breeding objectives are specific for each tea growing region in the country as the prevailing biotic and abiotic stresses do vary from region to region. Therefore, setting up breeding priorities in the developing region for specific cultivars has become a challenging task. In the face of reduced usage of pesticides in tea plantations due to stringent monitoring of pesticide residue in made tea, it has also become important to concentrate more on breeding cultivars that are resistant to major pests and diseases.

4.3.1 Early Attempts at Tea Improvement

In the early years, seed was introduced to Sri Lanka from various sources of seed *baries*, mainly from India and China. Those seeds were of mixed *jats* and hence high heterogeneity was present in the seedling populations. With the restriction on importing seeds from other countries, growers in Sri Lanka started establishing seed gardens in their tea plantations. Enterprising tea planters made initial selections of desirable seed material to raise seed bearers. The selection of vigorous seedlings in the nursery, mainly based on the morphological characteristics, followed by the establishment of seed gardens with those seedlings, marked the beginning of unplanned tea breeding (Kulasegaram, 1978). Hence, in the second phase of tea cultivation, the main source of planting materials were the seeds derived from those seed gardens with unknown parental types. The early plantations were therefore comprised of highly heterogeneous stands of seedling tea, which later formed the base population from which promising genotypes were selected. The emphasis of selection in the early days was mainly on the type of "*jat*".

The second phase of selection work commenced in 1920 when it was found that selection could be made for the yield. It was revealed that about 70% of the yield came from only 20% to 30% of the bushes (Richards, 1966). However, the scientific approach to tea cultivation including tea improvement began after the inception of the Tea Research Institute of Ceylon (now the Tea Research Institute of Sri Lanka) in 1925. Although some form of selection was practiced in an empirical way, a definite program with a rational method of selection was laid down only at the beginning of the 1940s and more concerted efforts at seedling selection were made only after the advent of an economical vegetative propagation technique in the late 1930s (Visser, 1958).

By 1946, considerable progress in clonal selection had been achieved through the exploitation of available natural variability in seedling populations (Tubbs, 1946). It was reported that the majority of the seedling selections were carried out on up country estates rather than in the low and mid country (Visser, 1958). Between 1938 and 1958, a total of 290 cultivars were established through mass selection followed by vegetative propagation (Visser, 1958). Over time, more than half of the cultivars were rejected, leaving only 55 cultivars of which 14 have been finally approved (Visser, 1958). After continuous scrutiny, the first approved list of cultivars for commercial planting was made available in 1968. Those cultivars contributed immensely to increased productivity in the tea plantations.

4.3.2 Conventional Breeding and Selection Program

The past approach, which mainly relied on mass selection of elite bushes from early plantations, was established with seeds obtained from randomly mated or open pollinated progenitors of unknown breeding value which had several limitations. It was realized that use of such seed material was not efficient in recombining important agronomic traits. Hence, controlled hybridization between known and complementary progenitors was performed to raise segregated populations in later years (Anandappa *et al.*, 1988). Accordingly, an organized tea breeding program was initiated at the Institute in 1961. Many advantages accrue from this method, compared to using open pollinated seed progenies and old seedling tea in the selection process. Hence, unlike in the very early years, the material used for developing new cultivars from the late 1970s onwards mainly comprised of either full-sib or half-sib families derived from proven parents through planned hybridization or from seeds derived from open pollination among proven vegetative propagation cultivars. As such, those materials contributed to improved trait association with breeding value.

At the inception of the plant breeding program at the Tea Research Institute of Sri Lanka, tea breeding objectives were mainly centered on developing high yielding cultivars. The cultivars were initially evaluated for yield attributes and the high yielding cultivars were then screened for other important characteristics such as made tea quality, pest and disease resistance and drought tolerance (Kulasegaram, 1978). By adopting that kind of strategy, the cultivars, which were identified as promising based on their yield potential, revealed in many cases undesirable characteristics such as susceptibility to major pests and diseases, poor quality and susceptibility to drought when subjected to different trait evaluations (Anon, 1976, 1994). As biotic and abiotic stresses can cause considerable yield loss and contribute to low productivity (Kulasegaram, 1978), the need to use a mixture of cultivars having different desirable attributes was emphasized to avert the problem (Sivapalan, 1986).

Up to now, TRISL has released cultivars of TRI 2000, TRI 3000, TRI 4000 series and estate selections. For the most popular TRI 2000 series, all selections were made at the TRISL from seedlings raised from a batch of open pollinated seeds introduced from a single seed bearer in 1937 (Richards, 1965). Exploiting the wide genetic variability present in the seedling tea fields by mass selection led to the generation of a number of distinct primitive cultivars. Tea planters in consultation with scientists at TRISL also made selections of desirable plant types from seedling populations on their estates. Some of those cultivars, commonly known as "Estate selections" are still very popular in plantations in various parts of the growing regions. The above two forms of selection, "Estate Selection" and "TRI cultivars" formed the basis for the recommendation of cultivars at the TRISL (Gunasekare & Anandappa, 2008).

The useful characteristics present in cultivars of the TRI 2000 series and some estate selections have been recognized and these cultivars were used in the breeding program to develop cultivars designated as the TRI 3000 and TRI 4000 series. Cultivars of the TRI 3000 series have been derived from selections made from open pollinated seed progenies of known parents whereas cultivars of the TRI 4000 series were the product of controlled hybridization (Anandappa, 1992). Hence, the TRI 3000 and TRI 4000 series cultivars were the result of many years of breeding and selection from progenies derived from parent cultivars whose characteristics of economic importance were known. Therefore, the TRI 3000 and TRI 4000 series cultivars have a combination of desirable characteristics as compared to their parents, i.e., the TRI 2000 series cultivars and the estate selections (Gunasekare & Anandappa, 2008).

Tea breeding and selection programs in general are multi-stage, multi-location testing and evaluation procedures, which span 20 to 25 years, thus requiring considerable investment in terms of land, labor, time and money. With the restriction of funds for R&D, it has become critical to justify the investment in breeding. Therefore, clear identification of breeding priorities and targets in keeping with the socio-economic status of the end-users is extremely important. Resistance to pests and diseases, for example, has received less attention in the past, especially since the application of agrochemicals was seen as an easy strategy to address the issue. However, owing to environmental concerns and legislation involved in maximum residue limits (MRLs) in made tea, it has become imperative to reduce the usage of pesticides in tea cultivation. Hence, the tea breeding strategies need to be focused on developing high yielding cultivars which also prove to have resistance to major pests and diseases. This would in turn help achieve maximum crop productivity and profitability in tea plantations.

4.3.2.1 Region Specific Tea Breeding Objectives

Accordingly, the present breeding approach is centered on developing cultivars to cope with the complex requirements determined by agro-ecology, biotic and abiotic stresses and consumer preferences of the end-product, based on quality in the international market. Tea breeding objectives are therefore primarily focused on high yield, high quality of processed tea, genetic resistance to major biotic and abiotic stresses prevailing in different tea growing regions and regional adaptability. Furthermore, tea in Sri Lanka is grown under widely varying soil and climatic conditions in the presence of various biotic and abiotic stresses. Hence, the region-specific breeding objectives have been identified and breeding strategies are targeted to incorporate and select the genotypes which combine the relevant desirable traits (Table 4.4).

Table 4.4 Region specific breeding objectives considered in the current breeding program of the TRISL

Tea growing region	Priority breeding objectives
Up country	Yield, made tea quality, resistance to blister blight, collar & branch canker (*Phomopsis theae* Petch), shot-hole borer, nematodes
Mid country	Yield, made tea quality, resistance to shot-hole borer, nematodes, drought
Low country	Yield, resistance to stem canker (*Macrophoma theicola* Petch), shot-hole borer, nematodes, low country live wood termite, drought

The tea industry being a labor-intensive enterprise, mechanization of harvesting operations is essential to meet future challenges on account of labor scarcity. In that current context, it has become essential to select cultivars that are amenable to mechanical harvesting, and this is being pursued in the current selection program using near-finished cultivars.

4.3.2.2 Stages of Tea Breeding and Selection

The tea breeding program at TRISL consists of 4 stages (Table 4.5), i.e., (1) generation of genetic variability, (2) selection of useful genotypes, (3) comparative tests to demonstrate the superiority of the selected genotypes and, (4) exposing promising cultivars to multi-regional sites for determination of stability and adaptability. Initially, genotypes are evaluated in small and medium scale experimental trials for preliminary evaluation. At each stage, progressive elimination of undesirable genotypes and an increase in the number of plants per genotype are adopted to generate reliable data to select the most outstanding genotypes possessing a combination of desirable traits. To accrue the comparative advantages, the test genotypes are always compared with existing proven cultivars which are used as standard cultivars. Genotypes exhibiting undesirable characteristics are eliminated relative to their overall performance and the promising cultivars are advanced into the next stage/phase in which the accessions are subjected to multi-regional evaluation. After concluding the regional trials, the promising accessions based on their yield potential and other key traits are selected to be included in large scale multi-location trials with the partnership of growers in their fields. Accordingly, superior cultivars are identified and recommended for commercial planting based on their regional performances. As such, elite genotypes undergo more extensive multi-location trials and commercial evaluation before release.

Table 4.5 Stages involved in breeding and evaluation of new tea cultivars at the TRISL

Evaluation stage	I	II	III	IV
Location	On-station	On-station	On-station	On-farm
No. of genotypes	>1,000 seedlings	100 – 200	15 – 25	8 – 10
No. of plants/genotypes	Single plant	7 – 10	16 – 24	250 – 300
No. of replicates	Single plant randomization	2 – 3	2 – 3	None *

* No replicates on the same site, but multi-location trials are considered as replicates.

In the current breeding approach, three types of source materials are used for selecting new cultivars, viz. (1) existing old seedling tea, (2) open pollinated seeds from bi- and poly-clonal seed gardens and, (3) progenies derived from controlled hybridization.

4.3.2.3 Controlled Hybridization

Controlled hybridization between known cultivars possessing important traits has been undertaken to create new progenies in order to incorporate desirable traits discussed above into existing proven cultivars. The progenitors for each cross are chosen so that the weaknesses of one are matched by the strengths of the other. In Sri Lanka, tea plants flower throughout the year with some seasonal and regional variations (Kulasegaram, 1978). However, in the past basic research carried out to understand the cultivar behavior in terms of floral biology and fecundity traits was scanty. Selection of parents for hybridization was mainly based on their desirable agronomic traits, without considering their reproductive capacity. A dearth of information on the reproductive biology of the crop has constrained the breeding success in the early breeding programs. This has prevented the strategic planning of a hybridization program to generate an adequate number of progenies for genetic studies.

In keeping with the immediate need to generate basic information, several studies were initiated recently to investigate the breeding systems in tea cultivars, stigma receptivity, pollen biology (Ariyarathna *et al.*, 2007), fecundity, flowering synchrony (Kumara *et al.*, 2008) among tea accessions in the germplasm to gain in depth knowledge to address the issues related to a controlled hybridization program. Although in Sri Lanka tea plants flower throughout the year, cultivar differences have been observed in the peak flowering period under local conditions (Kumara *et al.*, 2008). It was possible to categorize the tea germplasm accessions into four main groups based on flowering synchrony and the accessions capacity for profuse flowering and high fecundity (Kumara *et al.*, 2008). The stage at which the flower buds attain the optimum stigmatic receptivity was varied, based on the developmental stage of the bud and the accession (Fig. 4.10) (Ariyarathna *et al.*, 2007). Hence, this would have a great implication on the success of the hybridization program. The presence of pseudo-pollen was reported for the first time in tea in the study by Ariyarathna *et al.* (2007).

The above information complements the selection of appropriate male and female parents in pre-designing cross combinations. Hence, in the current breeding program, prior to conducting a hybridization program, selected parents are subjected to pre-evaluation of their reproductive ability by assessing the flower abundance, functional activity of stigma and pollen viability to identify appropriate male and female parents for the controlled program. By adopting such a strategic selection of parents, seed setting success has been increased by 35% in the current breeding program.

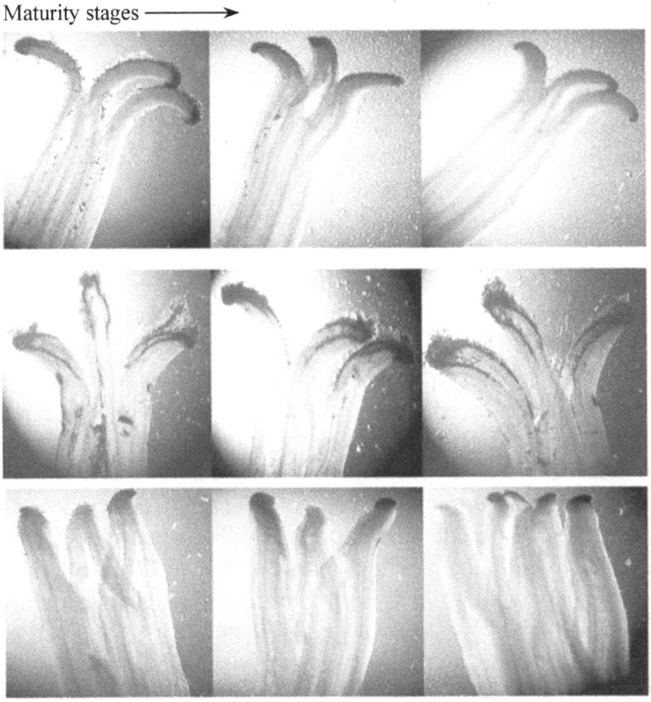

Fig. 4.10. Variability in stigmatic receptive areas revealed by esterase test in different developmental stages of flower buds in different tea cultivars

4.3.2.4 Biclonal and Polyclonal Seed Progenies as Alternative Planting Material

With reduced availability of suitable high potential tea growing areas, tea cultivation has been extended beyond the traditional tea growing boundaries. This creates another challenge for tea breeders, namely to develop cultivars for marginal conditions. Furthermore, with the revived interest in using seeds as planting material by the growers, TRISL has taken initiatives to produce genetically superior seedling material which would survive better under marginal conditions. Investigations were commenced in 1999 to evaluate the performances of biclonal and polyclonal seed progenies derived through natural pollination among proven cultivars established in isolated seed gardens.

Preliminary yield evaluation and uniformity within each biclonal and polyclonal progeny were studied in 4 regional trials. Of the 11 different biclonal and polyclonal seed progenies evaluated, 5 proved promising (Gunasekare *et al.*, 2004; Piyasundara *et al.*, 2003). To validate the potential of promising seed stocks, large scale commercial trials with the partnership of growers in marginal and drought prone areas have been established since 2004. According to the performance evaluated so far, it is apparent that 3 of them could be released for the industry in the near future for restricted use in low productive tea lands or drought prone areas.

4.3.3 Screening and Evaluation

Selection of genotypes that contain appropriate combinations of traits from a segregating progeny, to identify a few that may have potential as new cultivars, is a critical component of the selection and evaluation. At the same time, the procedure involved in screening and selection in tea has many limitations and needs substantial resources, especially time, land and labor. The selection criteria used at present are not amenable for early selection. This is further compounded by a dearth of information on heritability and genetic correlation of important traits, which is crucial for the success of any breeding effort aimed at improving the targeted attributes. The lack of information on those aspects is mainly due to cost implications and complexity associated with such long term investigations.

As stated before, the breeding objectives are targeted at yield, quality, regions specific biotic and abiotic problems and adaptability of new cultivars. Considering the emerging challenges in the international market and the problems encountered by growers in diverse environments with varying socio-economic status, certain modifications have been made when developing the new series of cultivars (TRI 5000 series) which are in the pipeline. The new strategy is based on giving equal attention to yield potential as well as to other important characteristics through a multi-disciplinary approach at early stages of the cultivar evaluation program. In developing the new series of cultivars, the emphasis was placed not only on high yield potential, but also on taking into account traits such as made tea quality, resistance to blister blight, canker, shot-hole borer and live wood termite, with a view to combining them into a commercially acceptable cultivar.

Depending on the screening and evaluation procedure involved as well as limitations encountered, all the priority traits identified cannot be evaluated simultaneously. Hence, trait evaluations have been segmented into different stages in the evaluation program while yield assessments are carried out throughout all stages (Table 4.6).

4.3.3.1 Yield

As the harvestable portion of tea consists of two leaves and a terminal bud, shoot or leaf characteristics form the basis of selection for the two most important traits, yield and quality. Initial yield assessments at the individual seedling level, Stage Ⅰ, are purely based on visual estimation, considering the size of leaf/bud, compactness of the plucking table and the frame size.

Table 4.6 Stages of evaluation cycle and targeted traits for evaluating genotypes at different stages of the cultivar development program at TRISL

Stage of evaluation	Traits evaluated	Time frame			
		1st to 5th Year	6th to 10th Year	11th to 15th Year	16th to 20th Year
Stage I	Preliminary yield evaluation Visual assessment for pests and diseases				
Stage II	Quantitative assessment of yield Systematic screening for diseases blister blight and stem canker Systematic screening for pest, shot-hole borer				
Stage III	Quantitative assessment of yield & yield components Systematic screening for diseases, blister blight and stem canker for confirmation Systematic screening for pest, shot-hole borer Evaluation for quality Screening for pest, live wood termite Separate screening experiments for poria and nematodes Evaluation for suitability for mechanical harvesting				
Stage IV	Yield & yield component assessments Evaluation of quality for final confirmation Systematic screening for major pests and diseases prevailing in the agro-ecological region concerned Growers feed-back on new cultivars through participatory appraisal Morphological characterization of cultivars using standardized descriptors and preparation of cultivar album Cost-benefit analysis Recommendation and release of new cultivars to the growers				

Yield evaluation criteria at the next and latter stages (Stages II & III) comprise a weekly yield record throughout the pruning cycle. Performances of test cultivars for yield are always judged by reference to standard cultivars, which often is a popular cultivar or locally adapted cultivar. The same standard cultivars hence are not considered as standard for trials at all locations.

Apart from the quantitative estimation of yield, the bush area (plucking surface area), number of shoots per unit area (shoot density), weight per shoot, ratio between active/dormant ("*banji*") shoots are also taken into account in selecting potentially high yielding cultivars. Those evaluation criteria are further supported with the fresh weight of pruning, recovery after pruning (qualitative assessment) and quantitative assessment of tipping weights (Ranathunga *et al.*, 2004).

4.3.3.2 Quality of Made Tea

Organoleptic tests of the made tea produced through a miniature manufacturing system, requiring approximately 500 g of fresh leaves, is adopted for testing the made tea quality of cultivars. As such, evaluation of cultivars for made tea quality can only be commenced at Stage III onwards of the evaluation phase where adequate harvestable fresh leaves (approximately 500 g) can be obtained for miniature manufacture.

The judgment of tea quality and assessment of its characteristics are generally undertaken by professional tea tasters. Three major components are considered in assessing the quality of made tea in cultivars. They are (1) appearance of black tea particles, (2) infusion/liquor characteristics and, (3) characteristics of the infused leaf. The infusion is evaluated and ranked for color, brightness, quality, strength, briskness and flavor. The quality of the made tea of cultivars is evaluated depending on the region where it is produced. For cultivars which are tested in the up country and mid country, prominence is given to infusion characteristics whereas in the low country it is primarily the appearance of the black tea that is important.

As it is a challenging task to evaluate a large number of genotypes at the early stages, owing to limitations on the green leaf required for manufacture, alternative criteria such as morphological markers for the preliminary selection of putative quality cultivars have been identified through multivariate statistical analysis (Ranathunga *et al*., 2008). It is planned to extend this study to identify chemical markers in the green leaf related to made tea quality, to develop a reliable screening criteria for quality.

4.3.3.3 Screening for Biotic and Abiotic Stress Sensitivity

Screening of genotypes in the evaluation program is conducted for blister blight, which is caused by a fungus, *Exobasidium vexans* Massee and canker by *Phomopsis theae* Petch at higher elevations as well as *Macrophoma theicola* Petch at lower elevations. Screening of new genotypes for all those 3 diseases are assessed in the field under natural infestation at different stages (Table 4.6). Screening cultivars for pests such as shot-hole borer (*Xyleborus fornicates* Eichhoff) and low country live wood termite (*Glyptotermes dialatus* Bugnion) is also carried out by natural infestation in the field.

Assessing the sensitivity of genotypes to another disease, which is commonly known as red root disease caused by a fungus, *Poria hypolateritia* Berk, is done under forced inoculation in the isolated pits, at later stages of the evaluation cycle. Screening of genotypes for nematode resistance is also done in isolated tanks filled with soil infested with nematodes. As such, especially designed screening experiments need to be conducted to evaluate the cultivars for the above two biotic stresses thus requiring additional planting materials to raise those screening experiments, which is one of the bottlenecks in the screening procedure.

4.3.3.4 Constraints Faced by Breeders in the Selection Program

Unavailability of reliable and early screening procedures to screen large numbers of genotypes at the very early stages of the evaluation cycle, preferably at the nursery stage, remains the great challenge for breeders in the cultivar development program. Hence, to reduce and replace the laborious, time consuming screening procedures, it is imperative to develop easy, rapid, inexpensive and early screening tools to save the resources as well as to retain useful genotypes without eliminating them at the early stages.

Laboratory bio-assays and chemical markers are being investigated to facilitate early screening of genotypes. A biochemical marker related to inherent disease resistance for blister blight has been identified recently and it was found that epicatechin was involved in the resistance mechanism (Punyasiri *et al.*, 2005). The total phenolic content in the tea cultivars susceptible to shot-hole borer was found to be high compared to resistant cultivars (Bombuwala, 2001) and the accumulation of a higher concentration of caffeine in the bark, after attack by the beetle was identified and was presumed to be the cause of an accumulation of caffeine in tea cultivars. This may be one of the factors which determine the resistance to shot-hole borer (Kumar *et al.*, 1995). Hence attempts to correlate the biochemical and molecular basis of resistance to biotic stress would be highly useful in establishing advanced breeding lines in a short time duration rather than using conventional screening methods. But validation and application of those markers as early screening criteria are yet to be explored.

4.3.4 Multi-Location Adaptability Trials and Recommended Tea Cultivars

Tea in Sri Lanka is grown under widely varying soil and climatic conditions with specific problems (biotic and abiotic stresses) and hence considerable genotypic differences in response to environmental changes were noted in tea (Wickramaratne, 1981b; Balasuriya, 1999). High productivity and stability in different environments have thus assumed greater importance because tea yields vary not only locally and seasonally, but also with genotype, elevation, climate and edaphic factors (Ranathunga & Gunasekare, 2008a). In the past breeding approach, a large number of advanced breeding lines were evaluated with the best existing commercial cultivars in the on-station regional trials. Often the concern at this stage was the lack of proper representation of the target environment by the selected environment where on-station trials were carried out.

Although in the past, the recommendation of cultivars for different planting districts was made based on the results of regional adaptability trials conducted in the up, mid and low country, those locations only broadly represent the major tea planting regions in Sri Lanka. A fundamental problem in evaluating new cultivars in the breeding program is the relationship between selection environment and target environment. The selection efficiency decreases as the selected environment becomes increasingly different from the target environment in terms of climate, soil and other factors. Wide variations in soil and climatic factors encountered even within regional boundaries contribute to varying effects on the growth and yield stability and adaptability of cultivars (Ranathunga & Gunasekare, 2008a). Hence, the present breeding approach is not to restrict selection of genotypes to those with only high yield potential and quality, but also to ensure their high adaptation to the target environment while considering their ability to withstand major stresses prevailing in the same environment.

After concluding the on-station regional trials, promising accessions that are selected based on their yield potential and other key traits are included in large scale multi-location adaptive trials. These trials are undertaken by scientists with the partnership of growers in their fields to select the few best genotypes for the target environment concerned (Gunasekare, 2008). Information generated at the adaptive trial stage enables the assessment of the stability and adaptability of selected cultivars in a range of diverse environments, covering different tea growing agro-ecological regions. In establishing adaptive trials, 6 – 8 promising new cultivars are planted in separate blocks with clear demarcations. A minimum of 2 standard commercial cultivars (one TRI cultivar and a popular estate cultivar for that particular site) are included for comparison. The size of the block is 250 – 300 plants per cultivar and without replication at a given site, for simplicity.

The objective of the scientist-grower collaboration in the final stage of evaluation of the potential cultivars is to validate the experimental findings and to get growers' feedback on the suitability and the acceptability of new cultivars. This will help the scientists to assess the commercial potential of the selected cultivars under estate-managed conditions instead of evaluating their performance under controlled experimental conditions. The results generated from the adaptive trials will also enable the plant breeder to consolidate the experimental findings on new cultivars, thereby creating more confidence in the performance of the cultivars prior to making a commercial recommendation and release. In addition, testing the selected cultivars in adaptive trials in growers' fields will provide an opportunity for tea growers to have early access to new potential cultivars before the cultivars are formally released or commercially available. This will increase the end-user acceptance of new releases.

At the time of concluding the adaptive trials, the cultivars demonstrating marked improvement over previously recommended cultivars (standards/controls included in the adaptive trials) will be selected as new elite cultivars and the rest of the cultivars will be eliminated from the program. Those adaptive trial sites are then converted into a nucleus of mother bush blocks to facilitate the dissemination of planting materials to the end-users once the TRISL makes the new cultivar recommendation.

The list of recommended cultivars is revised from time to time and new superior cultivars are added to the updated list. The new revised list of recommended cultivars contains 64 cultivars (Table 4.7), which have been recommended for various tea growing regions in Sri Lanka (Anon, 2002). The choice of cultivars to be used for planting is left to each individual estate/grower. However, growers are being advised to get an idea of the major pests and diseases prevalent in a particular estate in order to select the best set of cultivars which possess tolerance to those stress conditions (Gunasekare & Anandappa, 2008).

Table 4.7 National released tea cultivars in Sri Lanka

Cultivar	Year of release	Pedigree	Breeding method/ Source of selection	Recommended region *	Desirable attribute other than yield
TRI 4004	1994	TRI 777×TRI 2026	CH	Low	Shot-hole borer tolerant
TRI 4006	1994	TRI 2023×TRI 2026	CH	Up, Mid	Nematode tolerant
TRI 4014	1994	TRI 2023×TRI 2026	CH	Low	Drought tolerant
TRI 4024	1994	TRI 2023×TRI 2026	CH	Low	Shot hole borer tolerant
TRI 4034	1994	TRI 2023×TRI 2026	CH	Up	Shot hole borer tolerant
TRI 4042	1994	TRI 2023×TRI 2026	CH	Mid, Uva, Low	Low country live wood termite tolerant
TRI 4043	1994	TRI 2023×TRI 2026	CH	Low	Shot hole borer tolerant
TRI 4046	1994	TRI 2023×TRI 2026	CH	Mid, Uva	Drought tolerant
TRI 4047	1994	CY 9×ASM 4/10	CH	Mid, Low	Stem canker tolerant
TRI 4049	1994	CY 9×ASM 4/10	CH	Low	Low country live wood termite tolerant
TRI 4052	1994	CY 9×ASM 4/10	CH	Up, Uva, Low	Blister blight tolerant
TRI 4053	1994	CY 9×ASM 4/10	CH	Up, Mid, Uva, Low	Stem canker tolerant
TRI 4054	1994	CY 9×ASM 4/10	CH	Low	Shot hole borer tolerant
TRI 4055	1994	CY 9×ASM 4/10	CH	Low	Drought tolerant
TRI 4059	1994	TRI 2020×TRI 2023	CH	Low	Shot hole borer tolerant
TRI 4061	1994	TRI 2020×TRI 2023	CH	Low	Stem canker tolerant
TRI 4067	1994	CY 9×ASM 4/10	CH	Up	High quality made tea
TRI 4071	1994	N 2×TRI 2024	CH	Up, Mid, Uva	Drought tolerant
TRI 4078	1994	N 2×N 2	Self progeny	Up, Uva	Low country live wood termite tolerant
TRI 4079	1994	N 2×N 2	Self progeny	Up	High quality made tea
TRI 4085	1994	TRI 2024×DN	CH	Up	Shot hole borer tolerant
TRI 3013	1980	TRI 2024	OP	Up, Mid, Uva	Stem canker tolerant
TRI 3014	1980	TRI 2025	OP	Mid, Low	Drought & shot-hole borer tolerant

(To be continued)

(Table 4.7)

Cultivar	Year of release	Pedigree	Breeding method/ Source of selection	Recommended region *	Desirable attribute other than yield
TRI 3015	1980	TRI 2026	OP	Up, Mid, Uva	High yield
TRI 3016	1980	DT 95×ASM 4/10	CH	Up	Nematode tolerant
TRI 3017	1980	ASM 4/10×DT 95	CH	Mid, Uva	Nematode tolerant
TRI 3018	1980	TRI 2024×DT 1	CH	Up, Mid, Uva	High yield
TRI 3019	1980	DT 95×ASM 4/10	CH	Up, Mid, Uva	High yield
TRI 3020	1980	TRI 2025	OP	Up, Mid	Drought & nematode tolerant
TRI 3022	1980	TRI 2026×TRI 2023	CH	Uva, Low	Drought tolerant
TRI 3025	1980	TRI 2025	OP	Mid, Low	Low country live wood termite tolerant
TRI 3035	1980	TRI 2025	OP	Uva	Drought tolerant
TRI 3047	1980	ASM 4/10	OP	Low	Low country live wood termite tolerant
TRI 3051	1980	ASM 4/10	OP	Low	Drought tolerant
TRI 3052	1980	ASM 4/10	OP	Low	Drought tolerant
TRI 3055	1980	ASM 4/10	OP	Low	Stem canker tolerant
TRI 3069	1980	TRI 2025 4×	TET	Low	Low country live wood termite tolerant
TRI 3072	1980	TRI 2025	OP	Low	Drought & blister blight tolerant
TRI 3073	1980	TRI 2025	OP	Up	Blister blight tolerant
TRI 62/5	1978	ASM 4/10	OP/INT	Up, Mid, Uva, Low	Stem canker & nematode tolerant
TRI 62/6	1978	ASM 4/10	OP/INT	Low	Stem canker & shot-hole borer tolerant
TRI 62/9	1978	ASM 4/10	OP/INT	Mid, Uva, Low	High quality made tea
TRI 777	1958	Indo-China	OP/INT	Up	High quality made tea
TRI 2022	1958	ASM 4/10	OP/INT	Uva, Low	Stem canker tolerant
TRI 2023	1958	ASM 4/10	OP/INT	Up, Mid, Uva, Low	Shot-hole borer tolerant
TRI 2024	1958	ASM 4/10	OP/INT	Mid, Uva, Low	Nematode tolerant

(To be continued)

(Table 4.7)

Cultivar	Year of release	Pedigree	Breeding method/ Source of selection	Recommended region *	Desirable attribute other than yield
TRI 2025	1958	ASM 4/10	OP/INT	Up, Mid, Uva, Low	Drought & canker tolerant
TRI 2026	1958	ASM 4/10	OP/INT	Uva	Blister blight tolerant
TRI 2027	1958	ASM 4/10	OP/INT	Mid, Uva	Nematode, drought & canker tolerant
TRI 2043	1958	Indo-China	OP/INT	Up, Mid, Uva, Low	Blister blight tolerant, silver tip making quality
CH 13	1958	** Craighead	ES	Mid	Blister blight & drought tolerant
CY 9	1958	** Tangakelle	ES	Up, Mid	Drought tolerant
DG 39	1958	** Balangoda	ES	Mid, Uva, Low	Drought & shot hole borer tolerant
DG 7	1958	** Balangoda	ES	Mid, Uva, Low	Drought tolerant
DN	1958	** Dayagama	ES	Up, Mid, Low	Drought & shot hole borer tolerant
DT 1	1958	**Drayton	ES	Up	Blister blight tolerant
H 1/58	1958	** Hulandawa	ES	Low	Blister blight tolerant, high quality
K 145	1958	**Kirkoswald	ES	Up, Mid, Uva	Drought tolerant
KEN 16/3	1958	** Kenilworth	ES	Uva	Low country live wood termite
KP 204	1958	** Palmgarden	ES	Low	Canker & low country live wood termite tolerant
N 2	1958	** Norwood	ES	Up, Mid	Blister blight tolerant, high quality
NAY 3	1958	** Nayabedda	ES	Up, Uva	Blister blight tolerant
PK 2	1958	** Park	ES	Up	Canker & blister blight tolerant, high quality
S 106	1958	** Sirikandura	ES	Low	Drought & Canker tolerant

* Recommended region for planting, based on the elevation category;

Selections made from: CH—Controlled hybridized progenies (full-sib progenies); OP—Open pollinated progenies of bi-clonal and poly-clonal materials (half-sib progenies); ES—"Estate Selections" (Selections made from old seed tea populations existing on respective tea plantations/estates**); TET—Induced tetraploid; OP/INT—Selections made from open pollinated seeds imported/introduced from other countries

A new series of cultivars, the TRI 5000 series, which is in the pipeline, is at the final stage of evaluation in multi-location adaptive trials. This series is being evaluated at commercial level in growers' fields with the partnership of progressive growers in a wider range of agro-ecological regions (Gunasekare, 2008). It is intended to implement an agro-ecology based cultivar recommendation at the time of releasing the new series to facilitate growers in choosing the best possible cultivars for their specific localities more confidently. Hence, the new approach adopted in assessing the TRI 5000 series will enable delivery of grower acceptable cultivars for widely varying tea growing environments.

4.3.5 Dissemination of Planting Material to the Growers

Dissemination of planting material of the cultivars in the provisional list, which is the interim release, is mainly done through the plant breeding division, especially to establish adaptive trials on growers' fields. Once the cultivars are recommended for commercial planting, the dissemination of planting material is delegated to TRISL regional centers and to the two estates that comes under the purview of TRISL. The regional trials conducted on those sites are converted to serve as mother bush plots of the selected cultivars to obtain initial planting materials for dissemination to small scale growers, while on-farm commercial scale trials are used as nucleus plots for the dissemination of planting materials to large scale growers or to their own company estates.

4.3.6 Non-Conventional Approaches to Tea Breeding

As conventional tea breeding approaches require a long time to develop desirable tea cultivars, non-conventional methods such as polyploidy and mutation breeding were initiated at the TRISL, to hasten the on-going breeding program. Induction of polyploidy in tea was first attempted by Sebasthiampillai (1970) using colchicines treatment and few tetraploid cultivars have been developed. One such tetraploid cultivar (TRI 3069) has been recommended and released for commercial planting. Subsequently, crosses involving tetraploid cultivars were undertaken and the resultant progenies are being evaluated (Gunasekare & Ranathunga, 2003). Some anatomical and morphological markers have also been identified to screen polyploidy genotypes in tea (Ranathunga & Gunasekare, 2003; Thirukkumaren & Gunasekare, 2001).

To incorporate more desirable characteristics into existing proven genotypes without changing much of the desired traits, work on mutation breeding using ionizing radiation (γ-rays-Co^{60}) has been conducted. LD_{50} (lethal dose 50%) value was determined for nodal cuttings (Sarathchandra & Pieris, 2001) and putative mutants of the M_1 generation are being tested for their performance in the field.

Yield component evaluation of the M_1 generation showed increased shoot density and shoot growth rates in some irradiated plants compared to their control counterparts (Data unpublished).

Though both in the past and at present, new tea cultivars have been developed through conventional breeding methods with some integration of non-conventional breeding approaches discussed above, steps have been taken to explore the potential of integrating new biotechnological tools to supplement the conventional breeding program.

4.3.7 Application of Biotechnological Tools for Breeding

Research activities aimed at developing necessary protocols for cell and tissue culture as breeding tools were commenced at the TRISL in 1985. Initial attempts were geared towards developing micropropagation techniques (Arulpragasam & Lattif, 1986).

Shortening the breeding cycle or time taken for evaluation of new cultivars is an urgent need. To cut down the number of years spent on raising clonal material from the individual seedlings of the hybrid progenies generated from controlled hybridization, embryo culture techniques have been attempted. A reproducible *in vitro* protocol for plantlet formation from zygotic embryo axes with a high germination rate was developed (Fig. 4.11) (Kulasekera *et al.*, 2004). Direct somatic embryogenesis from cotyledon tissues and zygotic embryo axes were also developed to facilitate mass multiplication of hybrid seeds (Seran *et al.*, 2006b, 2007a). The rate of multiplication of zygotic embryo cultures proved that it is possible to use the protocol for raising clonal counterparts of the hybrid seed *in vitro*, directly to establish stage II trials in the field, rather than planting individual seedlings in the field followed by conventional vegetative propagation, which takes at least 5 – 6 years. By using embryo culture and micropropagation it was estimated to reduce 5 – 6 years of the evaluation cycle.

Fig. 4.11. Regeneration of plants from embryo cultures *in vitro*

Indirect organogenesis was established using a callus derived from cotyledon explants (Arulpragasam *et al.*, 1988) and stem explants (Gunasekare & Evans, 2000b) and somatic embryogenesis from a stem and leaf callus culture (Sarathchandra *et al.*, 1988; Seran, 2006c), to extend the technique of recovering somaclonal variants as additional sources of variation in the selection program.

An isolation protocol that can release an adequate yield of viable protoplasts from leaf mesophyll cells was accomplished and the competence of the isolated protoplasts was proved in culture to facilitate somatic hybridization (Gunasekare & Evans, 1998; 2001). Progress on work related to anther culture to produce homozygous lines was not successful, as it was unable to regenerate plants from anther callus.

Culture protocols necessary for rescuing inter-specific hybrids between *C. sinensis* (L.) O. Kuntze and *C. sasanqua* Thunb. are being developed. Continuous monitoring of crosses made between inter-specific hybrids revealed that fruit abscission takes place 3 months after pollination, and hence embryos need to be rescued *in vitro.* This response was found to be cultivar specific and TRI 3013 and TRI 3019 were identified as promising maternal parents of *C. sinensis* for inter-specific hybridization. A few inter-specific hybrids that have already been recovered through embryo rescue are being multiplied *in vitro* (Data unpublished).

Recombinant DNA technology is very attractive in a crop like tea where incorporation of a trait of interest into a proven genotype is not possible through conventional hybridization and selection methods. Since there is growing concern and resistance towards transgenic products among consumers, especially in the western world where the bulk of Sri Lankan tea is being exported, work in this area is being debated. However, reviewing the present status of transgenic work overseas and considering the financial constraints on research and development work at the institute, research on this line receives low priority in the institute's current research program.

4.3.8 Genomics to Aid Tea Breeding

Some molecular approaches have also been investigated to facilitate the tea breeding and evaluation program. Research directed towards marker-assisted selection (MAS) has been initiated at the TRISL. A mapping population (pseudo-test cross population) consisting of 300 individuals derived from parents TRI 2023 × TRI 2043 was established in the field (Gunasekare, 2007a). Using 148 individuals, a genetic linkage map of tea was constructed at TRISL using EST and genomic SSRs. A total of 193 polymorphic SSR primers consisting of 107 EST and 86 genomic SSRs were used to construct the map (Mewan *et al.*, 2007). Work in this area has been extended to find molecular markers linked to blister blight resistance using the same segregated population. The recent development of molecular markers in tea and their potential application in tea breeding is expected to offer the opportunity to develop reliable criteria for early selection and the fast screening of large numbers of genotypes.

4.4 Propagation and Extension System of New Cultivars

Propagation and extension of new cultivars are of paramount important in increasing the extent of tea with new cultivars. Hence, an easy and economical propagation method is required to ensure multiplication of new material with proven success to disseminate new planting materials to reach growers at a rapid pace to increase their acceptance.

4.4.1 Propagation Techniques

Tea can be propagated either by seeds or by cuttings. Seeds were the only source of planting material used until the economically viable vegetative propagation method was discovered and established at the TRISL (Kehl, 1950). Since then, vegetative propagation has been the most preferred method of propagation due to advantages that accrue from its uniformity and predictability.

4.4.1.1 Vegetative Propagation

Vegetative propagation using single node cuttings is the method adopted on a commercial scale. The most suitable cutting for propagation is the single node cutting consisting of a nodal leaf with an axillary bud and inter node with a 2.5 – 3.5 cm length. Those cuttings are obtained from mother bushes especially grown for the purpose of taking shoots. Cuttings are taken from the middle portion of the shoot discarding the tender apical region and woody basal portion. Approximately 3 – 4 single node cuttings can be taken from a properly grown up shoot. Single nodal cuttings are planted in polythene bags (size: 23 cm×10 cm; gauge: 150) filled with loamy soil having a pH in the range of 4.5 – 5.5 and are kept under shade until they are hardened. The soil is fumigated using a soil fumigant before planting cuttings. Aftercare operations in the nursery include watering, application of fertilizer (T 65), training of plants to encourage lateral branching and control of pests (mite, *Hemitarsonemus latus* Green and tea tortrix, *Homona coffearia* Nietner) and mainly blister blight diseases. Nursery plants will be ready for field planting after a nursery period of 8 – 9 months in the low country and 10 – 11 months in the up country.

In the nursery, different cultivars are evaluated for their rooting ability. Cultivars which do not show good rooting ability are discarded during the cultivar evaluation process and hence the rooting ability of a cultivar is taken as one of the criteria for evaluating the performance of cultivars during the nursery period.

4.4.1.2 Grafting

Though the grafting technique in tea has already been established at TRISL, it is not practiced widely in commercial plantations due to the skills and labor involved in adopting the method. Cleft grafting is the method practiced at present. Although grafting is undertaken with a view to combining desirable characteristics into a composite plant, with the development of new cultivars possessing more than one desirable trait in combination does not necessitate undertaking the tedious process involved in grafting.

4.4.1.3 *In Vitro* Techniques for Propagation

Initial work on tissue culture was mainly confined to micropropagation using stem nodal explants. Early attempts were geared towards establishing *in vitro* protocols for shoot multiplication using stem nodal explants (Arulpragasam & Lattif, 1986) and their subsequent *in vitro* rooting (Sarathchandra *et al.*, 1997; Gunasekare & Evans, 2000a). The protocol for shoot multiplication was patented in 1990. Seran *et al.* (2007b) investigated the germination and subsequent plant development of *in vitro* cultured zygotic embryos and embryonic axes in comparison to the conventional seed propagation of tea.

Though extensive research has been done on micropropagation, challenges yet remain regarding the commercial exploitation of the technique. To scale it up to commercial level and to be more cost-effective, micropropagation must compete with conventional propagation and should be able to produce plants at a lower cost within a reasonable time frame. As of now, micropropagation is still found to be cost prohibitive for the production of planting material on a commercial scale. Research work has therefore been directed towards developing a cost-effective micropropagation protocol using locally available and less expensive materials as alternative medium substitutes. In addition, instead of using expensive *in vitro* rooting techniques, a successful *ex vitro* rooting method has been developed requiring less time for the microshoots to produce a vigorous rooting system while achieving high survival rates during acclimatization (data unpublished) (Fig. 4.12). Until such time as an economically viable micropropagation protocol is developed, the presently developed protocol is being used to supplement various steps involved in propagating new material to cut down the number of years spent in developing new cultivars through conventional breeding and selection methods.

Fig. 4.12. Microshoots producing vigorous root system following rooting *ex vitro*

4.4.1.4 Tea Seeds and Seed Gardens

Bi-clonal and poly-clonal seed gardens with known and proven cultivar combinations have been established and maintained as isolated seed gardens to obtain seeds of improved quality. To ensure proper isolation from nearby tea fields, those seed gardens have been established mainly in the rubber estates. Establishment and maintenance of appropriate seed orchards with desired parental lines is a prerequisite for successful seed production. Selection of progenitor clones for seed gardens was based on the success rate of parents used in controlled hybridization and also considering their flowering synchrony. The progenitor clones are raised as vegetative propagation plants and planted with wide spacing (5 m apart). In bi-clonal seed gardens progenitor clones are planted in alternative rows, whereas in poly-clonal seed gardens they are planted according to special planting arrangements, such as square planting and double triangle planting (Gunasekare & Anandappa, 2008).

Tea seeds are recalcitrant and hence need to be used as soon as possible. The mature seeds collected from seed bearers are subjected to sinker-floater assessment by immersing seeds in water. Floating seeds are discarded and sinkers are sown on sand beds prepared with river sand. Germinating seedlings are then transferred to nursery bags as used for the propagation of single nodal cuttings. All the other aftercare operations remain the same as for vegetative propagation nursery plants.

4.4.2 Extension System in the Country

In the tea sector, the extension function is performed mainly by the Advisory and Extension Division of the Institute at the main station at Talawakelle and at other regional stations at Ratnapura, Kandy, Passara, Kottawa and Deniyaya. A decentralized advisory and extension service operating at TRISL offers a better system of disseminating newly developed technologies as well as close interaction with growers. Research findings related to technologies developed for tea are disseminated through formal meetings organized by the TRISL scientists and extension officers to the end-users, as well as for the officials of the Tea Small Holding Development Authority (TSHDA) who, in turn, offer their extension services to the smallholders in the sector.

4.4.3 Strategy for Promotion of New Cultivars

The adaptive trials that are conducted in multi-locations with the partnership of the extension division and the progressive growers on their plantations are used as one of the strategies for promoting new cultivars among the growers. The awareness created by conducting those trials on growers' fields allows them to be acquainted with new cultivars. The acceptance of these new cultivars is enhanced through this approach while facilitating the selection of highly desirable cultivars that fulfill growers' needs and suit their socio-economic conditions in a wide range of target environments. Growers are advised to establish and maintain their mother bushes of newly released cultivars to meet their requirements in a more systematic way (Gunasekare, 2004). Regional trials conducted for evaluating the cultivars are used as demonstration plots to familiarize the growers with new cultivars. Regular awareness and training programs in the form of workshops, discussions and regional scientific meetings are conducted by the scientists in close collaboration with advisory personnel to make the growers aware of the new cultivars as well as the cultivars already released.

4.4.4 Cultivar Specific Issues

There are no cultivar specific issues related to field establishment conditions regarding harvesting and pruning. All cultural practices remain the same for all the cultivars released so far to the industry. At present, harvesting of tea cultivars is mainly being done manually (Fig. 4.13) and only a few estates are adopting the tea harvesting shear developed by the TRISL on a pilot scale (Fig. 4.14). Cultivars amenable for mechanical harvesting are being tested in the current breeding program.

Fig. 4.13. Manual harvesting of tea shoots

Fig. 4.14. Mechanical harvesting using shear innovated by the TRISL

4.5 Future Trends and Goals for Breeding and Selection

The future aims and goals of the tea breeding program need to be centered on addressing the key challenges facing the tea industry. The ever increasing cost of production is the main threat to the sustainability of the local tea industry. In the face of emerging challenges in the tea industry, locally and internationally, breeding and crop improvement of tea need to be focused on diversified breeding objectives to cater for the needs of end-users, both growers and consumers. Furthermore, the generally high costs involved in breeding new tea cultivars should be highly targeted when addressing the current problems encountered by the end-users. Hence, it is necessary to further ensure that the plant breeding research undertaken is relevant to the end-users needs.

4.5.1 Cultivars with Expanded List of Traits

The tea industry being a labor-intensive enterprise, mechanization of harvesting operations is essential to meet future challenges in labor scarcity by developing new cultivars amenable to mechanical harvesting. With the escalating price of inorganic fertilizers, cultivars that use fertilizers highly efficiently need to be researched. At the same time, it is necessary to select cultivars suitable for organic farming.

4.5.2 Diversifying Breeding Objectives

Food and Agriculture Organization (FAO) of the United Nations projections to 2017 indicate that world green tea production is expected to grow at a considerably faster rate than black tea, 4.5% annually, compared to 1.9% for black tea (www.fao.org). Though breeding approaches at TRISL are not yet geared towards selecting cultivars suitable for green tea production, this aspect will be pursued in the future program to retain a competitive edge in the international market. Furthermore, to meet the changing demands of the consumers and to cater for health conscious consumers, it is necessary to take steps to breed cultivars with low caffeine content. Hence, incorporation of such new traits will be looked into in future breeding programs. Rapid introgression of these novel characteristics from *C. irrawadiensis* P. K. Barua or *C. taliensis* (W. W. Smith) Melchior needs to be exploited through marker-aided introgression breeding.

4.5.3 Handling Climate Change

In the face of global warming and climate change, understanding the genetic potential of cultivars with regard to stress tolerance at the DNA level would be a key step towards more efficient and targeted breeding. With reduced availability of suitable high potential tea growing areas, tea cultivation has been extended beyond the traditional tea growing boundaries. This creates another challenge for tea breeders, namely the development of cultivars for marginal conditions.

Climate change may also have a longer term effect on tea growing areas, possibly inducing a shift in the geographical area where tea is grown, and the associated biotic constraints (stresses), to a higher altitude, as is already seen with the pest, shot-hole borer, which is now becoming a prevalent pest at high altitudes. Hence, the development of new cultivars with improved multi-trait associations is necessary. Incorporation of these resistant traits in the tea breeding program is often hampered by the limitations of phenotypic selections on adult plants. Accurate identification of elite cultivars is restricted by environmentally dependent screening techniques for many biotic stresses. Hence, there is an urgent need to develop early and accurate screening techniques while integrating advanced biotechnological methods to increase the precision of selection.

The broad tea growing regions based on the elevation categories need to be further divided into smaller target environments based on biotic and abiotic stresses prevalent in those specific areas. Initiatives have already been taken to develop a spatial map indicating potential productivity levels as well as stresses prevailing in target environments and to recommend cultivars suitable for planting in those areas, to assist farmers in selecting the most appropriate cultivars to suit their needs and environment.

4.5.4 Complementary and Cost-Effective Strategies in Germplasm Management

The cost involved in plant genetic resource management activities is a matter of concern with the limited financial budget allocation for research activities at the institute. Hence, complementary and cost-effective strategies need to be formulated for sustaining the germplasm conservation and management program in the long run. Assembling a core germplasm, establishing a cryogenic germplasm bank for preserving the base collection and expanding the available collection have been identified as integral components of the future breeding program.

4.5.5 Application of Molecular Markers and Genomics

The application of advanced biotechnological tools in the analysis and manipulation of plant genomes provided practical approaches to enhance the efficiency of classical breeding and crop improvement programs, especially for perennial crops such as tea. These techniques are also useful in germplasm managerial aspects and genetic diversity analysis to identify parents for breeding programs.

4.5.5.1 Fingerprinting of Cultivars

Many of the improved tea cultivars are often morphologically similar and hence cause difficulties in discriminating among them using morphological traits alone. Hence, cultivar specific molecular markers need to be developed for fingerprinting of improved cultivars. Although plant genotyping using molecular techniques has an immediate practical application, this has not been fully explored in tea. Fingerprinting needs to be used as a tool to resolve problems in mis-identification of germplasm and certification of clonal cultivars in control quality as well as to facilitate plant variety protection laws, even before such a regulatory framework is in place in the country.

4.5.5.2 Cultivars with Enhanced Yield Stability

To date, the advancements made in tea breeding have been mainly centered on conventional breeding techniques, although many constraints are involved in applying these in today's context. The development of molecular markers has opened up new perspectives in enhancing the efficiency of conventional breeding approaches. Though wide arrays of molecular marker technologies are available, many of them are yet to be exploited for their practical applications. Particular emphasis needs to be placed on improving cultivar adaptation across multiple and varied environments to reduce the yield gap. MAS needs to be targeted on polygenic traits. Yield, arguably the most important trait, is highly influenced by the environment. Bridging the yield gap and increasing yield stability under different environmental/stress conditions are of great importance for the sustainability of the industry. Hence, genomic maps need to be targeted to analyze the stability of detected quantitative trait loci (QTLs), especially for yield and yield related traits, across diverse environments.

Adequate validation of molecular markers is needed to ensure their accuracy and applicability in the breeding program. Though the application of biochemical and molecular markers in conventional tea breeding programs is still at its exploratory phase, there is great scope for integrating them, to aim at reducing the time scale and increasing the efficiency of selection in the cultivar development program.

4.5.5.3 Return to Investment in New Breeding Technologies

Once novel enabling technologies become available, the return on investment in those technologies will often dictate the degree to which those innovations are integrated into the existing tea breeding program with success. Even if MAS can potentially be very useful for genetic improvement of long-term crops such as tea, the cost of application and investment in developing the technology must be considered, as with many new technologies. The total cost of molecular markers needs to be calculated based on development costs (costs involved in identifying molecular markers and detecting the association between the molecular marker and the trait of interest) and running costs (typing individuals for the appropriate markers in the selection program).

It is also a concern that developing new technologies requires a huge amount of investment, which stimulates a shift in funds at the expense of the conventional tea breeding program. Therefore, development of the MAS approach requires careful prioritization of traits and even specific genes for which markers are to be sought, to ensure the relative cost effectiveness of MAS vs. conventional or phenotypic selection methods. Furthermore, for the application of MAS to be worthwhile, the markers identified would be applicable in different genetic backgrounds, rather than confining its use to the progeny of the single cross from which it was developed.

4.5.6 Decentralized Approach in Cultivar Selection

As we move into the new century, the production strategies need to be oriented to meet world trends and demands. Consumers in the Middle East market prefer the distinct taste of low-grown tea produced in Sri Lanka. This led the rural people in the low country region to start growing tea on their smallholdings. Their current contribution to the national tea production is 70%. Unlike the large plantations which are managed by private companies, those smallholders have special needs. The majority of the smallholdings are in less favorable and marginal environments where modern tea cultivars have failed to meet their expectations. As many modern cultivars need high input to get optimum returns, smallholders failed to benefit from those new improved tea cultivars. Accordingly, future breeding perspectives need to be focused on adapting cultivars to the environment and to clients through a decentralized cultivar selection approach. By doing that it would be possible to tailor-make cultivars that can harness the full potential of each environment.

It is worth noting that the potential of the extensive application of genetic engineering techniques does not eliminate the need for breeding programs to cope with genotype×environment (G×E) interaction because almost no cultivar can assemble genes conferring superior performance in all environments within a relatively large region. Also, the possible selection for yield based on molecular markers may require preliminary definition of adaptation and yield stability targets, since a remarkable portion of useful markers are environment-specific. In a wide adaptation prospect, MAS may prove distinctly less effective than multi-environment, phenotypic selection for yield in the presence of relatively large G×E interaction (Cooper *et al.*, 1999), suggesting that participatory variety selection (PVS) is more useful and effective in developing cultivars for complex, diverse and risk-prone marginal areas while meeting the diverse socio-economic status of the farmers in the low country (Gunasekare, 2006).

Direct selection in the target environment (or in an environment identical to the production environment) is always the most efficient way of handling the situation described above. In this situation, farmer-driven trials become important to develop cultivars to suit their needs since targeting heterogeneous and remote environments is difficult to address through a centralized formal breeding program (Gunasekare, 2007b). Through this approach, it is proposed to evaluate a large number of advanced breeding lines with the participation of growers in their fields, while increasing the access and awareness of the potential cultivars from the initial stages of the plant breeding program (Gunasekare, 2006). Using this approach it would be possible to breed tea cultivars which suit the needs of tea growers in variable environments and perhaps be no less important than the opportunities offered by biotechnology.

A spatial map based on tea productivity levels is indeed a timely necessity to exploit the full potential of the target environment while identifying location specific cultivars for the area. A guideline will be formulated to assist the growers in selecting suitable planting material for diverse environments, including marginal areas which can capitalize on the use of improved biclonal and polyclonal seed material which can withstand marginal conditions.

4.5.7 Perspectives for the Future

Considering various facets of the issue and the opportunities discussed above, the future research and development of tea breeding need to be focused on integrating advanced biotechnological tools and decentralized participatory approaches into the conventional tea breeding program, so as to develop new tea cultivars better suited to the future demands of the industry and varied consumer preferences in the global market. Unless the breeding program and associated knowledge about pedigree, phenotypic and molecular markers are fully integrated, the response to selection realized in the conventional breeding program remains elusive. Therefore, the immediate challenge is to integrate these tools and new approaches as discussed above into the conventional tea breeding schemes without any further delay, to produce new tea cultivars that thrive better under changing environments and meet the ever-changing demands of end-users/consumers.

References

Anandappa TI (1973) Annual Report. Tea Research Institute of Sri Lanka, pp.38-39.
Anandappa TI (1992) New tea clones. Tea Bulletin, 12: 28-33.

Anandappa TI, Nanayakkara R, Solomon H R (1988) Seed setting ability of some Sri Lankan tea clones and their implication for tea breeding. In: Proceedings of Regional Tea (Scientific) Conference, 19-21 January, 1988, Colombo, Sri Lanka, pp.73-87.

Anon (1976) Comparative character ratings of recommended tea clones. TRI Advisory Circular, November, 1976, C-10.

Anon (1994) New tea clones for experimental planting. TRI Advisory Circulars, December, 1994, C-13.

Anon (2002) The suitability of tea clones for the different regions. TRI Advisory Circular, No - PN1, December, 2002.

Anon (2008) Statistical Pocket Book – Plantation Sector. Ministry of Plantation Industries, Colombo, Sri Lanka.

Ariyaratna HACK, Gunasekare MTK (2006) Genetic base of tea (*Camellia sinensis* L.) cultivars as revealed by pedigree analysis. Journal of Applied Genetics, 48(2): 125-128.

Ariyaratna HACK, Gunasekare MTK (2008) Pistil related morphological traits reflect genetic diversity of tea in Sri Lanka, In: Proceedings of 2nd Symposium on Plantation Crop Research. 16-17 October, 2008, Colombo, Sri Lanka, pp. 383-388.

Ariyaratna HACK, Kottawa Arachchi JD, Gunasekare MTK (2007) Floral biology and breeding system of tea (*Camellia sinensis* L.): Implications on the tea breeding program. Sri Lanka Journal of Tea Science, 72: 31-43.

Arulpragasam PV, Lattif R (1986) Studies on the tissue culture of tea. 1. Development of culture methods for multiplication of shoots. Sri Lanka Journal of Tea Science, 55: 44-47.

Arulpragasam PV, Lattif R, Seneviratna P (1988) Studies on the tissue culture of tea: 3 regeneration of plants from cotyledon callus culture. Sri Lanka Journal of Tea Science, 57: 20-23.

Balasuriya J (1999) Shoot population density and shoot weight of clonal tea at different altitudes in Sri Lanka. European Journal of Agronomy, 11: 123-130.

Bombuwala RMTP (2001) Biochemical interactions in shot-hole borer infestation of tea and studies of three microbial polysaccharides. Postgraduate Institute of Science News, 1.

Boukema IW, Hintum JL, Astley D (1997) Creation and composition of the *Brassica oleracea* core collection. Plant Genetic Resource Newsletter, 111: 29-32.

Central Bank (2007) Annual Report. Central Bank of Sri Lanka.

Cooper M, Podlich DW, Jenson NW, Chapman SC, Hammer GL (1999) Modeling plant breeding programs. Trends in Agronomy, 2: 33-64.

De Alwis KA, Panabokke CR (1972) Handbook of the soils of Sri Lanka (Ceylon). Journal of Soil Science Society of Ceylon, 2: 219-230.

Dissanayaka STB, Wijewardena JDH, Samrappuli L (1999) Management of the wet zone soils. In: Mapa RB, Somasiri S, Nagarajah S (eds.) Soils of the Wet Zone of Sri Lanka, pp.160-175.

Frankel, OH (1984) Genetic perspectives of germplasm conservation. In: Arber

WK, Limensee K, Peacock WJ, Starlinger P (eds.) Genetic Manipulation: Impact on Man and Society. London: Cambridge University Press, pp.161-170.

Fu YB (2000) Effectiveness of bulking procedures in measuring population pair-wise similarity with dominant and co-dominant genetic markers. Theoretical and Applied Genetics, 100: 1284-1289.

Goonatilake WAS, Priyantha C, Mewan KM, Gunasekare MTK (2006) Genetic diversity in tea (*Camellia sinensis* (L.) O. Kuntze) as revealed by RAPD-PCR markers. In: Proceedings of International Symposium on the Issues and Challenges of the 21st Century. 4-8 July, 2006, Sabaragamuwa University of Sri Lanka, p.28.

Gunasekare MTK (2004) Systematic establishment of mother bushes (multiplication plots) of tea cultivars. TRI Update, 9(1): 3-4.

Gunasekare MTK (2006) Adapting crop varieties to environments and clients through decentralized—participatory approach. The Journal of Agricultural Sciences, 2 (1): 34-45.

Gunasekare MTK (2007a) Application of molecular markers to the genetic improvement of *Camellia sinensis* L. (tea): A review. Journal of Horticultural Science & Biotechnology, 82(2): 161-169.

Gunasekare MTK (2007b) Current status and future directions in breeding tea. In: Gunasena HPM, Girihagama PC (eds.) Current Status and Future Directions of Plant Breeding Research in Sri Lanka. Sri Lanka Council for Agricultural Research Policy, pp.111-124.

Gunasekare MTK (2007c) Section on plant breeding. Tea Research Institute Technical Report, 2007.

Gunasekare MTK (2008) New series of improved tea cultivars: the potential. In: Proceedings of Experiments and Extension Forum of Tea Research Institute of Sri Lanka, (217): 25-29.

Gunasekare MTK, Evans PK (1998) Isolation of protoplasts from leaf tissue of tea (*Camellia sinensis* L.): Factors affecting protoplast yield and viability. Tropical Agricultural Research, 10: 1-11.

Gunasekare MTK, Evans PK (2000a) *In vitro* rooting of microshoots of tea (*Camellia sinensis* L.). Sri Lanka Journal of Tea Science, 66 (1/2): 5-15.

Gunasekare MTK, Evans PK (2000b) *In vitro* shoot organogenesis in callus derived from stem tissue of tea (*Camellia sinensis* L.). Sri Lanka Journal of Tea Science, 66 (1/2): 16-26.

Gunasekare MTK, Evans PK (2001) Isolation and culture of mesophyll protoplasts from tea (*Camellia sinensis* L.). Plant Tissue Culture, 11(1): 55-64.

Gunasekare MTK, Ranathunga MAB (2003) Polyploidy in tea (*Camellia sinensis* L.) and its application in tea breeding: A review. Sri Lanka Journal of Tea Science, 68(2): 14-26.

Gunasekare MTK, Kumara JBDAP (2005) Tea genetic resources in Sri Lanka: Genetic resources originating from estate selections. Sri Lanka Journal of Tea Science, 70(2): 69-81.

Gunasekare MTK, Pieris TUS (2006) Phenotypic variation in germplasm accessions of tea (*Camellia sinensis* L.) in Sri Lanka. Plant Genetic Resource

Newsletter (Rome), 146: 39-42.
Gunasekare MTK, Anandappa TI (2008) Planting materials. In: Zoysa AKN (eds.) Handbook on Tea. Tea Research Institute of Sri Lanka, Talawakelle, Sri Lanka, pp.34-49.
Gunasekare MTK, Arachchi JDK, Mudalige AK, Peiris TUS (2001) Morphological diversity of tea (*Camellia sinensis* L.) genotypes in Sri Lanka. In: Proceedings of 57th Annual Session of Sri Lanka Association for the Advancement of Science (SLAAS). Part I, p.83.
Gunasekare MTK, Ratnayake M, Ratnagoda BA (2003) Tea reserves: Preserving the old seedling tea. TRI Update, 8(1): 5-6.
Gunasekare MTK, Piyasundara JHN, Upali PD (2004) Improved seed progenies of tea (*Camellia sinensis* L.): A source of planting material, In: Zoysa AKN, Mohamed MTZ (eds.) Plantation Crop Research-Current Trends and Future Challenges. The Tea Research Institute of Sri Lanka, pp.103-108.
International Plant Genetic Resources Institute (IPGRI) (1997) Descriptors for Tea (*Camellia sinensis*)., Rome, Italy.
International Tea Committee (ITC) (2009, 2010) Annual Bulletin of Statistics. London.
Kehl FH (1950) Vegetative propagation of tea by nodal cuttings. Tea Quarterly, 21: 3-17.
Kulasegaram S (1978) Progress in tea breeding. Tropical Agriculture Research series, (11): 151-160.
Kulasekera KML, Gunasekare MTK, De Costa WAJM (2004) Effects of plant factors on *in vitro* seed germination of tea (*Camellia sinensis* L.). Proceeding of Peradeniya University Research Session, Sri Lanka, 9(10): 12.
Kumar NS, Hewavitharanage P, Adikaram NKB (1995) Attack on tea by *Xyleborus fornicatus:* inhibition of the symbiotic, *Monacrosporium ambrosium*, by caffeine. Phytochemistry, 40: 1113-1116.
Kumara DAP, Ariyarathna HACK, Ratnayake M, Kottawa Archchi JD, Gunasekare MTK (2008) Assessment of flowering synchrony in tea (*Camellia sinensis* L.) germplasm accessions in Sri Lanka: Implications to tea breeding program. In: Proceedings of National Symposium 2008. 23-24 October, 2008, University of Ruhuna, Sri Lanka, p.34.
Liyanage AC, Fernando WMU, Pathirana KPSK (1999) Determination of genetic diversity of tea cultivars in Sri Lanka using isozyme polymorphisms. Report on FAO/IAEA seminar, Philippines, 1999, pp.85-87.
Mewan KM, Liyanage AC. Jayamanne E, Gunasekare MTK, Karunanayake E (2005) Studying genetic relationship among tea (*Camellia sinensis* L.) cultivars in Sri Lanka using RAPD markers. Sri Lanka Journal of Tea Science, 70(1): 42-53.
Mewan KM, Saha MC, Konstatin C, Pang Y, Abeysinghe ISB, Dixon RA (2007) Construction of an EST-SSR based saturated genetic linkage map of tea (*Camellia sinensis* L.). In: Proceedings of the 3rd International Conference on O-Cha (tea) Culture and Science. 2-4 November, 2007, Shizuoka, Japan, p.52.
Panabokke CR, Kannangara CR (1975) The identification and demarcation of the

agro-ecological regions of Sri Lanka. In: Proceedings of Annual Sessions. Sri Lanka Association for the Advancement of Science, 31: 49.

Park YG, Kaundun SS, Zhyvoloup A (2002) Use of bulked genomic DNA-based RAPD methodology to asses the genetic diversity among abandoned Korean tea plantations. Genetic Resources and Crop Evolution, 49: 159-165.

Piyasundara JHN (2008) Systematic characterization of tea (*Camellia sinensis* L.) germplasm using morphological descriptors. Master of Philosophy Thesis, Postgraduate Institute of Agriculture, Peradeniya, Sri Lanka.

Piyasundara JHN, Gunasekare MTK (2008) Morphological comparison of tea (*Camellia sinensis* L.) germplasm originating from old seedling tea populations in different agro-ecological regions in Sri Lanka, In: Proceeding of National Symposium 2008. 23-24 October, 2008, University of Ruhuna, Sri Lanka, p.30.

Piyasundara JHN, Upali PD, Gunasekare MTK (2003) Preliminary yield evaluation of improved seed tea cultivars. Tea Bulletin, 18: 15-19.

Piyasundara JHN, Gunasekare MTK, Pieris TUS, Wickramasinghe IP (2006) Phenotypic diversity of Sri Lankan tea (*Camellia sinensis* L.) germplasm based on morphological descriptors. Tropical Agricultural Research, 18: 237-243

Piyasundara JHN, Gunasekare MTK, Wickramasinghe (2008) Characterization of tea germplasm in Sri Lanka using morphological descriptors. Proceedings of 2nd Symposium on Plantation Crop Research. Colombo, Sri Lanka, 16-17 October, 2008, pp.389-395.

Punyasiri PAN, Abeysinghe ISB, Kumar V (2005) Preformed and induced chemical resistance of tea leaf against *Exobasidium vexans* infection. Journal of Chemical Ecology, 31: 1315-1323.

Punyawardena BVR, Bandara TMJ, Munasinghe MAK, Jayaratna Banda N (2003) Agro-ecological regions in Sri Lanka. Map of Sri Lanka.

Ramaswamy MS (1960) Copper in Ceylon tea. Tea Quarterly, 31: 76-81.

Ranathunga MAB, Gunasekare MTK (2003) A comparative assessment of some morphological and anatomical attributes to identify markers for screening polyploidy genotypes of tea (*Camellia sinensis* L.). Sri Lanka Journal of Tea Science, 68(1): 12-19.

Ranathunga MAB, Gunasekare MTK (2008a) Genotype × environment interaction and yield stability of tea cultivars in Sri Lanka. In: Proceedings of National Symposium 2008. 23-24 October, 2008, University of Ruhuna, Sri Lanka, p.14.

Ranathunga MAB, Gunasekare MTK (2008b) Assembling of preliminary core collection of tea (*Camellia sinensis* (L.) O. Kuntze) genetic resources in Sri Lanka. Plant Genetic Resource Newsletter (Rome), 155: 41-45.

Ranathunga MAB, Piyasundara JHN, Paskarathevan R, Gunasekare MTK (2004) An alternative criterion for selecting high yielding cultivars of tea (*Camellia sinensis* L.). In: Proceedings of the 60th Annual Session of Sri Lanka Association for the Advancement of Science, p.49.

Ranathunga MAB, Gunasekare MTK, Ratnayaka M (2008) Morphological attributes for prediction of quality of made tea during early selection stages of tea breeding. In: Proceedings of 2nd Symposium on Plantation Crop Research. 16-17 October, 2008, Colombo, Sri Lanka, pp.82-90.

Richards AV (1965) The origin of the popular TRI clones. Tea Quarterly, 36: 183-187.

Richards AV (1966) The breeding, selection and propagation of tea. Tea Quarterly, 37: 154-160.

Richards AV (1967) Some observations on the performances of the popular TRI and estate clones. Tea Quarterly, 38: 245-248.

Richards AV, Sebastiampillai AR (1964) A note on the identification of some TRI clones. Tea Quarterly, 35: 168.

Sarathchandra TM, Pieris R (2001) Induction of mutations in tea (*Camellia sinensis* L.). In: Proceeding of the 57th Annual Session of Sri Lanka Association of the Advancement of Science. Part I, p.44.

Sarathchandra TM, Upali PD, Wijewardena RGA (1988) Somatic embryogenesis in stem and leaf callus culture. Sri Lanka Journal of Tea science, 57: 50-54.

Sarathchandra TM, Sarathchandra K, Hiriburegama K (1997) Root formation on *in vitro* micropropagated shoots of tea. Sri Lanka Journal of Tea science, 65: 5-10.

Sebasthiampillai AR (1970) A simple technique for the induction of polyploids in tea. Tea Quarterly, 46: 12-15.

Seran TH, Hiriburegama K, Gunasekare MTK (2005) Encapsulation of embryonic axes of *Camellia sinensis* L. (tea) and subsequent *in vitro* germination. Journal of Horticultural Science & Biotechnology, 80(1): 154-158.

Seran TH, Hiriburegama K, Gunasekare MTK (2006a) Short-term storage of encapsulated zygotic embryonic axes of tea at low temperature. Tropical Agricultural Research, 18: 358-366.

Seran TH, Hiriburegama K, Gunasekare MTK (2006b) Direct somatic embryogenesis from explants obtained from *in vitro* germinated embryonic axes of *Camellia sinensis* (L.) O. Kuntze. Journal of Horticultural Science & Biotechnology, 81(5): 883-890.

Seran TH, Hiriburegama K, Gunasekare MTK (2006c) Somatic embryogenesis from embryogenic leaf callus of tea (*Camellia sinensis* L.). Tropical Agricultural Research, 18: 367-375.

Seran TH, Hiriburegama K, Gunasekare MTK (2007a) Production of cotyledon-type somatic embryos directly from immature cotyledonary explants of *Camellia sinensis* L. Journal of Horticultural Science & Biotechnology, 82: 119-125.

Seran TH, Gunasekare MTK, Hirimburegama K (2007b) Germination and subsequent plant development of *in vitro* cultured zygotic embryos and embryonic axes in comparison to conventional seed propagation of tea (*Camellia sinensis* L.). Sri Lanka Journal of Tea Science, 71(2): 27-39.

Singh ID, Gunasekare MTK (2000) Conservation of tea genetic resources. TRI Update, 5(1): 1-2.

Singh ID, Shanmugarajah V, Gunasekare MTK, Ratnayake M, Sritharan U, Gunadasa SW (2003) Tea breeding in Sri Lanka (Chapter 3). In: Modder

WWD (eds.) Twentieth Century Tea Research in Sri Lanka. The Tea Research Institute of Sri Lanka, pp.37- 46.

Sivapalan P (1986) A strategy to adopt the appropriate genetic diversity for planting clonal tea. Sri Lanka Journal of Tea Science, 55(2): 53-57.

Sundaravathany A, Kumara AP, Gunasekare MTK (2005) Characterization of tea germplasm using reproductive traits. Journal of Jaffna Science Association, 12: 19.

Tea Research Institute of Sri Lanka (2008) Handbook on Tea. Talawakelle: Sri Lanka.

Thirukkumaren G, Gunasekare MTK (2001) Use of pollen morphology and physiology to differentiate ploidy levels of tea clones. Journal of Jaffna Science Association, 9(1): 6-7.

Tubbs FR (1939) The improvement of planting material. Tea Quarterly, 11: 38-49.

Tubbs FR (1946) Tea selection – the present position. Tea Quarterly, 17: 59-65.

Visser T (1958) The position of clonal selection in Ceylon. Tea Quarterly, 29: 154-159.

Walgama RS, Gunasekare MTK, Jayathilake MM, Kottawaarchchi JD, Liyanage DD (2008) Evaluation of tea (*Camellia sinensis* L.) germplasm for host-plant resistance to shot-hole borer, *Xyleborus fornicatus* Eichh (Coleoptera: Scolytidae). Sri Lanka Journal of Tea Science, 73: 29-38.

Wickramaratne MRT (1981a) Variations in some leaf characteristics in tea (*Camellia sinensis* L.) and their use in the identification of clones. Tea Quarterly, 50: 183-198.

Wickramaratne MRT (1981b) Genotype-environment interactions in tea (*Camellia sinensis* L.) and their implication in tea breeding and selection. Journal of Agricultural Sciences, 96: 471-478.

Wijeratne MA (2004) Tea industry in Sri Lanka. In: Proceedings of the International Conference on O-Cha (tea) Culture and Science. 4-6 November, 2004, Japan, Shizuoka, pp.51-54.

5

Tea Improvement in Kenya

Samson M. Kamunya[1*], Francis N. Wachira[1], Ram S. Pathak[2], Richard C. Muoki[1], Ram K. Sharma[3]**

[1] Tea Research Foundation of Kenya, P.O. Box 820, Kericho 20200, Kenya
E-mail: *skamunya@tearesearch.or.ke; **fwachira@tearesearch.or.ke
[2] Department of Crops, Horticulture and Soil Sciences, Egerton University, P.O. Box 536, Egerton, Njoro 20115, Kenya
E-mail: ramspathak@yahoo.com
[3] Division of Biotechnology, Institute of Himalayan Bioresource Technology, Post Box 6, Palampur, Himachal Pradesh 17061, India
E-mail: ramsharma@ihbt.res.in

Abstract: The chapter presents a detailed account of efforts at tea improvement in Kenya with achievements made and challenges to be surmounted since tea was first introduced into the country. Although tea was introduced into Kenya at the turn of the 20th century, concerted efforts at tea improvement could not take root until the early 1960s after the country gained self-rule from Britain. Owing to the heterogeneity of pioneer seedling populations that was accompanied by management constraints, early research efforts resulted in the development of whole single cuttings as propagation materials which, coupled with clonal selection, led to the release and commercialization of high yielding and quality clones. This resulted in rapid expansion of the Kenyan tea industry. To date, the Tea Research Foundation has released a total of 50 high yielding and good quality tea clones for commercial utilization, not just in Kenya alone but also in the entire East African region. The Kenyan tea industry, which almost solely involves the export of black CTC tea, is currently experiencing problems as a result of global annual over-production that is outstripping demand. To counter the declining revenue base of tea enterprises, attempts to undertake tea product diversification and value adding have been initiated. Furthermore, tea improvement activities integrating molecular markers and participatory clonal selection involving farmers and consumers are expected to fast-track the development and adoption of novel varieties within a relatively short period.

5.1 General Introduction to the Kenya Tea Industry

Tea plant (*Camellia sinensis* (L.) O. Kuntze) is the most widely consumed non-alcoholic beverage in the world and, consequently, the most important crop species in the genus *Camellia*, which has over 200 species reported so far (Chang & Bartholomew, 1984). Tea is popularly consumed either as a green (non-fermented), white, yellow or Oolong (semi-fermented) or black, dark (full-fermented) beverage with the process of manufacture for each type varying (Hampton, 1992; Takeo, 1992). Of the current global production of tea, about 78% comprises black tea, 20% green tea and 2% Oolong tea. While black tea is popular in Asia and some countries in the West, Oolong and green teas are largely consumed in China, Japan, Korea and a few countries in North Africa and the Middle-East, respectively (Basu, 2003). The manufacturing process of black tea, in particular, normally entails an oxidation of the polyphenolic compounds (fermentability) during which certain chemical changes take place (Hampton, 1992). Other tea products include the orthodox teas, decaffeinated tea, specialty and herbal teas and silvery tips (Banerjee, 1992b; Gill, 1992). Quality of each type of tea product is largely dependent upon the type of tea cultivar, which provides the raw material for its manufacture. Studies undertaken to evaluate quality of black tea by chemical means have revealed theaflavin (TF) and thearubigin (TR) to be major components that are largely responsible for briskness, brightness, strength and color of black tea (Roberts, 1958, 1962; Woods & Roberts, 1964; Owuor *et al.*, 1986, 2006). Thus, when tea liquor is termed as bright and brisk, usually such teas fetch high market value. Thus, quality improvement, in addition to other important attributes of tea, has been an essential component of tea breeding.

5.1.1 Introduction of Tea to Africa

The first tea plant was grown in the Cape of Good Hope in South Africa in 1687, but actual planting on the African continent was in the surrounding area of Durban in the second half of the 19th century (Anon, 1962). It was then planted near Blantyre, Malawi, in 1878 with good results. This elicited a commercial tea growing venture that commenced in 1891 at Malanje, Malawi. Successful tea growing in Malawi resulted in its spread to Mozambique, Zimbabwe and Tanzania during the 1920s. There are no reports, however, regarding the source of planting materials in these countries (Anon, 1962).

5.1.2 Introduction of Tea to Kenya

Tea was reportedly introduced into Kenya by the Caine brothers who imported dark-leafed "Manipuri" hybrid seed from Assam in 1904 and 1905 to establish a plantation at Limuru, Central Kenya (Matheson, 1950). In 1912, *C. sinensis* seed was imported from Sri Lanka to establish a plantation of tea with high quality and yield (Matheson, 1950). According to Matheson (1950), little interest existed over the next 12 years except for several small plantations which were established at Limuru in the East of Great Rift Valley, Kericho and Kaimosi in the West of Great Rift Valley (Fig. 5.1). However, advice given by the Howland brothers in 1924 on the use of quality seed from the light colored leaf var. *assamica* or Manipuri types for drought resistance stimulated serious planting by several companies. The planting expanded rapidly and by 1929 there were 2,162 ha of tea in Kenya (Greenway, 1945). By 1963, the acreage had increased to 21,448 ha and in 2009 the acreage stood at 158,394 ha (ITC, 2010).

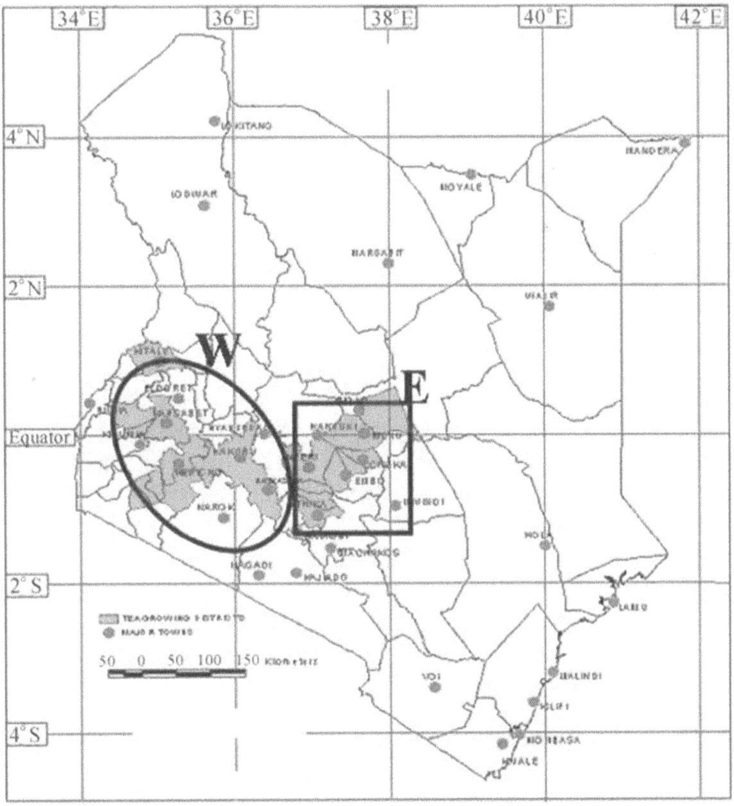

Fig. 5.1. Tea growing districts of Kenya (Map Source: Kenya Tea Development Agency (KTDA))

5.1.3 Economic Importance of Tea in Kenya

Tea is a major foreign exchange earner and a source of livelihood for millions of people in the tea growing world (Reeves *et al.*, 1987). Currently, tea is one of the leading cash crops in Kenya and makes a significant contribution to the economy. In the year 2010, the country produced 399,006 tonnes of made tea of which over 379,000 tonnes were exported, earning KSh 97 billion (US$ 1.23 billion) (Tea Board of Kenya (TBK), 2010) to register the highest export performance recorded by the industry. This represents about 26% of the total export earnings and about 4% of the gross domestic product (GDP). The major importers of Kenyan tea in 2010 were Pakistan, Egypt, the United Kingdom and Sudan, representing 22%, 21%, 14% and 5%, respectively (TBK, 2010). As the tea industry in Kenya is largely based in the rural areas where the vast majority of Kenyans live, over 62% of the crop is produced by the highly successful smallholder sub-sector, offering a direct source of livelihood to over 3 million people (about 10% of the total population) (TBK Statistics). Its contribution towards poverty eradication and infrastructural development in the rural areas has therefore been enormous. It also contributes to environmental conservation through enhanced water infiltration, reduced surface erosion and mitigation of global warming through carbon sequestration.

Internationally, Kenya ranked third in 2009 in annual tea production after China and India (ITC, 2010). However, a combination of a favorable tea growing environment and availability of improved clones puts Kenya in the lead in tea productivity (ITC, 2010). It is noteworthy that Kenya accounted for over 10% of world production and over 20% of the export share (FAO, 2010; ITC, 2010), making her the leading exporter of black tea. About 95% of Kenyan tea is exported as a generic product which is used to blend low quality teas from other countries. Currently, Kenya produces CTC black tea as the only product, for which the prices have declined owing to a global glut. Being the leading exporter of black tea on the world market, Kenya perhaps has borne the greatest impact of price fluctuations and stagnation, largely as a result of the recession in the world's leading economy, the USA. This calls for concerted action, the chief being the diversification and added value of tea products, not only to help Kenya maintain its position as a leading exporter but also to enhance foreign exchange earnings. This has been a major challenge that is being addressed in a multidisciplinary approach by researchers, processors, promoters and other stakeholders.

5.1.4 History of Tea Improvement in Kenya

Tea improvement efforts may have commenced with its first introduction into Kenya by white settler farmers at the turn of the 20th century (Anon, 1962). Seed comprising random open pollinated natural hybrids between the *C. sinensis* var.

assamica and var. *sinensis* cultivars was procured from the Assam region of North East India (Singh, 1979; Anon, 1962). The largely Manipuri hybrid seed was first planted in Limuru. The tea seedling populations arising from this planting became the source of seed for subsequent planting. However, the first commercial plantation could not be established until 1924. The early industry was dominated by colonial settlers who had the sole right to seed access. In 1960, the Special Crop Development Authority (SCDA) was founded to promote the cultivation of the crop within the smallholder agricultural subsector (M'Imwere, 1997). This was to later evolve into the Kenya Tea Development Authority (KTDA) whose major early task was to facilitate expansion of tea cultivation into native lands. The sector later saw rapid expansion and currently it accounts for about 55% of all tea produced in Kenya (ITC, 2010).

The pioneer tea plantations were established with planting material mainly from the descendants of the Manipuri hybrid seed (Wachira, 2002a). Reports indicate some hybrid seeds were introduced from the Mt Vernon Estate in Sri Lanka (Anon, 1962) although the proportion of this in the pioneer plantations is unknown. From the 1960s, supposed improved tea seed was also introduced from Uganda. These seeds were originally derived from germplasm introduced from Dangri Manipuri, Betjan Assam and Rajghur in India (Anon, 1962). However, in all cases of seed introduction, no data on collection and passport descriptors were kept (Wachira, 2002a). Scanty information on the actual seed source notwithstanding, it is generally acknowledged that its origins are in North East India. Early breeders therefore were able to select seed parents, which to them contained outstanding attributes. Thus, the Indian hybrid seed provenances from which pioneer seedling tea populations were established in Kenya were mixed random collections from local races (Anon, 1962; Cannell *et al.*, 1977; Singh, 1979). The early planters might have conducted mass selections whereby visual selection for *jats* akin to var. *assamica* in the seedling populations was practised on the basis of general vigor, plucking point density and large shoot size which later became the open-pollinated seed bearers (Green, 1966). This, however, resulted in slow progress in the yield and black tea quality improvement even though the later generations of seedling populations were much better than the ancestral pioneer stock. Lack of uniformity further compounded the problem as seedling populations were comprised of unique genotypes. Furthermore, initial selection was biased towards yield with little attention being accorded to other attributes such as black tea quality and resistance/tolerance to biotic and abiotic stress factors.

The attempt to raise more uniform progenies from selected seed parents based on arbitrary criteria and the ability to produce many seeds only led to the production of more variable populations as breeders were not fully in control of the pollination process. Visser (1969), for example, observed that good seed producers might not necessarily result in good clones in terms of tea production, especially where the trait of interest has low heritability.

Even though precise information of collection is unavailable, it is highly possible that the initial germplasm was obtained from restricted sources and therefore commercial plantations may have a narrow genetic base. This is evident through an appraisal carried out in 1999 (Wachira, 2002a), which revealed that clonal tea accounted for 38% and 80% of all tea in the estate and smallholder sectors, respectively, in Kenya. Nationally, the assessment found out that cultivation of clonal plants had greatly expanded to about 60%. Besides, the highly diverse pioneer seedling plantations were being uprooted for replacement with improved clones resulting not only in the loss of valuable genetic resources but also instant fixation of a few genotypes. The survey showed that despite the wide availability of a wide range of clonal cultivars to choose from, most growers cultivated only a narrow band, most of which was closely related genetically and therefore the growers were not accessing sufficient diversity at the farm level. Arising from the survey and other studies (Wachira *et al.*, 1997, 2001), deliberate efforts to introduce new germplasm, mainly to expand diversity and develop novel cultivars aimed at diversification of tea products, have been made (Tea Research Foundation of Kenya (TRFK), 2005). Thus, green tea cultivars from Japan and China have been introduced in the TRFK tea breeding program, which is currently being reviewed to make it co-evolve with emerging challenges in the global tea arena.

Perhaps what attributed to the successful development of the tea industry in Kenya are the tremendous efforts put into tea improvement research, and favourable climatic conditions and widespread popularity of the tea beverage. As the pioneer tea plantations comprised seedling tea populations, difficulties in management and quality maintenance arising from the heterogeneity in tea plantations elicited the research and development of vegetative propagation techniques. This resulted in the discovery of the single whole-leaf cutting of tea as a propagule and the release of clonal teas from the early 1960s (Goodchild, 1960). The early tea improvement efforts in Kenya saw the release of better yielding but poor to moderate black tea quality or high quality but low yielding clonal teas compared to the existing seedling teas. Although efforts to recombine the two traits in one genotype through hybridization have increased significantly over the last decade or so, a lack of prerequisite knowledge of inheritance patterns and the combining abilities of the desirable traits, like yield, quality and other secondary attributes, have resulted in slow progress in tea improvement. Emphasis has mostly been placed on good adaptations for varying growing conditions, yield and quality. Studies have also been initiated to understand the tea genetics better, based on molecular techniques (Wachira *et al.*, 1997; Hackett *et al.*, 2000; Wachira, 2002b), for the purpose of enhancing tea improvement. In addition, the ever changing global climate which at times leads to unprecedented weather phenomenon and which is worsened by the extension of tea growing in areas not traditionally meant for tea, have resulted in the emergence of new or increased virulence of endemic diseases and pests as well as prolonged drought periods. Lack of basic information on the defense mechanisms and stress tolerances of tea has circumvented attempts to develop pest or disease and abiotic stress resistant cultivars, and the development of ideotypes with an ideal architecture in terms of leaf arrangements (Yao *et al.*, 1987). Thus, the need exists to develop clonal teas with combined high yields and acceptable black to green tea quality that are also highly tolerant to abiotic and biotic stresses.

5.1.4.1 Early Tea Improvement Activities Elsewhere

The history of tea breeding dates back thousands of years in ancient China. Ever since the commencement of tea growing it was recognized that improvement of tea entails problems that are somewhat unique to the plant. This is so because firstly, unlike other woody perennials, in tea only a part of the total biomass constitutes the harvest and secondly the plant is highly heterogeneous and strongly, but not absolutely, a self-incompatible tree species (Rogers, 1975; Wachira & Kamunya, 2005a). It was noted that tea sets better with pollen from another bush, the average set of the plant with its own pollen being about one quarter of that obtained by cross pollination (Wight & Barua, 1939). Where selfing occurs, the seeds are smaller with reduced germinability or there are no seeds at all (Mamedov, 1961; Sebastiampillai, 1963). Consequently, the earlier breeding strategy relied on artificial pollination between plants that differed in some morphological features as a way of producing superior planting materials.

In the first authentic literature for tea, *Tea Classics* (*Cha Ching*) written by the Chinese tea scientist Lu Yu during the years 760 – 770 A.D., there were comprehensive descriptions of tea origin, cultivation, manufacture, drinking methods, history and culture etc. (Wu, 1987). A tea farmer in Fujian Province, China, successfully developed the first cuttings for the vegetative Propagation method in the 1780s. Shortly afterwards, a famous Oolong tea clone 'Tieguanyin' was successfully bred. Late in 1857, a famous green and black tea clone 'Fuding Dabaicha' was bred successfully. Currently, they are both predominant tea cultivars in Chinese tea gardens (China Tea Varieties Compilation Committee, 2001).

The early phase of tea breeding concentrated more on production of sufficient planting material rather than on high yield and quality, which were to be the main objectives of breeding later. In Assam and other parts of North East India, the emphasis was on mass-selection which involved random crosses between plants apparently varying in leaf shape, size, texture and growth features (Wight, 1956). Mass-selection, however, often failed not only to produce tea of high quality but also the uniform morphological attributes so essential for high yields and quality (Barua, 1963). However, it resulted in the development of several seed cultivars in Assam that were superior in yield and quality to *jats* which had been randomly planted earlier.

In Sri Lanka, mass-selection was restricted to choosing the outstanding mother bushes which could be propagated vegetatively to produce high yielding uniform stands (Visser & Kehl, 1958). Unlike in Assam, the emphasis in Sri Lanka was not so much on the selection of seed bearers (Visser, 1969). However, in African countries most tea populations were initially grown from open pollinated seeds (Cannell *et al.*, 1977). Owing to marked environmental heterogeneity (Hasselo,

1964) and lack of adequate genetic variability for mass-selection and the continuous exploitation of the same population for further improvement and expansion, very little or no progress could be realized (Green, 1971).

5.1.4.2 Criteria of Selection and Release in Early Tea Improvement Efforts

Tea is essentially a perennial plant that is highly outbreeding and self-incompatible (Rogers, 1975; Wachira & Kamunya, 2005a). Like any other woody perennial tree crop, tea has a lengthy juvenile period, poor juvenile-mature correlation, especially for growth characteristics and large plant size (TRFK, 1991). Following the development of vegetative propagation as the preferred method of producing uniform tea fields that are easy to manage for uniform tea quality and yield, initial releases were based on rootability, nursery growth, fast fermentability and field performance (Hainsworth, 1965). The best seedling fields were initially used as checks for yield and cup quality with the emerging clones being released upon outperforming these fields (Hainsworth, 1965; Green, 1969; Njuguna, 1985). Other criteria adopted included leaf color (Todd, 1955; Goodchild, 1960), with a light green color being thought to be associated with cup quality. Leaf pose particularly erect to semi-erect (Njuguna, 1989) and preference by pluckers for tea bushes with large and heavy shoots (Njuguna, 1987) were also considered important attributes for selection.

Selection for black tea quality has relied on the relationships between morphology and quality. Some workers, for example, have shown that pubescence has a positive correlation with quality, particularly that of orthodox tea (Wight & Barua, 1954; Venkataramani & Padmanbhan, 1964; Wu, 1964). Besides hairiness, different shades of leaf greenness are also considered to be expressions of quality as measured by liquor color, strength and overall aroma; light leafed cultivars being considered to produce high quality black tea, while dark-green and pale-green cultivars are associated with low quality (Wight *et al.*, 1963). Fermentability tests (Sanderson, 1963) and total polyphenol contents (Obanda *et al.*, 1997) are currently being used as potential indicators of black/green tea quality.

5.2 Collection, Appraisal and Utilization of Tea Germplasm

Past work (TRFK, 2001a; Wachira, 2002a) revealed that clonal tea accounted for about 60% of all tea in Kenya. However, in spite of the wide range of divergent clones to choose from, it has been found that farmers have chosen only a few clones for wide cultivation. This implied that farmers in Kenya were not effectively and efficiently utilizing the existing genetic resources (Wachira, 2002a). Both the tea plant (*C. sinensis*) and its relative species are important tea germplasms.

5.2.1 The Collection and Introduction of Tea Cultivars

A lot of diversity was, however, retained in pioneer seedling tea plantations most of which have been under the threat of being uprooted and replanted with clonal tea. In a study conducted in 2000 and 2001, analysis of molecular diversity revealed that the most widely cultivated clones accessed only a portion of the total available diversity in the pioneer plantations in Kenya (Wachira, 2002a). The existing gene pool in pioneer plantations in Kenya had therefore not been efficiently used. These plantations needed to be sampled for conservation before they were uprooted. Their sampling would contribute towards the establishment of a core collection of tea germplasm. Such a core collection is constructed in a way that it includes much of the available diversity with a minimum number of individuals. As this kind of activity requires active participation of major tea growers, the Foundation has been awaiting information from the stakeholders on their current acreage under seedling tea in their fields, the country of origin of the seedling material, year of introduction and current yield performance in kilograms of made tea per hectare prior to strategizing on how to undertake nationwide sampling of seedling tea for conservation.

Studies done in the past (Wachira *et al.*, 1995, 1997; Paul *et al.*, 1997) revealed that even though the Kenyan tea genetic resources were sufficiently diverse, there was a need to introduce unique tea germplasm preferably from China and Japan, which the country had hitherto not accessed. Thus, Material Transfer Agreements were entered between Kenya and Japan on the one hand and China and Kenya on the other, leading to the introduction of mostly var. *sinensis* tea genetic resources from Japan and China in 2001 (TRFK, 2002). The germplasm was comprised of seeds from popular green tea cultivars Yabukita and Yutakamidori from Japan and Chinese green and black tea cultivars, namely Yinghong 1, Hanlu, Xiuhong, Wulinghong and Yinghong 2. The cultivars have since been cloned (TRFK, 2006) and transplanted into field trials where they have established themselves well. Furthermore, in 2003, a germplasm exchange agreement with the Tea Research Institute of Tanzania (TRIT) was signed. This enabled TRFK to import clonal cuttings of 10 accessions of tea from Tanzania, namely TRIT 201/10, TRIT 201/16, TRIT 201/43, TRIT 201/44, TRIT 201/47, TRIT 201/50, TRIT 201/55, TRIT 201/73, TRIT 201/75, and TRIT 201/82. Additionally, more introductions were carried out in November 2004 comprising another batch of tea seed from China (TRFK, 2006). The germplasm was subjected to the requisite phytosanitary screening and, while under further observation in an open quarantine facility at TRFK, was scaled-up/cloned and transplanted in the field during the 2008 planting season (Kamunya & Muoki, 2008). The majority of these materials have established themselves well and efforts to have the appropriate raw material for the development of improved green tea cultivars are right on course.

5.2.2 The Relative Tea Species

Several species in the genus *Camellia* have been found to naturally hybridise freely. The prospect of using interspecific hybridisation in improving some traits in tea e.g. cold hardiness, drought tolerance, specific characteristics in chemical components, disease and pest resistance etc. therefore exists. This may be possible without impairing quality particularly if progenitors are chosen based on taxonomic and genetic proximity to tea. Scientists have often raised doubts about whether or not the existing tea populations in plantations have resulted from hybridisations between the three main varietal taxa only (i.e., var. *sinensis*, var. *assamica*, var. *assamica* ssp. *lasiocalyx*) or have also involved other *Camellia* species. Recent revisions of taxonomy within section *Thea* to which tea belongs have led to the identification and description of additional varieties of tea; namely var. *waldenae*, var. *dehungensis* and var. *publimba* (Chang & Bartholomew, 1984; Banerjee, 1992a). In Taiwan of China, a new *Camellia* species known *as C. buisanensis* Sasaki has been described (Su *et al.*, 2004). The contribution of these taxa to the genepool of cultivated tea in Kenya is unknown. Leaves from some other *Camellia* species e.g. *C. taliensis*, *C. grandibractiata*, *C. kwangsiensis*, *C. gymnogyna*, *C. crassicolumna*, *C. tachangensis, C. ptilophylla* and *C. irrawadiensis* are used as the source of a tea like beverage in parts of China which indicates that the breeding potential of additional underutilized *Camellia* species is very great (Chang and Bartholomew, 1984; Banerjee, 1992a). Because of the close morphological resemblance between many *Camellia* species, it is possible that several non-tea species and their hybrids with tea have gone undetected in our tea fields. For example, the presence of some tea bushes in seedling fields with leaf punctations, brick red leaf pigmentation (associated with the anthocyanin pigment and/or sasanquin) and others with coffee like aroma maybe indications of past interspecific introgression events (TRFK, 2004).

In an effort to access diversity from the secondary and tertiary gene pools of tea, several *Camellia* species were imported into the country (TRFK, 2001a). In 1999, these species which included eight horticultural cultivars of *C. japonica, C. brevistyla, C. sasanqua, C. irrawadiensis* and clones of the three tea varieties (var. *sinensis*, var. *assamica*, var. *assamica* ssp. *lasiocalyx*) were planted out in "TRFK *Camellia Gene Bank*", partly for conservation and partly for inter-specific hybridization (wide crosses). Other accessions included in the gene bank are *C. assimilis*, *C. oleifera*, *C. kissi*, *C. chrysantha*, *C. furfuraceae*, *C. brevistyla* and some teas from Taiwan of China (Taiwan yamacha TYC 43, TYC 57, TYC 78, TYC 80 and TYC 87) and Vietnam (Clone 3). Currently, the TRFK is conserving over 250 accessions of tea and related species.

5.3 Current Tea Breeding Strategies

Tea breeding essentially consists of three phases: generation of genetic variability, selection of useful genotypes and comparative tests to demonstrate the superiority of the selected genotypes. A third phase that involves exposing pre-released and promising clones to multiple sites (genotype-environment interaction) for stability and adaptability is always the final phase in plant improvement programs. Generally, the first two phases have been given adequate attention even though somehow confined to on-station trials in the two TRFK experimental stations. Simultaneous comparative testing of test genotypes has been hampered by lack of testing sites in different tea growing areas. The third phase has been undergoing rationalization as earlier efforts to expose promising clones to other sites were haphazardly done. Interested factories (as farmers' representatives) were asked to collect 500 free single-whole leaf cuttings for adaptability trials on their farms. Where technical follow-ups were undertaken, they were insufficient and in most cases there were none. Consequently, the performance of early releases with regard to different tea growing regions could not be documented. It is currently thought that involvement of farmers and other end-users right from selection of breeding materials to multilocational testing is the most cost-effective way of developing new elite cultivars with wider acceptability.

Tea breeding in East Africa commenced with the establishment of a breeding seed *barie* (orchard) at Rwebitaba Estate in Uganda in 1967 (Green, 1973). Two more polyclonal seed *baries* were established at Kangaita (TRFK, 1980) and Timbilil (TRFK, 1990) after the incorporation of TRFK in 1980. Additionally, the major tea companies in East Africa, notably James Finlay Kenya Ltd, Brooke Bond (Kenya), Eastern Produce Kenya Ltd and George Williamson (Kenya), initiated their own tea improvement activities by establishing seed *baries*. Successive breeding efforts at the TRFK have seen the expansion of the existing polyclonal seed *barie* as well as the establishment of new biclonal seed *baries* using elite commercial and promising cultivars. The various parental materials used in polyclonal and biclonal seed orchards in order to enhance hybridization operations are shown in Tables 5.1 and 5.2 (TRFK, 2006).

At Rwebitaba, parents consisting of elite Tea Research Institute of East Africa (TRIEA; currently defunct) clones were used as breeding stocks, with clone TRFK 6/8 as the common parent (Green, 1973). As a result, a total of 27 clones related to TRFK 6/8 were released for commercial utilization from this breeding program at Timbilil Estate, Kericho (Table 5.3). The clones constitute 60% of released clones to date and some of them have given comparable or better yields than TRFK 31/8 and AHP S15/10 (e.g. TRFK 303/577 and TRFK 303/1199), which are currently used as standard checks for high yield. Subsequent breeding and selection efforts have led to the injection into the tea industry of a further 22 clones to date. Thus, the Tea Research Foundation has released a total of 49 high yielding and good quality tea clones for commercial utilization, not just in Kenya alone but also in the entire East African region (Table 5.3). The improved cultivars have been published in the Regional Variety List for Kenya, Uganda and Tanzania (Association for Strengthening Agricultural Research in Eastern and Central Africa (ASARECA), 2004) and the Kenya National Crop Variety List (Kenya Plant Health Inspectorate Service (KEPHIS), 2004) for wider coverage. While only a small portion of all the released clones is currently being utilized by growers, the earlier released unutilized ones may have been rendered irrelevant by the more recently developed elite clones that have shown markedly better performance in a combination of preferred attributes.

Table 5.1 Breeding stocks and their expected genetic contribution in breeding program

High yield potential	High quality potential	Pest tolerance/ resistance	Drought tolerance	High soil pH tolerance	Cold tolerance	Genetic study
TRFK 31/8	TRFK 6/8[9]	TRFK7/9[3]	TRFCA SFS150	EPK TN14-3	EPK TN14-3	TRFK 12/2[1]
TRFK 303/577[8]	GW Ejulu-L	TRFK 57/15[3]	TRFK 303/577[8]	NDT Tai	TRFCA SFS150	TRFK K-Purple
TRFK 301/4	EPK TN 15-23	AHP SC31/37[3]			EPK C12	TRFK 31/30[2]
TRFK 301/5		AHP S15/10[3]			NRIT Yabukita[6]	TRFK 311/287[2]
EPK C12		EPK TN14-3[5]			NRIT Yutakamidori[6]	TRFK 382/1[7]
BBLK 35		TRFK 303/1199[3]				TRFK 382/2[7]
AHP S15/10[9]		TRFK 54/40[4]				TRFK 386/2[7]
AHP SC12/28[9]		TRFCA SFS150[3]				TRFK 371/1[7]
AHP SC31/37		AHP CG28U864[4]				TRFK 306[10]
AHP CG28V929[9]		TRFK 301/1[4]				
AHP CG28U864		TRFK L/16[4]				

[1] Non fermenter; [2] Tetraploid; [3] Resistant to red crevice mite; [4] Susceptible to Scales; [5] Preferred but highly tolerant to red crevice mite; [6] Green tea cultivar of low catechin content; [7] Triploid; [8] Susceptible to root knot nematodes; [9] Very susceptible to water stress; [10] anthocyanin tea cultivar

Table 5.2 Isolated biclonal seed garden progenitor clones and their attributes

Clones	Location	Attributes
1. TRFK 6/8 and AHP SC31/37	Timbilil	High quality and yield
2. TRFCA SFS150 and GW Ejulu-L	Timbilil	Drought tolerance, high yield and quality
3. TRFCA SFS150 and AHP CG28V929	Timbilil	Drought tolerance and high yield
4. TRFK 301/4 and EPK C12	Timbilil	High yield and cold tolerance
5. TRFK 31/30 and AHP SC12/28	Timbilil	Tetraploid and diploid
6. EPK TN14-3 and AHP CG28U864	Timbilil	High soil pH, cold and pest tolerance and high yield
8. GW Ejulu –L and TRFK 301/5	Kangaita	High quality and yield
8. TRFK 301/4 and AHP SC31/37	Kangaita	High yield
9. TRFK 311/287 and AHP S15/10	Kangaita	Tetraploid and diploid
10. TRFK 31/8 and NDT Tai	Kangaita	High yield and high soil pH tolerance
11. TRFK 12/2 and AHP SC12/28	Kangaita	Non-fermenter and high yield

NB: Timbilil (0°22′ S, 35°21′ E, 2180 m amsl) in Kericho District, West of Rift Valley and Kangaita (0.5° S, 37.3° E; 2100 m amsl) in Kirinyaga District, East of Rift Valley

Table 5.3 A list of released clones showing highest yields in Timbilil, Kericho, year of release, criteria of release and current status

Clone	Source of Seed	Yields kg mt/ (ha·year)	Year of release	Variety type	Criteria for release	Resistance to pests/diseases	Adaptability/ stability
TRFK 6/8	Kericho, Kenya	4,441	1964	var. *assamica*	Good rooting, nursery growth vigorous, fast fermenting, field growth in clonal plots	Moderately susceptible to frost	Average
TRFK 7/3	Ambangulu, Tanzania	4,592	1964	var. *assamica*	Good rooting, nursery growth vigorous, fast fermenting, field growth in clonal plots	Mod, susceptible to frost	Average, East
TRFK 7/14	Ambangulu, Tanzania	4,496	1964	var. *assamica*	Good rooting, nursery growth vigorous, fast fermenting, field growth in clonal plots	Susceptible to mites	Unknown/ commercial
TRFK 11/4	Kericho, Kenya	6,132	1964	var. *assamica*	Good rooting, nursery growth vigorous, fast fermenting, field growth in clonal plots	Moderate	Unknown/ commercial
TRFK 12/12	Kericho, Kenya	4,671	1964	var. *assamica*	Good rooting, nursery growth vigorous, fast fermenting, field growth in clonal plots	Susceptible to mites	Average
TRFK 12/19	Kericho, Kenya	4,686	1964	var. *assamica*	Good rooting, nursery growth vigorous, fast fermenting, field growth in clonal plots	moderate	Average
TRFK 31/8	Ambangulu, Tanzania	5,049	1964	var. *assamica*	Good rooting, nursery growth vigorous, fast fermenting, field growth in clonal plots	moderate	Average
TRFK 7/9	Ambangulu, Tanzania	5,246	1969	var. *sinensis*	Outyielded TRFK 6/8 by 9% with cup quality slightly poorer than TRFK 6/8 but better than seedling tea	moderate	Above average
TRFK 31/11	Ambangulu, Tanzania	3,753	1969	var. *assamica*	Yield 83% of TRFK 6/8 and cup quality better than TRFK 6/8	moderate	Unknown
TRFK 100/5	Rwebitaba, Uganda	5,238	1976	var. *assamica*	Yield comparable to TRFK 6/8 and cup quality slightly lower than TRFK 6/8 but better than seedling tea	tolerant	Unknown
TRFK 108/82	Rwebitaba, Uganda	5,329	1976	var. *sinensis*	Yield comparable to TRFK 6/8 and cup quality slightly lower than TRFK 6/8 but better than seedling tea	moderate	Unknown

(To be continued)

(Table 5.3)

Clone	Source of Seed	Yields kg mt/ (ha·year)	Year of release	Variety type	Criteria for release	Resistance to pests/diseases	Adaptability/ stability
TRFK 54/40	Kericho, Kenya	5,117	1986	var. *assamica*	Outyielded TRFK 6/8 by 37% by end of 2nd pruning cycle. Cup quality similar to TRFK 6/8	moderate	Unknown/ commercial
TRFK 303/178	OP of TRFK 6/8	5,722	1986	var. *assamica*	Outyielded TRFK 6/8 by 64% by end of 2nd pruning cycle. Cup quality better than seedling tea	Resistant	Unknown/ commercial
TRFK 303/216	OP of TRFK 6/8	5,383	1986	var. *assamica*	Outyielded TRFK 6/8 by 66% by end of 2nd pruning cycle. Cup quality better than seedling tea	Susceptible	Unknown/ commercial
TRFK 31/27	Ambangulu, Tanzania	4,100	1988	var. *assamica*	Outyielded TRFK 6/8 by 23% by end of 2nd pruning cycle	moderate	Average
TRFK 31/28	Ambangulu, Tanzania	3,180	1988	var. *assamica*	Outyielded TRFK 6/8 by 11% by end of 2nd pruning cycle	moderate	Unknown/ commercial
TRFK 31/29	Ambangulu, Tanzania	3,255	1988	var. *assamica*	Outyielded TRFK 6/8 by 17% by end of 2nd pruning cycle	moderate	Unknown
TRFK 55/55	Kericho, Kenya	3,066	1988	var. *assamica*	Outyielded TRFK 6/8 by 20% by end 2nd pruning cycle	moderate	Unknown
TRFK 55/56	Kericho, Kenya	3,147	1988	var. *assamica*	Outyielded TRFK 6/8 by 10% by end 2nd pruning cycle	susceptible	Unknown
TRFK 56/89	Kericho, Kenya	4580	1988	var. *sinensis*	Outyielded TRFK 6/8 by 23% by end of 2nd pruning cycle. Cup quality similar to TRFK 6/8		
TRFK 303/199	OP of TRFK 6/8	3,840	1988	var. *assamica*	Outyielded TRFK 6/8 by 14% by end of 2nd pruning cycle	resistant	Unknown/ commercial
TRFK 303/259	OP of TRFK 6/8	4,351	1988	var. *assamica*	Outyielded TRFK 6/8 by 12% by end of 2nd pruning cycle	moderate	Above average

(To be continued)

(Table 5.3)

Clone	Source of Seed	Yields kg mt/ (ha·year)	Year of release	Variety type	Criteria for release	Resistance to pests/diseases	Adaptability/ stability
TRFK 303/231	OP of TRFK 6/8	5,286	1989	var. *assamica*	Yield comparable to TRFK 31/8 and acceptable cup quality	Susceptible to Phomopsis	Unknown/ commercial
TRFK 303/348	OP of TRFK 6/8	3,405	1989	var. *assamica*	Yield comparable to TRFK 31/8 and acceptable cup quality		Unknown/ commercial
TRFK 303/352	OP of TRFK 6/8	3,340	1989	var. *assamica*	Yield comparable to TRFK 31/8 and acceptable cup quality	resistant	Unknown/ commercial
TRFK 303/366	OP of TRFK 6/8	3,369	1989	var. *assamica*	Yield comparable to TRFK 31/8 and acceptable cup quality		Unknown/ commercial
TRFK 303/388	OP of TRFK 6/8	3,376	1989	var. *assamica*	Yield comparable to TRFK 31/8 and acceptable cup quality		Unknown/ commercial
TRFK 303/577	OP of TRFK 6/8	7,817	1989	var. *sinensis*	Outyielded TRFK 6/8 by 54% and TRFK 31/8 by 22% by end of 2nd pruning cycle. Cup Quality better than seedling tea (St. 48)	Susceptible to root knot nematodes	Average
TRFK 303/745	OP of TRFK 6/8	3,523	1989	var. *assamica*	Yield comparable to TRFK 31/8 and acceptable cup quality		Unknown/ commercial
TRFK 303/791	OP of TRFK 6/8	3,927	1989	var. *assamica*	Yield comparable to TRFK 31/8 and acceptable cup quality		
TRFK 303/978	OP of TRFK 6/8	3,920	1989	var. *assamica*	Yield comparable to TRFK 31/8 and acceptable cup quality		Unknown/ commercial
TRFK 303/999	OP of TRFK 6/8	3,945	1989	var. *assamica*	Outyielded TRFK 6/8 by 26% and TRFK 31/8 by 9% by end of 2nd pruning cycle. Cup Quality better than seedling tea (St. 48)	resistant	Average

(To be continued)

(Table 5.3)

Clone	Source of Seed	Yields kg mt/(ha·year)	Year of release	Variety type	Criteria for release	Resistance to pests/diseases	Adaptability/stability
TRFK 303/1199	OP of TRFK 6/8	5,569	1989	var. *assamica*	Outyielded TRFK 6/8 by 78% and TRFK 31/8 by 15% by end of 2nd pruning cycle. Cup quality similar to TRFK 6/8	Susceptible to root knot nematodes	Below average
TRFK 303/186	OP of TRFK 6/8	3,798	1994	var. *assamica*	Yields significantly and consistently above TRFK 6/8 and comparable to 31/8	moderate	Unknown/commercial
TRFK 303/152	OP of TRFK 6/8	3,591	1994	var. *assamica*	Yields significantly and consistently above TRFK 6/8 and comparable to 31/8	resistant	Unknown/commercial
TRFK 303/179	OP of TRFK 6/8	3,591	1994	var. *assamica*	Yields significantly and consistently above TRFK 6/8 and comparable to 31/8	moderate	Unknown/commercial
TRFK 303/35	OP of TRFK 6/8	3,073	1994	var. *assamica*	Yields significantly and consistently above TRFK 6/8 and comparable to 31/8	resistant	
TRFK 303/156	OP of TRFK 6/8	4,410	1994	var. *assamica*	Yields significantly and consistently above TRFK 6/8 and comparable to 31/8	resistant	Unknown/commercial
TRFK 337/3	TRFK 6/8× TRFK 31/11	4,104	1995	var. *assamica*	Yield comparable TRFK 31/8 and acceptable cup quality	resistant	Unknown/commercial
TRFK 337/138	TRFK 6/8× TRFK 31/11	4,097	1995	var. *assamica*	Yield comparable TRFK 31/8 and acceptable cup quality	moderate	Unknown/commercial
TRFK 338/13	TRFK 31/11× TRFK 6/8	4,097	1995	var. *assamica*	Yield comparable TRFK 31/8 and acceptable cup quality	moderate	Unknown/commercial
TRFK 347/26	OP of TRFK 6/8	3,269	1995	var. *assamica*	Yield comparable TRFK 31/8 and acceptable cup quality	resistant	Unknown/commercial
TRFK 347/314	OP of TRFK 6/8	3,362	1995	var. *assamica*	Yield comparable to TRFK 31/8 and acceptable cup quality		Unknown/commercial

(To be continued)

(Table 5.3)

Clone	Source of Seed	Yields kg mt/ (ha·year)	Year of release	Variety type	Criteria for release	Resistance to pests/diseases	Adaptability/ stability
TRFK 347/336	OP of TRFK 6/8	3,260	1995	var. *assamica*	Yield comparable TRFK 31/8 and acceptable cup quality	moderate	Unknown/ commercial
TRFK 347/573	OP of TRFK 6/8	3,225	1995	var. *assamica*	Yield comparable TRFK 31/8 and acceptable cup quality	resistant	Unknown / commercial
TRFK 301/4	Reunion	4,864	2001	var. *assamica* ssp. *lasiocalyx*	Yield and cup quality comparable to TRFK 31/8 and TRFK 6/8, respectively	Susceptible to root knot nematodes	Average
TRFK 301/5	Reunion	5,909	2001	var. *assamica* ssp. *lasiocalyx*	Yield and cup quality comparable to TRFK 31/8 and TRFK 6/8, respectively	Resistant	Average
TRFK 371/3	OP of AHP S15/10	6,000	2008	var. *assamica*	Outyielded TRFK 31/8 by over 68% by end of 2nd pruning cycle. Cup quality comparable to TRFK 6/8, high levels of total polyphenols, tolerant to root knot nematode and suitable for green tea manufacture	Tolerant to root knot nematode and mites	Above average
TRFK 430/90	TRFCA SFS150× EPK TN14-3	6,000	2008	var. *assamica*	Outyielded TRFK 31/8 by 68% by end of 2nd pruning cycle. Cup quality comparable to TRFK 6/8, high levels of total polyphenols, and suitable for mechanical plucking	Tolerant to root knot nematode and mites	Average
TRFK 306	OP of TRFK 91/1	4,000	2011	Var. *assamica*	Yield comparable to TRFK 31/8; purple (anthocyanin-rich) pigment	Highly tolerant	Wide adaptability; released for processing of high value medicinal product

Presently, generated knowledge on the design and composition of current seed orchards indicate that their current setup may not be appropriate. They have been established in the middle of commercial tea plantations using progenitor materials whose combining abilities and genetic worth are unknown. Even though these plantations are maintained in the vegetative phase, some clones flower profusely even under the plucking table with accruing pollen grains being transferred to the breeding materials by foraging insects (Muoki *et al*., 2007). This might have led to production of inferior genotypes as some of the adjoining commercial plantations are still comprised of low yielding seedling accessions. As pollination in tea is predominantly entomophilous, cross pollination may be enhanced by the establishment of bee cages in newly designed seed orchards that are strategically placed, and surrounded by strips of other multi-storeyed vegetation such as broad leafed indigenous trees, which would effectively act as buffer zones against extraneous pollen, besides conserving the environment.

5.3.1 Breeding and Clonal Selection

In the recent past, breeding programs have been intensified and expanded to include the improvement of more than one economic trait in a single clone. Fruits borne from such efforts are expected to be harvested soon. Hybridization programs take advantage of the existence and/or creation of tremendous genetic variability in desirable traits involving choosing and crossing of disparate parents possessing desirable traits. Numerous crosses employing complementary mating design based on existing information have been undertaken since the inception of the rationalized breeding program. Seeds resulting from such crosses are collected alongside open pollinated ones upon maturity and used in the formation of basal populations for future selection. The majority of seedlings and clones that are at different stages of testing, either in progeny tests or replicated clonal field trials, are either half-sibs (open pollinated) or full-sibs. The response of the progenies in relation to yield and black tea quality and other secondary traits as well as estimation of their parental combining abilities will help in determining the best mating design for fruitful tea improvement.

Recent studies show that although Kenya's tea germplasm is predominantly of the (*C. sinensis* var. *assamica*) type, it is highly diverse though many of the clones are genealogically related (Wachira *et al*., 2001). It has been thought that the risks of having a population with a narrow genetic base may be high and huge losses can be encountered in the event of the onset of biotic and abiotic stresses. There has been a thrust in the breeding strategy to buffer the existing germplasm against the emergence of such risk factors by deliberately crossing disparate parents through intra-specific or inter-specific hybridization. This strategy has been aimed at broadening the genetic base, as well as introgressing new genes controlling useful traits that were otherwise not present in the base population. There are numerous genotypes under different stages of investigation, which are a result of direct crosses among var. *sinensis*, var. *assamica* ssp. *lasiocalyx* and var. *assamica*. A few elite clones like TRFK 301/4 and TRFK 301/5 (var. *assamica* ssp. *lasiocalyx*) have been released to the industry for on-farm diversification (Mamati *et al.*, 2001). Two new clones, namely TRFK 430/90 and TRFK 371/3, with combined optimum yield and black tea quality as the primary traits, and which are able to survive under diverse abiotic and biotic stress factors, have recently been released to the industry (Kamunya & Wachira, 2006).

Offspring that are established in progeny trials are evaluated for a period of seven years followed by selection for desirable traits (usually yield and black tea quality). Two to five percent of the best progeny are selected and advanced to the next phase of testing (clonal field trials) in which all the entries are replicated. The seven-year period consists of two years of formation of a plucking table, four years of assessment of yield and response to invasion by pests/disease and drought effects and one year assessment of recovery from pruning. The progeny trials are usually not replicated but are planted as hedges owing to variable numbers of offspring per family obtainable for evaluation. Thus, it has not been possible to divorce the effects of the environment from that of the genes while determining the performance of the progenies. It is normally assumed that the superior performance or response of the plant to the trait under investigation is genetic in origin. Unfortunately, this is not always true as the possibility of the existence of favorable or hostile microhabitats within a test site would affect a superior or inferior genotype either negatively or positively. Methods of counteracting such variability prior to progeny testing, possibly through micropropagation at the earliest opportunity (say at nursery stage or two years in hedges), would enable the establishment of replicated clonal progeny trials. This would ensure that only superior genotypes with the traits of interest are selected from one stage to the next. Studies already undertaken have shown that only a small portion of the seedling population is selected for further evaluation in clonal field trials. The occurrence of superior genotypes in such a population may sometimes be as low as 0.0025% (Wight, 1958). It is estimated that one seedling plant in 200 to 300 has high yield or good quality, i.e., one seedling per 40,000 to 100,000 may therefore combine both yield and black tea quality (Wight, 1958, 1961; TRIEA, 1966; Green, 1966; Kulasegaram, 1978), a low probability indeed and an indication of the monumental task involving the development of new cultivars with a combination of polygenic traits. Massive tracts of land are needed every year for evaluation of newly generated genotypes for attributes of interest. The present situation, whereby high potential land suitable for tea is generally lacking, may therefore have a negative impact on the progress of tea improvement. Currently, there are over 28 progeny trials with over 15,000 offspring at the TRFK tea improvement program (TRFK, 2006). The selected bushes are appropriately labeled and allowed to produce elongated shoots (whips) from which single node whole leaf cuttings are collected and vegetatively propagated in the nursery, usually for a minimum period of 8 months. Vegetative propagation is, at the moment, the most cost-effective and rapid means of multiplying tea plants for either commercial use or further evaluation. This is normally the first stage through which selected seedling teas are cloned.

The clonal field trials (CFTs) are always replicated with test clones being tested alongside parental clones and with commercial standard checks for yield, quality and other secondary attributes. This is normally done in order to ensure that newly developed clones are competitively selected and judicious judgments based on data are used to release new clones to the industry. The clonal field trials are usually evaluated for desirable traits, mainly yield and black tea quality, for ten years, which is equivalent to two standard pruning cycles. Until recently, clonal field trials were only evaluated at one site at the TRFK Headquarters on the Timbilil Estate (Table 5.3). Since the inauguration of a TRFK sub-station to the east of the Rift Valley, a number of CFTs have been established at two sites, thereby enabling genotype × environment interaction parameters to be estimated. Over 17 clonal field trials at different stages of development, with over 260 different clones, are currently in existence at the TRFK tea improvement program (TRFK, 2006).

Tea is a perennial crop that takes long to attain optimum yield potential (21 to 30 years) (Gazi, 1978), but it is normally assumed that reliable judgment on the performance of a good clone can be made after 8 years. However, before new clones are released to the industry, they have to be subjected to extra testing, in which case they are exposed to different environments in evaluations referred to as clonal adaptability trials. Plant genotypes are known to respond and perform differently at different localities owing to variations in environmental factors (soils, climate, elevation, geographic location, pests and diseases). Environmental variations affecting clonal response in yield (Ng'etich & Stephens, 2001; Ng'etich *et al.*, 2001; Wachira *et al.*, 2002) and black tea quality (Owuor & Othieno, 1987; Owuor *et al.*, 1988, 1990) have been observed in Kenya, justifying the need to test potential clones on multilocational sites. Genotypes that are entered for genotype-environment interactions (G×E) are tested for adaptability and stability for all traits of interest. Such evaluation programs have been initiated for all released, pre-released and promising potential clones in the west and east of the Rift Valley in collaboration with various stakeholders in the tea industry (TRFK, 2006). It is envisaged that a tea map indicating different agro-ecological zones favorable to specific clones will eventually be developed and made available to farmers. It may however be noted that such a program will take long to bear fruit due to longevity of the tea plant, and even where certain clones may show early take-off, their ability to sustain their performance will call for longer periods of observation. The tea growing environment is highly variable across time as well as space, leading to significant and unpredictable G × E interactions. For example, a particular tree genotype may grow well in wet years but be a poor competitor in dry years, while the neighboring plant might have the opposite response (Bradshaw, 1998).

5.3.1.1 Breeding for Potential in Combined Optimal Yield and Tea Quality

The need to develop tea cultivars with optimum potential combining yield and black/green tea quality has recently become the single most important breeding objective largely due to factors that are not unique to tea. Being a universal most popular beverage, its cultivation requirements pose a serious limitation to its expansion in areas that are unsuitable for its optimal growth. Moreover, in Kenya its expansion and/or sustainability to new areas that are suitable for tea production is faced by stiff competition from human settlement and other competing enterprises. Secondly, increased world tea production and static absorption call for high quality diversified tea products. The various clones that are currently being used to combine high yield and quality due to their recognized possession of these attributes are depicted in Tables 5.1 and 5.2. It would be worth noting that both yield and tea quality are complex traits which are controlled by many genes. For example, while in black tea theaflavins and thearubigins are the most important attributes affecting tea quality (Hara, 2001), in green tea the composition of various green catechins becomes the overriding factor. The amount of theaflavins is closely related to the commercial value of black tea (Owuor, 1992). From a health point of view, elucidating the most important components contributing to medicinal properties of the end-product, their genetic background and correlation with yield components, would be most useful in guiding breeding work. Such knowledge is currently inadequate but studies attempting to address it have been commenced.

5.3.1.2 Polyploidy Breeding

There has been a deliberate effort to introduce polyploidy breeding in tea improvement following the discovery of naturally occurring polyploids which contain more than the basic dosage of chromosomes ($n = 15$) (Wachira & Kiplangat, 1991; Wachira, 1994b). For example, thirty-eight triploids ($3n$) and two tetraploids ($4n$) have been identified among seedling populations at Timbilil Estate (Table 5.4) and some of them have been incorporated into the breeding program in polyclonal and biclonal seed *baries* (TRFK, 2002). Observations made so far indicate that polyploids, in spite of their high growth vigour, have given consistently lower yields than diploids (Wachira & Ng'etich, 1999), but further investigations into their possible utilization in genetic studies is continuing (TRFK, 2001a).

Table 5.4 Identified polyploid clones, their chromosome number and ancestry

Clone number	Ploidy-Chromosome No.	Original seed source
TRFK 311/287	$4n = 60$	TRFK 6/8 × TRFK 31/11 - hand-pollinated from TRIEA Uganda
TRFK 31/30	$4n = 60$	Seed from Ambangulu Estate, Tanzania
TRFK 52/1	$3n = 45$	"Between" seed from Kanywankoko Estate, Uganda
TRFK 77/1	$3n = 45$	Open-pollinated seed fom Mimosa Estate, Malawi
Dimbolil 3	$3n = 45$	Seedling from James Finlay (K) Ltd, Dimbolil Estate
TRFK 77/2	$3n = 45$	Open-pollinated seed from Koiwa Estate
TRFK 383/1	$3n = 45$	Open-pollinated BB35 × BB5, Koiwa Estate
TRFK 331/2	$3n = 45$	Seed from Chemosit Estate *barie*, Unilever tea
TRFK 378/1	$3n = 45$	Open-pollinated BB35 × BB7, Koiwa Estate
TRFK 412/1	$3n = 45$	Open-pollinated BB21 × BB5, Koiwa Estate
TRFK 371/1	$3n = 45$	Open-pollinated seed of AHP S15/10 from Chepgoiben Estate, James Finlay (K) Ltd
TRFK 400/1	$3n = 45$	Seed from Chepgoiben Estate *barie*
TRFK 389/1	$3n = 45$	Open-pollinated BB35 × BB152, Koiwa Estate
TRFK 392/1	$3n = 45$	Open-pollinated BB7 × BB35, Koiwa Estate
TRFK 394/1	$3n = 45$	Open-pollinated BB2 × BB35, Koiwa Estate
TRFK 395/1	$3n = 45$	Open-pollinated BB5 × BB35, Koiwa Estate
TRFK 54/49	$3n = 45$	Seed from Ambangulu Estate, Tanzania
TRFK 386/1	$3n = 45$	Seed from Sotik Tea Co.
TRFK 381/1	$3n = 45$	Seed from BB5 × BB2
TRFK 84/1	$3n = 45$	Mixed seed from Congo, Toro and Entebbe, Unganda
TRFK 84/2	$3n = 45$	Mixed seed from Congo, Toro and Entebbe, Unganda
TRFK 85/1	$3n = 45$	Seed collected from clones from Kakonde Estate
TRFK 382/2	$3n = 45$	Seeds from BB5 × BB35
TRFK 382/1	$3n = 45$	Seeds from BB5 × BB35
TRFK 386/2	$3n = 45$	Seed from Sotik Tea Co.
TRFK 76/3	$3n = 45$	Seed from Ramjat, Lugeri Estate, Malawi
TRFK 76/1	$3n = 45$	Seed from Ramjat, Lugeri Estate, Malawi
TRFK 76/2	$3n = 45$	Seed from Ramjat, Lugeri Estate, Malawi
TRFK 75/1	$3n = 45$	Commercial seed from Lugeri Estate, Malawi
TRFK 31/36	$3n = 45$	Seed from Ambangulu Estate, Tanzania
TRFK 31/38	$3n = 45$	Seed from Ambangulu Estate, Tanzania
TRFK 31/39	$3n = 45$	Seed from Ambangulu Estate, Tanzania
TRFK 31/40	$3n = 45$	Seed from Ambangulu Estate, Tanzania
TRFK 31/41	$3n = 45$	Seed from Ambangulu Estate, Tanzania
TRFK 18/7	$3n = 45$	Seed from Ambangulu Estate, Tanzania
TRFK 18/27	$3n = 45$	Seed from Ambangulu Estate, Tanzania
TRFK 18/26	$3n = 45$	Seed from Ambangulu Estate, Tanzania
TRFK 18/28	$3n = 45$	Seed from Ambangulu Estate, Tanzania
TRFK 54/50	$3n = 45$	Seed from Ambangulu Estate, Tanzania
TRFK 550/1	$3n = 45$	Open-pollinated seed from polyclonal mixture, Timbilil Estate, Kericho

Polyploids tested so far are of restricted genetic background, most having been derived from low to medium yielding progenitors like BBK 2, BBK 5, BBK 7 and BBK 35 (TRFK, 1999). However, indications exist of the undertaking of selections for diversification and/or for further improvement into secondary polyploids from the ongoing plant improvement program (TRFK, 2001a). Hybridization efforts between triploids and diploids have often not borne good results as they yielded poor seed set of which some exhibited low viability. This has been attributed to complications arising from the pairing of chromosomes during the process of fertilization. Suitable techniques may need to be devised to be able to effectively utilize polyploidy breeding in plant improvement programs.

5.3.1.3 Breeding for Pest Resistance

Tea mites, especially the red crevice mite (*Brevipalpus phoenicis*), have been reported to cause yield losses estimated to vary between 14% and 30% with heavy infestation (Sudoi, 1995; 1996). Similarly, scale insects (*Aspidiotus* species) can cause yield losses of about 10% (Sudoi *et al.*, 1996). Chemical control measures are costly and hazardous to health and the environment. As Kenyan tea is credited for being pesticide residues free, it is imperative to safeguard this status which is in conformity with the stipulations of the European Union on pesticide residue limits. This is achieved by the development of cost-effective and environmentally friendly pest control measures such as breeding for pest resistance.

Clones TRFK 7/9, TRFK 57/15, AHP SC31/37, TRFK 303/1199, TRFCA SFS150 and AHP S15/10 have been found to be relatively tolerant to the red crevice mite. Clone EPK TN14-3 has been noted to be less susceptible to scale insects though it is moderately preferred by the red crevice mite (Sudoi *et al.*, 1996). These clones are among the cultivars that are being used in the breeding program for pest resistance.

Incidences of root knot nematodes damage on clone TRFK 303/577, a popular cultivar that is high yielding and drought tolerant especially in areas where it is planted on former coffee land, are now widespread (Otieno *et al.*, 2002). A recent study revealed widespread clonal variation in response to infection by root knot nematodes in farmers' fields in Kenya, indicating scope for enhanced improvement in resistance/tolerance to the pest arising from participatory tea improvement initiatives (Kamunya *et al.*, 2008).

5.3.1.4 Breeding for Tolerance to Environmental Stress

Due to global climatic changes, the frequency, longevity and severity of drought have increased especially in the traditional tea growing areas. Similarly, stress associated with low temperatures as well as high soil pH, have been serious constraints on tea production in some parts west of the Rift. Additionally, tea farming is increasingly being extended into the non-traditional tea-growing areas that were formerly considered marginal and therefore unsuitable for the plant. Therefore, the need for consolidation of inherent tolerance to drought in the tea plant is imperative for sustainable tea production (Nagarajah & Ratnasurya, 1981). In the breeding program, clones with tolerance attributes to various abiotic stress factors, such as clones TRFCA SFS150, TRFK 303/577, EPK TN14-3 and EPK D99/10 (Table 5.1), have already been identified and introduced into the breeding program.

5.3.1.5 Inter-specific Hybridization

Inter-specific hybridization (species introgression) has lately been initiated with the aim of improving upon the vegetative and hardiness characteristics of tea by crossing tea (*C. sinensis*) with some closely allied species, particularly in India and Japan (Wachira, 1994a). It has been demonstrated that tea can easily be crossed with 10 different species (Bezbaruah, 1987). Hybridization work has successfully been carried out with *C. irrawadiensis, C. taliensis, C. japonica* and *C. kissi* (Ackerman, 1970, 1973; Bezbaruah, 1974, 1987). Two species, *C. irrawadiensis* (Wilson's Camellia) and *C. taliensis* (Forest's Camellia), have merited special attention as they lack caffeine. Their liquors however lack the quality of tea. To date, no interspecific hybrids have produced commercially acceptable tea of good quality.

A program of inter-specific hybridization is being implemented in Kenya at the TRFK, the focus being to produce diversified tea products from the resulting hybrids (TRFK, 1998). Numerous intra-specific crosses aimed at diversification of tea products have also been carried out at the TRFK with Japanese germplasm 'Yabukita' and 'Yutakamidori' (TRFK, 2001a), which are popular commercial cultivars for green tea in Japan. Offspring from these crosses are still young, with analytical investigations expected to cast light on their quality. The catechin levels to be determined will form the basis for identification of desirable green tea genotypes in Kenya. Owing to possibilities of natural hybridization, it is unclear if teas presently being cultivated are original cultivars (Visser, 1969). It is widely held that the present cultivars have emanated from hybridization from the three main taxa as well as other *Camellia* species (Banerjee, 1992a). Thus, breeders have a wide field of choice when it comes to choosing which traits to target in their breeding program.

5.3.2 Combining Abilities

The parental genetic values are expressed in terms of combining abilities. The two types of combining abilities that are of special interest to plant breeders are general and specific combining abilities. General combining ability (GCA) is defined as the average performance of the progeny of an individual when it is mated to a number of other individuals in the population (Falconer, 1989). Although GCA may be expressed in absolute units, it is usually more convenient and meaningful to express them as deviations from the overall mean. Thus a parent with GCA of 0 has an average general combining ability. A positive GCA indicates a parent that produces above average progeny, whereas a parent with a negative GCA produces progeny that perform below average for the population.

Specific combining ability (SCA) on the other hand refers to the average performance of the progeny of a cross between two specific parents that are different from what would be expected on the basis of their general combining abilities alone. It can either be positive or negative. SCA always refers to a specific cross and never to a particular parent (Falconer, 1989). While GCA is a measure of the additive genic action, the SCA is assumed to be a deviation from additivity (i.e., non-additive genic action). Scanty information exists on combining ability for perennial crops. For instance, combining ability studies for cocoa (*Theobroma cacao* L.) showed SCA effects to be greater than those of GCA for yield (Dias & Kageyama, 1995). However, a separate study involving diallel crossings had earlier revealed GCA to be more important than SCA for the same trait (Berry & Cilas, 1994).

Information on combining abilities in tea had until recently been lacking owing to which tea breeding work at the TRFK was not taking cognizance of the type of combining abilities of parents involved (Kamunya *et al.*, 2009). Current studies have led to the availability and analysis of diallel crosses with revelation of heterotic patterns based on parental combining abilities assisting in choosing the appropriate materials, design and structure of seed *baries* and mating designs for future breeding programs (Kamunya *et al.*, 2007, 2009).

5.3.3 Registration for Plant Breeders' Rights and Adoption of New Cultivars

The longevity of tea breeding demands that the process of registration for protection of improved cultivars begins at an appropriate stage in order not to delay the accessibility by growers of the new cultivars for commercial use. The Distinctness, Uniformity and Stability (DUS) test is normally superimposed on clonal adaptability trials that are normally undertaken in farmers' fields in some chosen unique sites within the tea growing region (Fig. 5.2). The test is conducted in collaboration with the Plant Variety Protection (PVP) office of the Kenya Plant Health Inspectorate Service (KEPHIS). Morphological characterization for DUS tests are carried out according to the International Union for the Protection of New Varieties of Plants (UPOV) guidelines for tea (UPOV TG/238/1) that was prepared by Dr. Liang Chen of the Tea Research Institute of the Chinese Academy of Agricultural Sciences as the leading expert (UPOV, 2008), and the International Plant Genetic Resources Institute descriptors for tea (IPGRI, 1997). About one year ahead of potential release, the Plant Breeder files for an application for a Grant of

Plant Breeders Rights with KEPHIS attaching yield and morphological data of the earmarked clones for release. The PVP office then constitutes the National Performance Technical Committee (NPTC) that evaluates the data collected and visits all the sites where the clones are being evaluated. Once satisfied that the clones/cultivars qualify for commercial use, the PVP constitutes another committee namely the National Variety Release Committee (NVRC) that convenes at the Ministry of Agriculture headquarters and scrutinises all the relevant information on all varieties that have passed the criteria for release. The NVRC gives the final recommendations to the Minister for Agriculture who officially releases all the improved cultivars through a special Kenya Gazette notice.

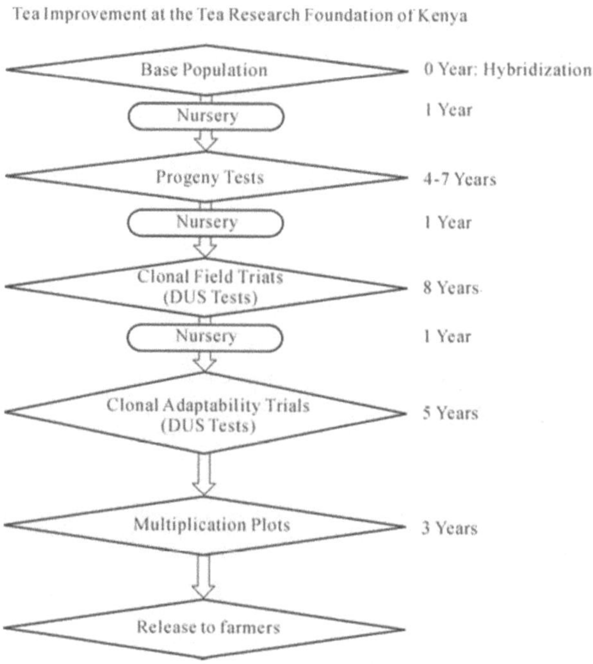

Fig. 5.2. Schematic presentation of tea breeding process at the Tea Research Foundation of Kenya

In the case of tea, the propagating materials are normally issued in the form of single-whole cuttings to growers through the factories with licensed nurseries for production of high quality vegetatively propagated (VP) material. Rooted VP materials are also sourced from the TRFK experimental stations' nurseries situated in the tea growing regions east and west of the Great Rift Valley. The TRFK Advisory Office and KTDA extension services are tasked to create awareness through field days, national and regional agricultural shows, field visits and through the media. In order to maintain quality, sustain productivity and ease management of tea fields, new plantations in Kenya are being established using clonal teas. Owing to the late entry of smallholders into tea farming just when vegetative propagation had been adopted (M'Imwere, 1997), about 80% of the sub-sector currently comprises clonal tea. The converse is true for the large estate sub-sector, which formed the bulk of the pioneer seedling tea population with the clonal tea proportion, though gradually rising, presently constituting about 40%.

5.4 Future Strategies and Opportunities

The dwindling revenue base from tea enterprises occasioned by the increased cost of production, the glut of black tea on the world market and the appreciation of the Kenya Shilling against the US Dollar, means that the availability of the requisite raw material for the diversification of tea products needs to be given serious consideration if the tea business is to remain relevant. The appropriate raw material is nothing else but the cultivar which is the final technological output of any plant improvement program. Owing to the prolonged breeding cycle of tea that is tied to its perenniality, careful consideration of the progenitor clones harboring the desired traits has to be ensured. If such parents are lacking in the national improvement program, efforts to source them from other countries or farming communities through mutually negotiated material transfer agreements could be the most cost-effective option. Additionally, any technology that may lead to a reduced cost of production would not only endear itself to the farmers but would also lead to a raised income for poor farmers and a significant reduction in poverty levels. Thus, breeding for cultivars that are suitable for mechanical plucking has now gained considerable importance (Kamunya & Wachira, 2006; TRFK, 2008) (Fig. 5.3). The same approach has also been adopted by other countries (Apostolides *et al*., 2006). Similarly, strategies addressing elite green tea cultivars, value-addition by screening existing germplasm and the undertaking of further breeding for high levels of antioxidants and low levels of caffeine aimed at accessing niche markets are also being instituted.

Fig. 5.3. Mechanical tea plucking trial in one of the experimental sites in Kenya

Efforts geared towards selection for black tea quality made use of fermentability based on the chloroform test (Sanderson, 1963). Fast fermenting clones were assumed to produce high quality black tea but that was not always the case. The slow fermenting ones were always discarded and this might have led to inadvertent rejection of potential clones for other tea products such as green tea or tea with high levels of antioxidant like total polyphenols, which were not considered important then. Thus, where the breeding program is earmarked for diversifying tea products, rationalized breeding activities call for a total overhaul of breeding objectives. While revising the breeding objectives, addressing traits that are not demand driven would be costly and an untenable venture. Thus, farmers' and consumers' needs come to the fore while formulating the objectives. For example, where consumers like specific tastes, changes impacting positively on breeding and relevant selection criteria for elite cultivars must be considered. In Malawi, at the Limbe tea auction, buyers look for two distinct types of tea: one is red coppery tea, while the other is yellow tea (Apostolides *et al.*, 2006). Selection of these attributes requires the establishment of rapid and reliable selection criteria and methods.

In green tea, the polyphenolic fraction is mostly composed of catechins. There are six components of catechins, the most important being epigallocatechin gallate (EGCG). These catechins have recognised antioxidative properties owing to which they are finding new applications in the confectionery, food, cosmetic and pharmaceutical industries with market opportunities for such cultivars increasing daily (Hara, 2001).

There has been tremendous improvement in Kenyan tea over the years with notable replacement of seedling teas with high yielding and better quality clones. Kenyan tea has gained a reputation in the world for high black tea quality and yields. However, it has not yet been possible to produce clones with combined optimum yield and black tea quality but, as previously mentioned, one of the current objectives in breeding programs is geared at coming up with such a clone. Careful parental choices, judicious breeding strategies as well as understanding the tea genetics, are key prerequisites to attaining quicker genetic progress while maintaining a broad genetic base. A combination of earlier identified morphophysiological and recently developed molecular markers for early selection of potential clones are expected to shorten the breeding cycle and allow easier and more accurate choice of disparate stocks possessing high yield and black tea quality (Sanderson, 1963; Wachira, 1994a). In an attempt to come up with clones that will result in quantum leaps in yield and quality as well as tolerance to biological and environmental stresses, the following highlighted strategies are either being explored or plans are under way to try them out.

5.4.1 Tissue Culture

Wachira (1990) reviewed the applicability of tissue and cell culture in tea improvement but only observed its usefulness in assisting research rather than being an alternative method of propagation, due to its cost implications. However, the case of the successful multiplication of an elite clone in Sri Lanka with up to 30,000 to 35,000 plantlets produced from 50 nodal explants in one year has been reported (Arulpragasam, 1990). Tissue culture utilization in developing pure lines of tea through generation of double haploids is currently under investigation (TRFK, 2001b), with useful information to help elucidate tea genetics expected to be generated soon.

5.4.2 Mutation Breeding

The objective of mutation breeding is to induce desirable mutations to enhance both quality and yield. However, in the absence of information on the precise location of loci of the genes responsible for these characteristics, preliminary investigations were restricted mostly to irradiating cuttings, pollen grains and seeds, anticipating that some of the treated plants would do better than those untreated (Singh & Sharma, 1982). The mutagens used include X-rays and γ-rays and ethyl methane sulphonate as a chemical mutagen. However, these treatments failed not only to produce superior mutants, but those treated had reduced vigor, stunted growth and a lesser amount of foliage and number of branches (Singh & Sharma, 1982). According to Sharma and Ranganathan (1985), use of irradiated pollens caused fruit drop. In the TRFK tea improvement program, mutagens have been incorporated in soma-clonal cultured materials with the aim of developing new genotypes that are able to tolerate stress factors (TRFK, 2001b). The mutagens used include colchicine, hydroxyquinoline and sulfanimide but since investigations are still at the earliest stage, no useful results have been reported. Some plantlets have however been raised from this investigation and have been transplanted in the field for further evaluation.

It has also been reported that tea clones differ in their responses to γ-radiation: clones from var. *sinensis* and var. *assamica* origin are generally more tolerant to γ-radiation than those from var. *assamica* ssp. *lasiocalyx* origin. However, 2 krad appears to be the upper limit for survival (Singh, 1980). The significance of these findings is not clear at present, but the apparent genetic variation in the response to mutagens suggests scope for further exploitation of this strategy in broadening the genetic diversity of tea.

5.4.3 Genetic Transformation in Tea

Foreign genes have been introduced in several woody crops including the rubber tree by using *Agrobacterium tumefaciens* Ti plasmid (Horsch *et al.*, 1985;

Venkatachalam *et al.*, 2006). Although transgenic technology has immense potential for genetic improvement of tea (Mondal *et al.*, 2004), it was not considered important until 2000, prior to which reports in tea research regarding this is non-existent. This may have been caused by initial challenges in developing a protocol for gene transfer. Mondal *et al.* (2001) were able to optimize transformation conditions and production of transgenic tea via *A. tumefaciens*. While research interest in genetic transformation in tea was highly motivated by the need to develop blister blight resistant cultivars in India, such a degree of interest has not attracted equivalent attention in Kenya largely due to the lack of capacity to perform such work. Furthermore, the assurance needed by Kenyan tea consumers abroad that the tea product is not derived from genetically modified cultivars strongly negates the need to conduct this type of research. Owing to global climatic change that is currently characterized by severe incidences of prolonged drought and frost, not to mention the increased susceptibility to pests and diseases when plants get stressed and the new challenges emerging in traditional tea growing areas of Kenya, there is a need to review the tea improvement effort and move away from largely conventional to current biotechnological inputs, including genetic transformation for faster genetic improvement of tea.

5.4.4 Breeding for Medicinal Tea

Tea is increasingly becoming recognized as a health drink with research in its pharmacological properties focusing on the possible components that make up tea and the effect at a given concentration (Chen, 1995). Therefore, tea breeding is currently targeting the selection and raising of populations with high functional components such as catechins, flavanols, theanine, β-carotene, 2-amino-5-(N-ethylcarboxyamido) pentanoic acid and polysaccharides (Chen, 1995). The health effects of tea have been extensively studied (Anon, 2003a, 2003b). It has been linked to lowered heart disease and cancer risk (Hara, 2001; Weisburger, 2006) through the action of flavonoids, a type of antioxidant, relieving some allergy symptoms and recently to boosting the body's immune system (Anon, 2003a, 2003b; Basu, 2003).

Tea derives its pharmacological properties largely from its polyphenols content. The polyphenols include catechins: catechin (C), epicatechin (EC), gallocatechin (GC), epigallocatechin (EGC), epigallocatechin gallate (EGCG) and epicatechin gallate (ECG). These biochemicals are formed through intermediary glucose metabolism comprising the pentose, shikimate and the prephenelate pathways. They play a prominent role in green and black tea quality (Nakagawa & Torii, 1965; Obanda *et al.*, 1992) and have been reported to have important medical properties, which include the ability to reduce serum cholesterol levels (Ohtsura, 1991), alleviation of hypertension and vascular disorders (Matsubara *et al.*, 1985; Yilddizogle-Ari *et al*., 1991), prevention of breast and prostate cancer (Ohtsura, 1991) and inhibition of inflammation (Maeda, 1989; Sugiyama, 1995; Yamada, 1995). EGCG has recently been found to boost the body's immune system by warding off human immunodeficiency virus (HIV) (Shearer, 2003) and inhibit HIV reverse transcriptase activity (Nakane & Ono, 1990). Epidemiological studies have also demonstrated that catechins in tea inhibit diabetes, including hyperglycemia, by reducing elevated sorbitol, decreasing protein glycosylation, decreasing lipid peroxidation and by inhibiting diabetic cataracts (Vinson *et al*., 2001). Epidemiological studies done in Europe revealed that drinkers of black tea had a lower incidence of heart diseases (Weisburger, 2006). These findings were attributed to tea polyphenols acting as effective antioxidants, inhibiting the oxidation of low-density lipoprotein (LDL) cholesterol caused by reactive oxygen species which leads to antherogenesis. Studies involving constituents of the polyphenolic fraction of green and black tea (theaflavins and thearubigins) showed that polyphenols reduced the mutagenic capacity of different types of carcinogens (Weisburger, 2006) as well as displaying powerful antibacterial action (Hara, 2001). Other investigators reported that tea and tea polyphenols decrease the rate of growth of tumor cells through mechanisms involving alterations in gene expression (Ohta *et al*., 2004). Additionally, tea polyphenols increase the rate of apoptosis (cell death) of tumor cells leading to their elimination (Hara, 2006). As tea polyphenols have antiviral and antibacterial properties, regular tea drinkers have been found to have healthier intestinal bacterial flora than those that drink less or no tea (Ichikawa *et al*., 2004; Hara, 2001). According to Hara (2006), catechins have been found to inhibit the growth of food-borne pathogenic bacteria, while they do not have adverse effects on beneficial bacteria such as *Bifidobacterium/Lactobacillus*. Furthermore, tea catechins, particularly EGCG, were confirmed to interact with the influenza virus in a way that rendered the virus non-infective to the cells (Hara, 2006). Owing to the suppressive power of tea polyphenols over reactive oxygen species formed during normal cellular metabolism which causes premature ageing, various studies found that a regular daily intake of five or more cups of tea facilitates a healthier ageing process (Weisburger, 2006).

In India, tea is classified as a health food, akin to Rasayanas known to ancient Indians (Dhawan, 2006). Rasayanas is a general term encompassing a group of health foods and herbal based drugs which, if regularly used, convey the concept of attainment of positive health, increased resistance to diseases and assured longevity (Dhawan, 2006). Rasayanas re-establish youth, strengthen life and the brain function and provide the capability to counteract diseases.

Because of its rich phenolic composition, tea is increasingly being put to other uses in products other than foods and drinks. For example, numerous environment-friendly industrial cleaning agents, deodorizers and antimicrobial agents have been formulated using tea (Yayabe, 2001). Owing to the health promotive and disease preventive properties of tea, some extracts of green tea, known as green tea polyphenols (GTP) are finding commercial application in several substances such as deodorants, antioxidative agent for food and cosmetics and an anti-tooth-decay agent (Basu, 2003).

Other polyphenols with potential pharmaceutical properties are the anthocyanins (purple pigmentation: Fig. 5.4) (Walker, 1975; Clifford, 2000) and flavanoid pigments (Dufresne & Farnworth, 2001). Tea has one of the highest total flavanoid contents of all plants at 15% of the leaf by dry weight and is also the major source of flavanoids in the UK diet, providing 80% of dietary flavonoids for the population as a whole (UK Tea Trade Technical Committee).

Fig. 5.4. Anthocyanin-rich tea clone at the foreground of a clonal field trial

As great variation has been revealed in the expression of these important biochemicals in different tea germplasm (Takeda, 1994; Magoma *et al.*, 2001), large scale efforts in tea product diversification demand that the inheritance pattern be properly understood and potential molecular markers identified to hasten development of superior cultivars of tea that are not only high yielding but also of high pharmacological value. Thus, breeding strategies geared towards meeting anticipated future demand for clones with more health attributes have been initiated and would be further strengthened through germplasm exchange.

Studies involving animals and humans have shown that both green and black teas are equally beneficial in their pharmacological properties (Hara, 2001; Karori *et al.*, 2008). A study carried out to compare total polyphenols of some selected Kenyan teas with teas from other countries revealed that Kenyan tea germplasm had 7% to 27% more total polyphenols than germplasm from China and Japan, which is traditionally used for green, Oolong and pouching tea manufacture and the extraction of total polyphenols (Wachira & Kamunya, 2005b). Owing to their potent antioxidant properties, total polyphenols can be used in marketing and bargaining for premium prices for Kenyan tea.

5.4.5 Quantitative Trait Loci (QTLs) Mapping

Most attributes of agricultural importance frequently manipulated by plant breeders (e.g. size, shape, yield, quality, tolerance to abiotic and sometimes biotic stresses) display a quantitative mode of inheritance and normally exhibit continuous variation (Collard *et al.*, 2005). Continuous variation in a phenotype can be explained by the independent actions of many distinct genetic factors, each having small effects on the overall phenotype (Thompson & Thoday, 1974). These polygenes differ little from genes affecting simply inherited traits in which they cannot be monitored directly by conventional methods, but a biometrical approach which partitions the total variation into genetic and non-genetic components has been devised (Jinks, 1981). Procedures for estimating the number of effective genes controlling a quantitative trait (Mather & Jinks, 1982; Becker, 1984) and the theoretical basis for interpreting the association of marker loci with QTLs have been developed (Tanksley *et al.*, 1982). Molecular markers have also been used for studying quantitative inheritance (Tanksley, 1993; Kearsey *et al.*, 2003; Elberse *et al.*, 2004).

The principal behind detecting a linkage between a marker and QTL involves evaluation of progeny for a characteristic of interest and for their genotypes at marker loci at regular intervals (say 10 to 20 cM) throughout the genome. Any associations discovered between the segregating marker and trait is owed to linkage. One-way analysis of variance on marker—genotype classes is the simplest way to look for QTLs (Soller *et al.*, 1976). Other more complex methods are those that involve identification of two linked markers flanking a QTL on either side (Lander & Botstein, 1989) or using a logarithm of odds (LOD) and joint effects of several QTLs on a trait using stepwise multiple regression (Cowen, 1989; Stam, 1991).

Linkage between genetic markers and quantitative traits of economic importance has now been documented in a number of plants. For example, in the tomato, selection for specific QTLs in segregating progeny has led to the development of insect resistant lines (Niehuis *et al.*, 1987) and cold tolerance (Vallejos & Tanksley, 1983). Elberse *et al.* (2004) detected a number of QTLs affecting growth related traits in wild barley. Saintagne *et al.* (2004) demonstrated the applicability of QTL variation in discriminating between some species of oak. QTLs associated with wood property traits in pine have been identified and verified (Brown *et al.*, 2003). Work on cacao detected yield stable QTLs in a study that spanned 15 years (Crouzillat *et al.*, 2000). In tea, Wachira (1996) attempted to demonstrate the existence of various markers that significantly associated with QTLs influencing the 9 traits measured. However, as the study involved only single-tree seedling progeny that was not replicated the precise measurement of a phenotypic trait in the cross for each quantitative character could not be established. Replicated field trials now available, coupled with more suitable models for resolving QTLs, are assisting identification of complex agronomic traits based on molecular markers.

5.4.6 From Conventional to Molecular Breeding

A "quantitative trait" describes a characteristic for which the observed variation is due to the segregation of several genes and where, for each gene, the effects of the allelic differences on the phenotype are generally small compared with the effects of the environment (Kearsey & Pooni, 1996). Genetic mapping of QTL involves identifying and determining the degree of association between the continuous traits and sets of genetic markers.

The ability to assess complex phenotypes such as yield, quality, drought tolerance and susceptibility to pests and diseases in tea at the seedling stage using genetic markers would greatly accelerate new cultivar development. In addition to the selection of advantageous traits, markers linked to complex traits could be used to select against negative characteristics and could even be used to select the combination of parents that would give rise to progeny with the desired genotype.

An essential requisite for accurate QTL identification in any plant species is a saturated genetic map covering the entire genome. If certain regions of the genome are not adequately represented by genetic markers, the QTL located in such regions will not be reliably mapped, because it will be difficult to determine if the QTL has a genuinely small phenotypic effect, or is merely weakly linked to flanking markers (Lander & Botstein, 1989).

An experiment entailing QTL mapping using molecular and phenotypic traits that had been measured on a pseudo-test cross comprising 42 clonal progeny (Stock 463) has recently been conducted at the TRFK (Kamunya & Muoki, 2009; Kamunya *et al.*, 2010). Over 250 RAPD, 15 SSR, 12 ISSR and 96 AFLP primers were screened using the bulked segregant analysis (BSA) technique in order to identify possible markers associated with yield, quality and stress related traits in Kenyan tea. The various bulks utilized in the study were formulated based on phenotypic data that had been collected during the trial period. Polymorphic and informative primers were further utilized in complete genotyping of the mapping population. The primers generated a total of 267 (50 RAPD, 7 ISSR, 11 SSR and 199 AFLP) informative markers in the mapping population upon complete genotyping. Of the 267 markers generated for Stock 463, 149 dominant markers showed a 1:1 segregation ratio (backcross) and were used to construct a linkage map of tea. The map consisted of 30 linkage groups out of which 19 resulted from maternal alleles and 11 are paternal. The 30 linkage groups spanned 1,411.5 cM with a mean interval of 14.7 cM between loci. Single-point genome-wide regression analysis detected a total of 64 QTLs controlling various traits across the two sites. Seventy-five percent of these QTLs were mapped into specific chromosomes, while the rest remained unlinked. Of these, QTLs for yield at the Timbilil site (YLD-T), yield at the Kangaita site (YLD-K), drought tolerance at the Kangaita site (DT-K) and pubescence (PUB) were localised at 2 cM, 2.7 cM, 3 cM and 1.4 cM from markers OPG-07-2800, E-AGC/M-CAG-725, OPT-18-2500 and OPO-02-650, respectively. The effects due to genotype-environment interactions were such that not a single marker could be associated with the same trait across environments. However, some markers portrayed pleiotropic effects by associating with more than one trait. Considering the long period required to develop elite tea cultivars (about 20 years), the study demonstrated that adoption of marker-assisted selection (MAS) could increase breeding efficiency by reducing the developmental interval for elite cultivars inhabiting novel traits. Identification of putative QTLs tightly linked to markers in this study, augmented by known genetic parameters, demonstrates the great potential of integrating molecular markers in tea breeding.

5.4.7 Participatory Crop Improvement: Involvement of Farmers in Tea Improvement

Participatory Crop Improvement (PCI) emerged in the past decade as an alternative plant breeding approach for developing countries in response to the recognition that conventional breeding of the formal institutions had brought little significant crop improvement to small scale farmers in agro-ecologically and socio-economically marginal and variable environments (Virk *et al.*, 2003; Witcombe *et al.*, 2003). A major reason for this is the fact that Formal Crop Improvement (FCI) in developing countries concentrated on cereals and cash crops, such as tea, in favorable high input agricultural systems. It was expected that at least some of the materials, which were developed for high input production systems, would also be successful in low input environments. However, farming systems in marginal environments are too different from those in the more favorable production areas (Lipton & Longhurst, 1989).

In developing countries, FCI programs are largely carried out on-station under well-controlled conditions, thus reducing environmental variation and increasing heritability and expected genetic gain. However, the majority of small scale farmers operate in environments in which variable complex stresses have a dominating effect on crop performance (Banziger *et al.*, 1997). The importance of adaptation to variable and risky low-input rain fed conditions, secondary crop uses and cultural preferences have received little or no attention. As breeding work focuses more on breeding for high yields and adaptation, the need to explain other characteristics of importance to small scale farmers call for closer attention. The need to produce a stable cultivar demands that selection be carried out in marginal environments preferably with farmers. Generally, if a breeder wishes to produce a strain or cultivar that should perform well in a particular environment, selection should be carried out in that environment (Kearsey & Pooni, 1996). On the other hand, in order to produce a stable cultivar, selection should be carried out in a poor environment.

The common usage of relatively high input levels to minimize abiotic and biotic variation and to target moderate to high input agriculture in FCI reduces the ratio of environmental variance vs. genotypic variance in comparison with the use of lower input levels. It also increases the discrepancy between on-station and on-farm conditions (Ceccarelli *et al.*, 1992). If environments are sufficiently different, G×E interactions can result in different ranking of evaluated germplasm representing the so-called crossover effect. Thus, products from FCI programs are not necessarily adapted to the marginal environments. On-station selection, therefore, does not in such a case result in the most productive materials for the specific conditions in the farmer's fields.

Differences in selection criteria contribute to diversity in materials selected by breeders and farmers. While farmers pay more attention to yield stability and characteristics of quality and secondary uses, the breeder may be paying attention to high yields (Thiele *et al.*, 1997). Because the criteria other than yield appear to be very important factors in cultivar adoptions and rejection, it is logical to include them explicitly in the analysis of G×E interaction and related issues in the context of PCI. Another drawback of the FCI system is its slow release of relatively few genetically homogeneous genotypes. It takes an FCI program 12 to 18 years to develop a new cultivar. A breeding program easily works with thousands of heterogeneous or homogeneous entries in different selection stages, of which only a fraction reach the on-farm testing phase, of which in turn only a few cultivars are released. Many materials which could potentially have been valuable for other conditions and preferences are eliminated in the process. The released cultivars are usually genetically uniform, which is not a necessity for small scale farmers. On the contrary, materials that contain some genetic diversity may be more suitable for variable and heterogeneous environments, providing them with an increased buffering capacity and potential to adapt. The rights of cultivar registration and plant variety protection add to the time needed for release and involve costs that form an additional drawback in responding to the needs for diversity within and between crop cultivars.

5.4.7.1 Why PCI? Strategy and Justification

PCI aims to link formal and local systems of crop improvement, combining the complementary capacities and expertise, seeking to combine the improvement in productivity with the supply of agrobiodiversity needed by farmers (Hardon, 1995). The PCI-strategy is to insert useful genetic diversity into the local systems and build on farmers' capacity for seed selection and exchange. Rather than trying to improve the impact of conventional breeding programs that generate a limited number of genetically uniform cultivars, the idea is to flush out into farmers' fields a larger number of materials, representing a wider range of genetic diversity. PCI builds on the recognition of farmers' capacity to select what best fits their environment and improved development of local crop adaptation through farmers' cultivar and seed selection. It relies on farmers' seed production and exchange to maintain and diffuse cultivars.

The main advantage of PCI over conventional breeding is that it involves farmers in developing, adapting and adopting new cultivars, setting breeding goals and selecting parents according to their requirements. The level of participation, however, varies with the nature and objectives of the project and availability of resources. It develops a spirit between different organizations and farmers of working closely together and appreciating each other's capabilities and contributions. The strengths and capabilities of different stakeholders are fully utilized in an integrated form.

A common functional distinction within PCI is Participatory Variety Selection (PVS), which is the selection between advanced or genetically stable populations and lines and Participatory Plant Breeding (PPB), which is selection within segregating populations (Witcombe *et al.*, 1996). In PVS, farmers are given cultivars (finished products from plant breeding) for testing in their own fields. After a successful PVS program, the cultivars preferred by farmers can be used as parents in a breeding program where farmers participate as active collaborators. This involves breeding and selection to create new cultivars and is called PPB. However, the distinction between PVS and PPB is not always clear. In the case of cross-pollinating populations, selection between populations (PVS) is usually combined with within-population selection (PPB). On-farm evaluation allows weighing of preferences and needs by the end-user of the products, and enables exploitation of G×E interaction through seeking location-specific adaptation to the complex and variable environment.

5.4.7.2 Success Cases of PCI

The success of the reported cases so far confirms the positive impact that participatory approaches can have on crop improvement in a marginal environment (Sperling *et al.*, 2001). The success of these cases is largely based on the fact that, through collaboration with farmers and on-farm selection in the target area, selection criteria and characteristics that were not given sufficient weight in the selection in FCI are now identified and incorporated in material. The farmers' willingness and capacity to invest time and resources in selection and participation with breeders will depend strongly on the benefits they derive from it. The benefits are access to materials with increased yield, yield stability or other improvements, status, knowledge and increased capacities (empowerment) and benefits from seed exchange. The latter benefit however assumes that locally selected materials have a wide agro-ecological adaptation and are attractive to a larger group of farmers (Sperling *et al.*, 2001).

The farmers' empowerment is considered as an important social benefit from PCI. The type of participation is presumably influencing the empowerment impact of the farmers. If farmers are only consulted and are not given a decision in the identification of material, the setting of selection criteria and selection itself, there is no true participation or an empowerment benefit. Empowerment or the capacity of farmers to work on improving their own is recognized as an important condition for sustainable agricultural development.

5.4.7.3 Situation of Tea Improvement in Relation to PCI in Kenya

Studies conducted in the recent past have revealed that the performance of tea clones relative to each other vary considerably with environments so that clones which are superior in one environment are not correspondingly superior elsewhere (Ng'etich *et al.*, 2001; Wachira *et al.*, 2002). Such genetic variation in response to environmental changes and in adaptation has not been adequately studied in tea. Earlier tea improvement efforts relied on limited involvement of the farmers in selection. Tea cultivars were developed in one site, usually at the Foundation headquarters in Timbilil Estate with all the stages in the selection process being evaluated at the site. The elite cultivars were then released to all the farmers in the country in order initially for them to test their adaptability to the local niche environments. They would subsequently be expected to adopt those which would be suitable to their conditions and preferences. Owing to the perennial nature of the tea growth cycle, the farmers eventually adopted a few of the released cultivars. Indeed one clone, TRFK 6/8, endeared itself to the farmers owing to its high black tea quality. It has been adopted by more than 60% of small scale farmers. Another clone, TRFK 31/8, a high yielding clone, has also been widely adopted by farmers.

The current tea improvement program has fully embraced the concept of modified PCI by involving the farmers in clonal adaptability studies through, in part, their factory tea extension agents and partly through the smallholder management agency, the Kenya Tea Development Agency (KTDA). Some farmers have volunteered their fields as testing sites and even their resources to maintain and collect data from the trials (Fig. 5.5). Extension officers supervise data collection activities on a weekly basis and later forward them to the plant breeder at TRFK.

Fig. 5.5. Involving small scale farmers in testing and selection of elite clones is a routine exercise in TRFK tea improvement program

To ensure that quality and reliable data are collected, training sessions on clonal identification, plot labeling, data recording and reporting are held on-farm. The initial data recording is done by a team comprising TRFK technical staff, KTDA tea extension staff and the farmer. Subsequently, the Tea Extension Coordinators and their assistants conduct supervision and provide linkage between the farmer and TRFK staff. The breeder and his technical staff carry out follow up visits on a quarterly basis to hold talks with parties involved, which include finding out the problems the farmers are facing and how best to solve them. Currently, a total of 10 participatory clonal adaptability trials are being carried out in different tea growing regions of Kenya. One such trial is being conducted on a collaboration basis between the TRFK and the Tea Research Institute of Tanzania (TRIT) with additional clones from TRIT.

A successful example of PCI in Kenya can be demonstrated by three clonal trials planted in 2003 in selected smallholder farms with the aim of testing for tolerance/resistance to root knot nematode of some selected released and promising clones (Kamunya *et al.*, 2008). Three farms, situated in the Kirinyaga district on the eastern side of the Great Rift Valley, were chosen on the basis of harboring high levels of nematode populations and their owners' willingness to surrender them for experimentation until sufficient data were collected. A combined total of 58 clones, i.e., 8 released clones, 25 (9 released and 16 promising clones) and another 25 released clones were screened for nematode tolerance/resistance in the three sites, respectively. Evaluation carried out over a two-year period from 2004 revealed that clones TRFK 31/8, TRFCA SFS 150, TRFK 301/5, EPK TN 14-3, AHP PMC 61, AHP SC 31/37, TRFK 54/40, TRFK 56/89, TRFK 100/5, TRFK12/12, TRFK 7/3, TRFK 337/3, TRFK 371/3 (a promising clone), TRFK 55/55, TRFK 338/13, TRFK 55/56, TRFK 31/28, TRFK 12/19, TRFK 430/4, TRFK 481/200, TRFK 378/1, TRFK 430/52, TRFK 375/5, TRFK 337/137, TRFK 347/573, TRFK 400/4, TRFK 430/7, TRFK 480/318, TRFK 371/8, TRFK 400/7, TRFK 400/10, and TRFK 371/6 were found to be highly tolerant to nematodes. Conversely, clones TRFK 303/577, TRFK 6/8, TRFK 303/1199, TRFK 301/4, TRFK 303/216, TRFK 57/15, TRFK 31/29, KTDA KAG 4, EPK D 99/10 and BBK 21 were discovered to be highly susceptible to nematodes. Yield data collection showed that the resistant/tolerant cultivars gave significantly higher yields than the susceptible cultivars (TRFK, 2006).

Similar trials have been set up in the Trans Nzoia, Mt Elgon, Gucha, Nandi and Meru North districts. Preliminary results show remarkable variability in clonal performance and preferences by the farmers (TRFK, 2006). With on-farm clonal trials and selection in the target area, selection criteria and characteristics that were initially not given sufficient weight in the selection under FCI, are now being identified with farmers' participation and incorporated in new materials ahead of release for large scale commercial utilization.

5.4.7.4 The role of Farmers and Benefits

From earlier discussion on success cases of PCI, the role of farmers emerged to be:

(i) Identification of traits, which are considered to be minor for the FCI to address;

(ii) Recognition of traits that correspond to the farmers' preferences;

(iii) Identification of characteristics that better suit their farming systems.

The farmers' willingness and capacity to invest time and resources in selection and participation with breeders will depend strongly on the benefits they derive from it. Benefits are access to materials with increased yield, high quality, stability or other improvements, status, knowledge and increased capacities (empowerment).

5.5 Conclusions

The role of tea improvement efforts in increased acreage, production and productivity of Kenyan tea germplasm cannot be overemphasized. To date a total of 50 high yielding and good quality tea clones have been released for commercial utilization, not just in Kenya but in the entire East African region. The resulting enhanced profitability and rapid expansion of the Kenyan tea industry that has been realized over the last 100 years has made tea one of the most popular enterprises particularly in the rural areas. The Kenyan tea industry which almost solely entails the sale of CTC black tea is, however, currently experiencing problems as a result of global annual production that is outstripping demand by about 1%. This has further been aggravated by the increased cost of production that threatens to reverse the gains realized over the last century. To counter the declining trend in the revenue base from the tea enterprise, efforts to undertake value-addition and product diversification have been initiated. Suitable improved clones with the potential for processing high catechins green tea and "silvery tips" (white tea) have already been selected, while a purple tea cultivar with potential pharmacological value was pre-released in 2011. Other improved clones are in different stages of testing for caffeine deficiency and high value tea seed oil. Furthermore, tea improvement activities integrating biotechnological tools and participatory clonal selection are expected to fast-track the development and adoption of novel cultivars within a relatively shorter period.

Acknowledgements

This review was supported by the Tea Board of Kenya and Council for Science and Industrial Research of India.

References

Ackerman WL (1970) Inter-specific Hybridization of *Camellia.* Americana Camellia Year Book, pp.65-79.

Ackerman WL (1973) Species compatibility relationships within the genus *Camellia*. Journal of Heredity, 64: 356-358.

Anon (1962) Historical notes on tea introduction in Africa. In: Tea Estates in Africa (Compiled by Wilson, Smithett & Co). London: Mabey & Fitzclarence Ltd, 6-9.

Anon (2003a) A spot of tea may be a shot in the arm for body's defenses. The Kansas City Star, Washington.

Anon (2003b) Five cups of tea keep the doctor away. The Guardian. 22 April, 2003.

Apostolides Z, Nyirenda HE, Mphangwe NIK (2006) Review of tea (*Camellia sinensis*) breeding and selection in Southern Africa. International Journal of Tea Science, 5(1&2): 13-19.

Arulpragasam PV (1990) Micropropagation of tea. In: Achievements and Future Prospects. International Conference on Tea Research: Global Perspective. 11-12 January, 1990, Culcutta, pp.1-5.

Association for Strengthening Agricultural Research in Eastern and Central Africa (ASARECA) (2004) Regional Variety List for Kenya, Uganda and Tanzania. The Secretariat, Seed Regional Working Group, Kenya, p.100.

Banerjee B (1992a) Botanical classification of tea. In: Willson KC & Clifford MN (eds.) Tea: Cultivation to Consumption. London: Chapman and Hall, pp.39-52.

Banerjee B (1992b) Selection and breeding of tea. In: Willson KC & Clifford MN (eds.) Tea: Cultivation to Consumption. London: Chapman and Hall, pp.53-81.

Banziger M. Betran FJ, Laffite HR (1997) Efficiency of high nitrogen selection environments for improving maize for low nitrogen target environments. Crop Science, 37: 1103-1109.

Barth S, Busimi AK, Friedrich HU, Melchinger AE (2003) Heterosis for biomass yield and related traits in five hybrids of *Arabidopsis thaliana* L. Heynh. Heredity, 91: 36-42.

Barua DN (1963) Selection of vegetative clone. Tea Encyclopaedia, 163: 32-88.

Basu B (2003) Drink tea and keep healthy. International Journal of Tea Science, 2(3): 5-7.

Becker WA (1984) Manual of Quantitative Genetics (4th Edition). Pullman: Academic Enterprises, p.188.

Berry D, Cilas C (1994) Genetic study of the behavior to black pod disease of cocoa (*Theobroma cacao* L.) obtained by diallel crossings. Agronomie, 14(9): 599-609.

Bezbaruah HP (1974) Tea breeding—a review. Indian Journal of Genetics and Plant Breeding (S), 34(A): 90-100.

Bezbaruah HP (1987) Use of interspecific hybrids in tea breeding. Two and A Bud, 34: 1-4.

Bradshaw HD (Junior) (1998) Case history in genetics of long-lived plants: Molecular approaches to domestication of a fast-growing forest tree: *Populus*. In: Paterson AH (eds.) Molecular Dissection of Complex Traits. New York: CRC Press, pp.219-228.

Brown GR, Bassoni DL, Gill GP, Fontana JR, Wheeler NC, Megraw RA, Davis MF, Sewell MM, Tuskan GA, Neale DB (2003) Identification of quantitative trait loci influencing wood property traits in loblolly pines (*Pinus taeda* L.). III. QTL verification and candidate gene mapping. Genetics, 164: 1537-1546.

Cannell MGR, Njuguna CK, Ford ED (1977) Variation in yield among competing individuals within mixed genotypes of tea: A selection problem. Journal of Applied Ecology, 14: 969-985.

Ceccarelli S, Grando S, Hamblin J (1992) Relationship between barley grain yield measured in low- and high-yielding environments. Euphytica, 64: 49-58.

Chang HT, Bartholomew B (1984) *Camellias*. Portland: Timber Press.

Chen ZM (1995) Tea in 21st Century. In: Proceedings of 1995 International Tea-Quality-Human Health Symposium. 7-10 November, 1995, Shanghai, China, pp.3-6.

China Tea Varieties Compilation Committee (2001) China Tea Varieties. Shanghai: Shanghai Scientific and Technical Publishers, p.9 (in Chinese).

Clifford MN (2000) Anthocyanins—nature, occurrence and dietary burden. Journal of the Science of Food and Agriculture, 80: 1063-1072.

Collard BCY, Jahufer MZZ, Bronwer JB, Pang ECK (2005) An introduction to markers, quantitative trait loci (QTL) mapping and marker-assisted selection for crop improvement: The basic concepts. Euphytica, 142: 169-196.

Cowen NM (1989) Multiple linear regression analysis of RFLP data sets used in mapping QTLs. In: Helentjaris T and Burr B (eds.) Development and Applications of Molecular Markers to Problems in Plant Genetics. New York: Cold Spring Harbor Laboratory Press.

Crouzillat D, Menard B, Mora A, Phillips W, Petiard V (2000) Quantitative trait loci analysis in *Theobroma cacao* using molecular markers: Yield QTL detection and stability over 15 years. Euphytica, 114: 13-23.

Dhawan BN (2006) Tea as a Rasayana. In: Jain NK, Siddiqi MA, Weisburger JH (eds.) Protective Effects of Tea on Human Health. CAB International, pp.6-15.

Dias LAS, Kageyama PY (1995) Combining ability for cacao (*Theobroma cacao* L.) yield components under southern Bahia conditions. Theoretical and Applied Genetics, 90: 534-541.

Dufresne CJ, Farnworth ER (2001) A review of latest research findings on the health promotion properties of tea. Journal of Nutritional Biochemistry, 12: 404-421.

Elberse IAM, Vanhala TK, Turin JHB, Stam P, van Damme JMM, van Tienderen PH (2004) Quantitative trait loci affecting growth-related traits in wild barley (*Hordeum spontaneum*) grown under different levels of nutrient supply. Heredity, 93: 22-33.

Falconer DS (1989) Introduction of Quantitative Genetic (3rd Edition). New York:

John Wiley and Sons, Inc., p.438.
Food and Agriculture Organization (FAO) (2010) http://faostat.fao.org/.
Gazi MS (1978) Distributional pattern of yield and vacancy of tea in Bangladesh. Tea Journal, 14(2): 19-22.
Gill M (1992) Speciality and herbal teas. In: Willson KC & Clifford MN (eds.) Tea: Cultivation to Consumption. London: Chapman and Hall, pp.513-528.
Goodchild NA (1960) Vegetative propagation. TRIE Pamphlet, (17): 60-69.
Green MJ (1966) Clonal selection in seedling stump nurseries. Tea in East Africa, 6(4): 11-12.
Green MJ (1969) New clonal release from the TRI. Tea, 10(1): 15.
Green MJ (1971) An evaluation of some criteria used in selecting large yielding tea clones. Journal of Agricultural Science, 76: 143-156.
Green MJ (1973) TRI breeding schemes. Tea in East Africa, 13(2): 5.
Greenway PJ (1945) Origins of some East African food plants. Part V. East African Journal of Sciences, 11: 56-63.
Hackett CA, Wachira FN, Paul S, Powell W, Waugh R (2000) Construction of a genetic linkage map for *Camellia sinensis* (tea). Heredity, 85: 346-355.
Hainsworth E (1965) The 1965 clonal release. Tea, 6(1): 15.
Hampton MG (1992) Production of black tea. In: Willson KC & Clifford MN (eds.) Tea: Cultivation to Consumption. London: Chapman and Hall, pp.459-511.
Hara Y (2001) Green Tea: Health Benefits and Applications. New York: Marcel Dekker.
Hara Y (2006) Prophylactic functions of tea catechins. In: Jain NK, Siddiqi MA, Weisburger JH (eds.) Protective Effects of Tea on Human Health. CAB International, pp.16-24.
Hardon J (1995) Participatory Plant Breeding. The outcome of a workshop on participatory plant breeding, 26-29 July, 1995. Plant Genetic Resources, IPGRI Rome.
Hasselo HN (1964) Productivity gradient in sloping tea land in Ceylon. Tea Quarterly, 35: 307-317.
Horsch RB, Fry JE, Hoffmann NL, Eichholz D, Rogers SG, Fraley RT (1985) A simple and general method for transferring genes into plants. Science, 227: 1229-1231.
Ichikawa H, Kunii M, Isemura M (2004) Mechanism of apoptosis induction selective for cancer cells by EGCG. In: Proceedings of 2004 International Conference on O-Cha (tea) Culture and Science. 4-6 November, 2004, Shizuoka, Japan, pp.458-459.
International Plant Genetic Resources Institute (IPGRI) (1997) Descriptors for Tea (*Camellia sinensis*). Rome, Italy.
International Tea Committee (ITC) (2010) Annual Bulletin of Statistics, London.
International Union for the Protection of New Varieties of Plants (UPOV) (2008) TG/238/1: Guidelines for the Conduct of Tests for Distinctness, Uniformity and Stability of Tea (*Camellia sinensis* (L). O. Kuntze). Geneva.
Jinks JL (1981) The genetic framework of plant breeding. Philosophical

Transactions of the Royal Society (London), B. 292: 407-419.

Kamunya SM, Wachira FN (2006) Two new clones (TRFK 371/3 and TRFK 430/90) released for commercial use. Tea, 27(1&2): 4-16.

Kamunya SM, Muoki RC (2009) Botanical and Genetical Investigations. Proceedings of Annual Technical Conference. 13-14 March, 2009, Tea Research Foundation of Kenya Training Centre, Kericho, Kenya, pp.6-28.

Kamunya SM, Muoki RC, Wachira FN, Pathak RS (2007) Inheritance of yield, drought tolerance and quality traits in tea (*Camellia sinensis* (L.) O. Kuntze). Tea, 28(2): 20-29.

Kamunya SM, Wachira FN, Langát J, Otieno W, Sudoi V (2008) Integrated management of root knot nematode (*Meloidogyne* spp.) in tea. International Journal of Pest Management, 54(2): 129-136.

Kamunya SM, Wachira FN, Pathak RS, Muoki RC, Wanyoko JK, Ronno WK, Sharma RK (2009) Quantitative genetic parameters in tea (*Camellia sinensis* (L.) O. Kuntze): I. Combining abilities for yield, drought tolerance and quality traits. African Journal of Plant Science, 3(5): 93-101.

Kamunya SM, Wachira FN, Pathak RS, Korir R, Sharma V, Kumar R, Bhardwaj P, Muoki RC, Ahuja PS, Sharma RK (2010). Genomic mapping and testing for quantitative trait loci in tea (*Camellia sinensis* (L.) O. Kuntze). Tree Genetics and Genomes, 6: 915-929.

Karori SM, Ngure RM, Wachira FN, Wanyoko JK, Mwangi JN (2008) Different types of tea products attenuate inflammation induced by *Trypanosoma brucei* infected mice. Parasitology International, 57: 325-333.

Kearsey MJ, Pooni HS (1996) The Genetical Analysis of Quantitative Traits. New York: Stanley Thornes (Publishers) Ltd., p.381.

Kearsey MJ, Pooni HS, Syed NH (2003) Genetics of quantitative traits in *Arabidopsis thaliana*. Heredity, 91: 456-464.

Kenya Plant Health Inspectorate Service (KEPHIS) (2004) Kenya National Crop Variety List, 45.

Kulasegaram S (1978) Progress in tea breeding. Proceedings of Symposium on Methods of Crop Breeding. October, 1977, Tropical Agricultural Research Service, Tokyo, Japan, 11: 151-160.

Lander ES, Botstein D (1989) Mapping mendelian factors underlying quantitative traits using RFLP linkage maps. Genetics, 121: 743-756.

Lipton R, Longhurst M (1989) New Seeds and Poor People. London: Unwin and Hyman, p.473.

M'Imwere ZK (1997) Tea production in the smallholder sector in Kenya: Achievements, problems and prospects. Tea, 18(2): 75-86.

Maeda Y (1989) Inhibitory effects of tea extracts on histamine release from mast cells. Food Hygiene and Safety Science Journal, 30(4): 295-299.

Magoma GN, Wachira FN, Obanda M (2001) The pharmacological potential and catechin diversity inherent in Kenyan tea. Tea, 22(2): 83-93.

Mamati GE, Wachira FN, Njuguna CK (2001) 2001 released clones. Tea, 22(1): 6-7.

Mamedov MA (1961) Tea selection in Azarbajdzan. Agrobiologia, 1: 62-67.

Mather K, Jinks JL (1982) Biometric Genetics. London: Chapman and Hall.

Matheson JK (1950) Tea. East African Agriculture. Matheson & Bovill, E.W. OUP. pp.198-206.

Matsubara Y, Kumamoto H, Lizuka Y, Murakami T, Okamoto K, Miyake H, Yokoi K (1985) Structure and hypotensive effect of flavonoid glycosides in *Citrus unshiu* peelings. Agricultural and Biological Chemistry, 49: 909-914.

Medina-Filhno HP (1980) Linkage of *Aps-1*, Mi and other markers on chromosome 6. Report of Tomato Genetics Cooperative, 30: 26-28.

Mondal TK, Bhattacharya A, Ahuja PS, Chand PK (2001) Transgenic tea (*Camellia sinensis* (L.) O. Kuntze cv. Kangra Jat) plants obtained by *Agrobacterium*-mediated transformation of somatic embryos. Plant Cell Reports, 20: 712-720.

Mondal TK, Bhattacharya A, Laxmikumaran M, Ahuja PS (2004) Recent advances of tea (*Camellia sinensis*) biotechnology. Plant Cell, Tissue and Organ Culture, 76: 195-254.

Muoki RC, Wachira FN, Pathak RS, Kamunya SM (2007) Assessment of the mating system of *Camellia sinensis* in biclonal seed orchards based on PCR markers. Journal of Horticultural Science & Biotechnology, 82(5): 733-738.

Nagarajah S, Ratnasurya (1981) Clonal variability in root growth and drought resistance in tea (*Camellia sinensis*). Plant and Soil, 60: 153-155.

Nakagawa M, Torii H (1965) Studies of flavanols in tea. 4. Enzyme oxidation of flavanols. Agricultural and Biological Chemistry, 20: 278-284.

Nakane H, Ono K (1990) Differential inhibitory effects of some catechins derivatives on the activities of HIV reverse transcriptase and cellular deoxyribonucleic acid and RNA polymerase. Biochemistry, 29(11): 2841-2845.

Ng'etich WK, Stephens W (2001) Responses of tea to environment in Kenya. 1. Genotype × environment interactions for total dry matter production and yield. Experimental Agriculture, 37: 333-342.

Ng'etich WK, Stephens W, Othieno CO (2001) Responses of tea to environment in Kenya: Yield and yield distribution. Experimental Agriculture, 37: 361-372.

Niehuis J, Helentjaris T, Slocum M, Ruggero B, Schaefer A (1987) Restriction fragment length polymorphism analysis of loci associated with insect resistance in tomato. Crop Science, pp.797-803.

Njuguna CK (1985) 1986 TRFK released clones. Tea, 6(2): 4-5.

Njuguna CK (1987) 1988 TRFK released clones. Tea, 8(2): 40-42.

Njuguna CK (1989a) TRFK released clones. Tea, 10(1): 5-6.

Njuguna CK (1989b) Yield performance of TRFK new released clones from the breeding program. Tea, 10(2): 64-72.

Obanda M, Owuor PO, Njuguna CK (1992) The impact of clonal variation of total polyphenols content and polyphenol oxidase activity of fresh tea shoots on plain black tea quality parameters. Tea, 13(2): 129-132.

Obanda M, Owuor PO, Taylor SJ (1997) Flavanol composition and caffeine content of green leaf as quality indicators of Kenyan black teas. Journal of the Science of Food and Agriculture, 74: 209-215.

Ohta T, Kobayashi T, Kondo T, Hara Y, Kaji K (2004) Suppression of tumor angiogenesis by EGCG—Comparison with the effect of propolis. In: Proceedings of 2004 International Conference on O-Cha (tea) Culture and Science. 4-6 November, 2004, Shizuoka, Japan, pp.481-482.

Ohtsura M (1991) Biochemical examination of chronic tea consumption in the rabbit. Journal of Japanese Food Industry, 38(7): 52-54.

Otieno W, Sudoi V, Wachira F, Mamati GE, Chalo R (2002) A report on outbreak of root knot nematodes on tea in Kerugoya and Imenti. TRFK Quarterly Bulletin, 7(3): 6-8.

Owuor PO (1992) Comparison of gas chromatographic volatile profiling methods for assessing the flavour quality of Kenya black teas. Journal of the Science of Food and Agriculture, 59: 189-197.

Owuor PO, Othieno CO (1987) Environmental effects on tea quality. I. Locational effects on chemical composition and quality of clonal teas. Proceedings of International Symposium on the Chemistry of Tropical Natural Products. on 24-28 August 1987 Moi University, Kenya.

Owuor PO, Reeves SG, Wanyoko JK (1986) Correlation of theaflavins content and valuations of Kenyan teas. Journal of the Science of Food and Agriculture, 37: 507-513.

Owuor PO, Tsushida T, Horita H, Murai T (1988) Effects of geographical area of production on the composition of the volatile flavour compounds in Kenyan clonal black CTC teas. Experimental Agriculture, 24(2): 227-235.

Owuor PO, Obaga SMO, Othieno CO (1990) The effect of altitude on the chemical composition of black tea. Journal of the Science of Food and Agriculture, 50: 9-17.

Owuor PO, Obanda M, Apostolides Z, Wright LP, Nyirenda HE, Mphangwe NIK (2006) The relationship between the chemical plain black tea quality parameters and black tea color, brightness and sensory evaluation. Food Chemistry, 97: 644-653.

Paul S, Wachira FN, Powell W, Waugh R (1997) Diversity and genetic differentiation among populations of Indian and Kenyan tea (*Camellia sinensis* (L.) O. Kuntze) revealed by AFLP markers. Theoretical and Applied Genetics, 94(2): 255-263.

Roberts EAH (1958) Chemistry of tea manufacturing of N.E. India. Journal of the Science of Food and Agriculture, 9: 381-390.

Roberts EAH (1962) Economic importance of flavanoid compounds. The Chemistry of Flavanoid Compounds, Pergamon, New York, pp.468-510.

Reeves SG, Owuor PO, Othieno CO (1987) Biochemistry of black tea manufacture in Kenya. Tropical Science, 27: 121-123.

Rogers SS (1975) Preliminary observations on pollen tube incompatibility in some tea clones. Tea Quarterly, 45: 463-470.

Saiki RK, Gelford DH, Stoffel A, Scharf SJ, Higuchi R, Horn GT, Mullis KB, Erlich HA (1988) Primer directed enzymatic amplification of DNA with a thermostable DNA polymerase. Science 239: 487-491.

Saintagne C, Bodenes C, Barreneche Pot D, Plomion C, Kremer A (2004)

Distribution of genomic regions differentiating oak species assessed by QTL detection. Heredity, 92: 20-30.
Sanderson GW (1963) The chloroform test: A study of its suitability as a means of rapidly evaluating fermenting properties of clones. Tea Quarterly, 34: 193-196.
Sebastiampillai AR (1963) Report on plant breeding. Annual Report, Tea Research Institute of Ceylon, pp.87-89.
Sharma VS, Ranganathan V (1985) The world of tea today. Outlook on Agriculture, 14: 35-40.
Shearer W (2003) Green tea extract may fight HIV. BBC News.
Singh ID (1979) Indian tea germplasm and its contribution to the world's tea industry. Two and A Bud, 26(1): 23-26.
Singh ID (1980) Non-conventional approaches to the breeding of tea in North East India. Two and A Bud, 27: 3-6.
Singh ID, Sharma PC (1982) Studies in radiation breeding in tea plants. Proceedings of the 4th Annual Symposium on Plantation Crops, pp.1-19.
Soller M, Brody T, Genizi A (1976) On the power of experimental designs for the detection of linkage between marker loci and quantitative loci in crosses between inbred lines. Theoretical and Applied Genetics, 47: 35-39.
Sperling L, Ashby JA, Smith ME, Weltzien E, McGuire S (2001) A framework for analyzing participatory plant breeding approaches and results. Euphytica, 122: 439-450.
Stam P (1991) Some aspects of QTL analysis. In: Pesek J, Hartmann J (eds.) Biometrics in plant breeding. Proceedings of the 8th Meeting of the Eucarpia Section, Biometrics in Plant Breeding. 1-6 July, 1991, Brno, Czechoslovakia.
Su MH, Yang SZ, Hsieh CF (2004) The identity of *Camellia buisanensis* Sasaki (Theaceae). Taiwania, 49(3): 201-208.
Sudoi V (1995) Effects of spraying petroleum oil on the control of scale insects *Apidiotus* sp. and their effects on natural enemies. Tea, 16: 119-123.
Sudoi V (1996) Influence of soil applied nitrogen (NPKS 25: 5: 5: 5) on *Brevipalpus phoenicis* Geikjkes (Acari: Tenuipalpidae) mite incidence and damage symptoms on tea. PhD Thesis, Moi University, Eldoret, Kenya.
Sudoi V, Khaemba BM, Wanjala FME (1996) Screening of tea clones for their resistance to *Brevipalpus phoenicis* Geikjkes (Acari: Tenuipalpidae) attack. Journal of Plantation Crops, 24: 291-295.
Sugiyama K (1995) Anti-allergic effects of tea. In: Proceedings of the 3rd International Green Tea Seminar. 1 September, 1995, Seoul, Korea, pp.59-64.
Takeda Y (1994) Difference in caffeine and tannin contents between tea cultivars, and application to tea breeding. Japan Agriculture Research Quarterly, 28(2): 117-123.
Takeo T (1992) Green and semi-fermented teas. In: Willson KC & Clifford MN (eds.) Tea: Cultivation to Consumption. London: Chapman and Hall, pp.413-457.
Tanksley SD (1993) Mapping polygenes. Annual Review of Genetics, 27: 205-233.
Tanksley SD, Medina-Filho H, Rick SM (1982) Use of naturally-occurring

enzyme variation to detect and map genes controlling quantitative traits in an interspecific backcross of tomato. Heredity, 49: 11-25.

Tea Board of Kenya (TBK) (2010) Annual Report and Accounts 2009/2010.

Thiele G, Gardner G, Torrez R, Gabriel J (1997) Farmer involvement in selection of new varieties: Potatoes in Bolivia. Experimental Agriculture, 33: 1-16.

Thompson JN, Thoday JM (1974) A definition and standard nomenclature for "polygenic loci". Heredity, 33: 430-437.

Todd JR (1955) Green leaf and yellow leaf. TRIEA Pamphlet, 12: 23-29.

TRFK (1980) Breeding. Annual Technical Report. Tea Research Foundation of Kenya (TRFK), Tea Board of Kenya, p.30.

TRFK (1990) Annual Technical Report. Tea Research Foundation of Kenya (TRFK), Tea Board of Kenya, p.25.

TRFK (1991) Annual Technical Report. Tea Research Foundation of Kenya (TRFK), Tea Board of Kenya, pp.15-25.

TRFK (1998) Annual Technical Report. Tea Research Foundation of Kenya (TRFK), Tea Board of Kenya, pp.38-41.

TRFK (1999) Annual Technical Report. Tea Research Foundation of Kenya (TRFK), Tea Board of Kenya, p.41.

TRFK (2001a) Annual Technical Report. Tea Research Foundation of Kenya (TRFK), Tea Board of Kenya, pp.22-25.

TRFK (2001b) Annual Technical Report. Tea Research Foundation of Kenya (TRFK), Tea Board of Kenya, p.40.

TRFK (2002) Annual Technical Report. Tea Research Foundation of Kenya (TRFK), Tea Board of Kenya, pp.20-28.

TRFK (2004) Annual Technical Report. Tea Research Foundation of Kenya (TRFK), Tea Board of Kenya.

TRFK (2005) Annual Technical Report. Tea Research Foundation of Kenya (TRFK), Tea Board of Kenya.

TRFK (2006) Annual Technical Report. Tea Research Foundation of Kenya (TRFK), Tea Board of Kenya.

TRFK (2008) Annual Technical Report. Tea Research Foundation of Kenya (TRFK), Tea Board of Kenya.

TRIEA (1966) Tea estate practice. Tea Research Institute of East Africa (TRIEA).

Vallejos CE, Tanksley SD (1983) Segregation of isozyme markers and cold tolerance in an interspecific backcross of tomato. Theoretical and Applied Genetics, 66: 241-247.

Venkatachalam P, Jayashree R, Rekha K, Sushmakumari S, Sobha S, Jayasree PK, Kala RG, Thulaseedharan A (2006) Rubber Tree (*Hevea brasiliensis* Muell. Arg). Methods in Molecular Biology, 344: 153-164.

Venkataramani KS, Padmanabhan TS (1964) A preliminary assessment of the relationship between certain leaf characteristics and cup quality. Annual Report United Plant Association of South India, Science Department, Tea Section, pp.50-63.

Vinson JA, Wu N, Teufel K, Zhang J (2001) Beneficial effects of green and black tea on animal models of antherosclerosis and diabetes. In: Proceedings of 2001

International Conference on O-Cha (tea) Culture and Science (Session III). 5-8 October, 2001, Shizuoka, Japan, p.367.

Virk DS, Singh DN, Kumar R, Prasad SC, Gangwar JS, Witcombe JR (2003) Collaborative and consultative participatory breeding of rice for the rainfed uplands of eastern India. Euphytica, 132: 95-108.

Visser T (1969) Tea (*Camellia sinensis*) (L.) O. Kuntze. In outlines of perennial crop Breeding in the Tropics. Miscellaneous paper 4, 459-493. (Ferwerda FP, Wit F eds.) Wageningen - The Netherlands.

Visser T, Kehl FH (1958) Selection and vegetative propagation of tea. Tea Quarterly, 29: 76-86.

Wachira FN (1990) Biotechnology: An assessment of its applicability in the improvement of Kenyan tea clones-tissue and cell culture. Tea, 11(1): 34-41.

Wachira FN (1994a) Clonal yield performance of some cambod teas (Shan tea), *C. sinensis* var. *assamica* ssp. *lasiocalyx* (Planchon ex-Watt). Tea, 15(2): 70-73.

Wachira FN (1994b) Triploidy in tea (*Camellia sinensis*): Effect on yield and yield attributes. Journal of Horticultural Science & Biotechnology, 69(1): 53-60.

Wachira FN (1996) Development of molecular markers in *Camellia*. PhD Thesis, Dundee University, Scotland, p.222.

Wachira FN (2002a) Genetic diversity and characterisation of Kenyan tea germplasm. A tea genetic diversity (TGD) project. TGD final project document, Kericho, Kenya.

Wachira FN (2002b) Genetic mapping of tea: A review of achievements and opportunities. Tea, 23(2): 91-102.

Wachira FN, Kiplangat JK (1991) Newly identified Kenyan polyploid tea strains. Tea, 12(1): 10-13.

Wachira FN, Ng'etich W (1999) Dry-matter production and partition in diploid, triploid and tetraploid tea. Journal of Horticultural Science & Biotechnology, 74: 507-512.

Wachira FN, Kamunya SM (2005a) Pseudo-self incompatibility in tea clones (*Camellia sinensis* (L.) O. Kuntze). Journal of Horticultural Science & Biotechnology, 80(6): 716-720.

Wachira FN, Kamunya SM (2005b) Kenyan teas are rich in antioxidants. Tea, 26(2): 81-89.

Wachira FN, Waugh R, Hackett CA, Powell W (1995) Detection of genetic diversity in tea (*Camellia sinensis*) using RAPD markers. Genome, 38: 201-210.

Wachira FN, Powell W, Waugh R (1997) An assessment of genetic diversity among *Camellia sinensis* L. (cultivated tea) and its wild relatives based on randomly amplified polymorphic DNA and organelle-specific STS. Heredity, 78: 603-611.

Wachira FN, Tanaka J, Takeda Y (2001) Genetic variation and differentiation in tea (*Camellia sinensis*) germplasm revealed by RAPD and AFLP variation. Journal of Horticultural Science & Biotechnology, 76: 557-563.

Wachira FN, Ng'etich W, Omolo J, Mamati G (2002) Genotype × environment interaction for tea yields. Euphytica, 127: 289-296.

Walker JRL (1975) The biology of plant phenolics. The Institute of Biology's Studies in Biology, (54): 23-32.

Weisburger JH (2006) Tea is health-promoting beverage in lowering the risk of premature killing chronic diseases: A review. In: Jain NK, Siddiqi MA, Weisburger JH (eds.) Protective Effects of Tea on Human Health. CAB International, pp.1-5.

Wight W (1956) Commercial selection and breeding of tea in India. World Crops, 8: 263-268.

Wight W (1958) The agrotype in tea taxonomy. Nature, 181: 893-895.

Wight W (1961) Combiners for tea breeding. Two and A Bud, 8(3): 3-5.

Wight W, Barua DN (1939) The tea plant industry: Some general principles. Tropical Agriculture (Ceylon), 93: 4-13.

Wight W, Barua DN (1954) Morphological basis of quality in tea. Nature, 173: 630-631.

Wight W, Gilchrist RCHH, Wight J (1963) Note on the colour and quality of tea leaf. Empire Journal of Experimental Agriculture, 31: 124-126.

Witcombe JR Joshi A, Joshi KD, Sthapit BR (1996) Farmer participatory crop improvement. I. Variety selection and plant breeding methods and their impact on biodiversity. Experimental Agriculture, 32: 445-460.

Witcombe JR, Joshi A, Goyal SN (2003) Participatory plant breeding in maize: A case study from Gujarat, India. Euphytica, 130: 413-422.

Woods DJ, Roberts EAH (1964) The chemical basis of quality in tea. Journal of the Science of Food and Agriculture, 15:19-25.

Wu CT (1964) Studies on the hereditary variation and morphology of pubescence on young tea shoots. Journal of the Agricultural Association of China, 47(1): 22.

Wu JN (1987) Review on 'Cha Ching'. Beijing: Agriculture Publish Press, pp.168-206.

Yamada K (1995) Immune regulatory function of food components and the development of anti-allergic food. Journal of Food Science and Technology, 42(11): 952-958.

Yao GK, Wu X, Xu N (1987) Analysis of the 'optimum type' structure of tea plant. Proceedings of International Tea Quality-Human Health Symposium, China, pp.32-36.

Yayabe F (2001) Industrial application of tea extracts. In: Proceedings of 2001 International Conference on O-Cha (tea) Culture and Science. 5-8 October, 2001, Shizuoka, Japan, pp.80-83.

Yilddizogle-Ari N, Atlan VM, Altinkurt O, Ozturk Y (1991) Pharmacological effects of rutin. Phytotherapy Research, 5: 19-23.

6

Japanese Tea Breeding History and the Future Perspective

Junichi Tanaka

NARO Institute of Crop Science
2-1-18, Kannondai, Tsukuba, Ibaraki 305-8518, Japan
E-mail: tanajun@affrc.go.jp

Abstract: Because almost all tea production is steamed green tea in Japan, this is the main target of breeding. Japanese tea breeding is very old and dates back as far as the 19th century, when breeders in the private sector selected elite clones from the original Japanese tea plants. However, plantings of one clonal green tea cultivar, Yabukita, bred privately by Sugiyama Hikosaburo in the early 20th century, now occupy about 80% of the land under Japanese tea cultivation. Recently, marker-assisted selection (MAS) and generational acceleration has been put to practical use. A move towards an era of genomic selection, sharing of information on DNA markers—especially single nucleotide polymorphisms (SNPs)—may well be desirable.

6.1 The Japanese Tea Industry and Issues in Breeding

To understand the issues of importance to the Japanese tea industry and breeding, we need to know the background to the Japanese tea industry. In this section, the history and the social and meteorological environment of the Japanese tea industry and breeding are described.

6.1.1 History of Japanese Tea

The first record of the introduction of tea to Japan was when two Japanese monks, Kukai and Saicho, brought seeds from China in the 9th century. However, this does not preclude the possibility that tea was introduced earlier. In the 13th century, the dissemination of tea drinking led Eisai, the Buddhist priest attributed to bringing green tea from China to Japan, to write *Kissa-Youjyou-ki* and to evangelize tea drinking in Japan. In the 16th century, Sen-no Rikyu promoted the drinking of powdered green tea, and the tea-drinking became popular with the aristocracy and the samurai. Among common people, the consumption of *hiboshi-bancha*, a tea that was harvested, boiled and then sun dried, spread gradually. In the 18th century, Sohen Nagatani established the process of steaming green tea: new young leaves were harvested, steamed, kneaded and then dried. This new type of green tea was received favorably and was the origin of the Japanese green tea that we know today. In the late 19th century, as a result of promotion by the Japanese government, tea production quickly expanded to become a major national export industry, much like the raw silk industry. Around that time, tea exports consisted of roasted green tea and a small amount of fermented tea, and the major export market was North America. However, quality control problems arose and tea exports slowed. After World War II, the Japanese government re-promoted tea as a major export product, but the tea was not priced competitively as compared with the products of other countries. By the time Japanese trade liberation took full effect in 1971, the export of Japanese tea had dramatically declined, but domestic consumption of green tea had increased. Currently, the Japanese tea farmers produce approximately 90,000 tonnes/year mainly as Japanese green tea from approximately 48,000 ha of tea gardens. The population of Japan is about 130 million, and the consumption of Japanese green tea per capita is about 700 g. Aside from this, a Japanese consumes on average 150 g of Oolong tea and 150 g of black tea per year, and almost all these kinds of tea are imported. As a recent tendency, the consumption of Japanese green tea has decreased.

6.1.2 The Original Japanese Tea

Because almost all Japanese tea cultivars are derived from the original Japanese tea, research into the introduction of this original tea was needed to understand Japanese tea breeding. The original Japanese tea plant was a bush with small leaves. It was classified as *Camellia sinensis* var. *sinensis* and was grown in the forests of western Japan. It is debatable whether this original Japanese tea plant was indigenous to Japan before recorded human history or whether it was introduced from abroad. Matsushita (2002) concluded in his book that tea was not native to Japan, because tea plants have not been found in Japan's pristine natural forests. Moreover, DNA marker analysis (Matsumoto *et al.*, 1994, Yamaguchi & Tanaka, 1999) has revealed that the genetic diversity of the original Japanese tea was smaller than that of China tea. This suggests that humans brought small tea populations of tea plants from China, and these populations then spread throughout western Japan.

6.1.3 Japanese Tea Breeding History

Japanese tea breeding is very old and dates back as far as the 19th century, when breeders in the private sector selected elite clones from the original Japanese tea plants. In 1932, the Japanese government began organized cross-breeding and since that time many black and green tea cultivars have been bred. However, plantings of one clonal green tea cultivar, Yabukita, bred privately by Sugiyama Hikosaburo in the early 20th century, now occupy about 80% of the land under Japanese tea cultivation. The planting of Yabukita spread mainly after the 1970s. In the 1970s, the tea harvesting system changed from hand-plucking to machine harvesting. Tea plants grown from seedlings do not have synchronous bud break and harvest times and are therefore unsuitable for machine harvesting. Clonal cultivars can be harvested uniformly by machine and retain their high quality. There are other clonal cultivars, but Yabukita is beyond comparison in terms of tea quality and yield. "Yabukita" has therefore become well known as a brand name, rather than a cultivar name. Although some high quality cultivars have recently been bred, they are still unlikely to challenge the name Yabukita has made for itself, and if breeders are to produce superior cultivars they will need to revisit the basic principles of tea breeding. Currently, the total clonal ratio in Japanese tea gardens is about 92.1%.

6.1.4 The Major Japanese Cultivars

Around 98% of domestic tea products are steamed green teas, including *gyokuro* and *genmai-cha* (teas mixed with roasted rice). The main target of breeding is therefore green tea, although breeding for other end-products, such as powdered green tea, is being conducted at the Kyoto Prefectural Agricultural Experiment Station. Until the 1970s, there were also black tea breeding programs. Japan's major cultivars are listed in Table 6.1. And Japanese registered cultivars are listed in Table 6.2.

6.1.5 Japan's Major Tea-Growing Districts and Their Climate

Japan has 47 prefectures. Tea is grown from Akita Prefecture (around latitude 40°N) to Okinawa Prefecture (around latitude 26° N). On the eastern side of Japan, Shizuoka Prefecture (at about latitude 35° N) grows about 40% of the total Japanese tea, and Kagoshima Prefecture (at about latitude 31° N) a further 20%

Table 6.1 Major tea cultivars of Japan

Cultivar	History	Usage	Features
Yabukita	Selected from original Japanese tea	Japanese green tea	Leading cultivar, high quality, mid yield, cultivated
Yutakamidori	Selected from seedlings of Asatsuyu	Japanese green tea	Early budding and harvesting, high yield, cultivated mainly in Kagoshima Pref.
Kanayamidori	S-6×Yabukita	Japanese green tea	Unique aroma, high yield, not tolerant to bad soil conditions
Sayamakaori	Selected from seedlings of Yabukita	Japanese green tea	High yield, mulberry scale resistance, upright leaves
Saemidori	Yabukita×Asatsuyu	Japanese green tea	High quality, early budding and harvesting
Okumidori	Selected from seedlings of Yabukita	Japanese green tea	High quality, late budding and harvesting
Meiryoku	Yabukita×Z-1	Japanese green tea	High yield
Asatsuyu	Selected from original Japanese tea	Japanese green tea	High quality, low yield
Asahi	Selected from original Japanese tea	Powdered green tea	Thin leaves, excellent quality as powdered tea but low yield
Samidori	Selected from original Japanese tea	Powdered green tea	High quality as powdered tea

Table 6.2 Japanese registered tea cultivars

Cultivar	Registered number	Registered year	History	Main usage	Bred station
Benihomare	No. 1	1953	Selected from Indian tea seedlings	Black tea	Kanaya
Asatsuyu	No. 2	1953	Selected from original Japanese tea	Japanese green tea	Kanaya
Miyoshi	No. 3	1953	Selected from original Japanese tea	Japanese green tea	Kanaya
Tamamidori	No. 4	1953	Selected from original Japanese tea	Pan fried green tea	Kanaya
Sayamamidiri	No. 5	1953	Selected from seedlings of Yabukita	Japanese green tea	Saitama Pref.
Yabukita	No. 6	1953	Selected from original Japanese tea	Japanese green tea	Sugiyama Hikosaburo Resistered by Shizuoka Pref.
Makinoharawase	No. 7	1953	Selected from original Japanese tea	Japanese green tea	Shizuoka Pref.
Koyanishi	No. 8	1953	Selected from original Japanese tea	Japanese green tea	Sugiyama Hikosaburo Resistered by Shizuoka Pref.
Rokurou	No. 9	1953	Selected from original Japanese tea	Japanese green tea	Sugiyama Hikosaburo Resistered by Shizuoka Pref.
Yamatomidori	No. 10	1953	Selected from original Japanese tea	Japanese green tea	Nara Pref.
Takachiho	No. 11	1953	Selected from original Japanese tea	Pan fried green tea	Miyazaki Pref.
Indo	No. 12	1953	Selected from var. *assamica* seedlings	Black tea	Kagoshima Pref.
Hatsumomiji	No. 13	1953	Ai2×Nka05	Black tea	Kagoshima Pref.
Benitachiwase	No. 14	1953	Ai2×Nka01	Black tea	Kagoshima Pref.
Akane	No. 15	1953	Ai2×Nka03	Black tea	Kagoshima Pref.
Natsumidori	No. 16	1954	Selected from original Japanese tea	Japanese green tea	Kanaya

(To be continued)

(Table 6.2)

Cultivar	Registered number	Registered year	History	Main usage	Bred station
Yaeho	No. 17	1954	Selected from original Japanese tea	Japanese green tea	Shizuoka Pref.
Asagiri	No. 18	1954	Selected from original Japanese tea	Japanese green tea	Kyoto Pref.
Kyomidori	No. 19	1954	Selected from original Japanese tea	Japanese green tea, powdered green tea	Kyoto Pref.
Hatsumidori	No. 20	1954	Selected from original Japanese tea	Japanese green tea	Kagoshima Pref.
Benikaori	No. 21	1954	Ai21×Nka03	Black tea	Kagoshima Pref.
Benifuji	No. 22	1960	Benihomare×C19	Black tea	Kanaya
Himemidori	No. 23	1960	Selected from original Japanese tea	Japanese green tea	Hainuzuka
Izumi	No. 24	1960	Selected from seedlings of Benihomare	Pan fried green tea	Hainuzuka
Satsumabeni	No. 25	1960	Nka03×Ai18	Black tea	Kagoshima Pref.
Okumusashi	No. 26	1962	Sayamamidori×Yamatomidori	Japanese green tea	Saitama Pref.
Yamanami	No. 27	1965	Selected from Chinese tea seedlings	Pan fried green tea	Miyazaki Pref.
Benihikari	No. 28	1969	Benikaori×MakuraCn1	Black tea	Makurazaki
Unkai	No. 29	1970	Takachiho×MiyaF1-9-4-48	Pan fried green tea	Miyazaki Pref.
Kanayamidori	No. 30	1970	S6×Yabukita	Japanese green tea	Kanaya
Sayamakaori	No. 31	1971	Selected from seedlings of Yabukita	Japanese green tea	Saitama Pref.
Okumidori	No. 32	1974	Selected from seedlings of Yabukita	Japanese green tea	Kanaya
Toyoka	No. 33	1976	Sayamamidori×Yabukita	Japanese green tea	Saitama Pref.
Okuyutaka	No. 34	1983	Selected from seedlings of Yutakamidori	Japanese green tea	Kanaya
Meiryoku	No. 35	1986	Selected from seedlings of Yabukita	Japanese green tea	Kanaya
Fukumidori	No. 36	1986	Yabukita×Sai23-F1-107	Japanese green tea	Saitama Pref.

(To be continued)

(Table 6.2)

Cultivar	Registered number	Registered year	History	Main usage	Bred station
Shunmei	No. 37	1988	Selected from seedlings of Yutakamidori	Japanese green tea	Kanaya
Minekaori	No. 38	1988	Yabukita×Unkai	Pan fried green tea	Miyazaki Pref.
Minamikaori	No. 39	1988	Yabukita×MiyaA11	Japanese green tea	Miyazaki Pref.
Saemidori	No. 40	1990	Yabukita×Asatsuyu	Japanese green tea	Makurazaki
Fushun	No. 41	1991	Z1×Kanayamidori	Japanese green tea	Kanaya
Minamisayaka	No. 42	1991	MiyaA-6×NN27	Japanese green tea	Miyazaki Pref.
Hokumei	No. 43	1992	Sayamamidori×Sai5507	Japanese green tea	Saitama Pref.
Benifuki	No. 44	1993	Benihomare×MakuraCd86	Black tea	Makurazaki
Ryofu	No. 45	1997	Houryoku×Yabukita	Japanese green tea	Kanaya
Musashikaori	No. 46	1997	Yabukita×Sai27-F1-73	Japanese green tea	Saitama Pref.
Sakimidori	No. 47	1997	NN27×ME52	Japanese green tea	Miyazaki Pref.
Harumidori	No. 48	2000	Kanayamidori×Yabukita	Japanese green tea	Makurazaki
Sofu	No. 49	2002	Yabukita×Shizu-Inzatsu131	Japanese green tea	Kanaya
Sainomidori	No. 50	2003	Selected from seedlings of Sayamakaori	Japanese green tea	Saitama Pref.
Harumoegi	No. 51	2003	NN27×ME52	Japanese green tea	Miyazaki Pref.
Miyamakaori	No. 52	2003	Kyoken283×Saitama-1gou	Japanese green tea	Miyazaki Pref.
Yumewakaba	No. 53	2006	Yabukita×Saitama-9gou	Japanese green tea	Saitama Pref.
Yumekaori	No. 54	2006	Sayamakaori×Miyazaki-8gou	Japanese green tea	Miyazaki Pref.

of the total crop. Fig. 6.1 shows the share of tea cultivation area by prefecture. These areas—especially Shizuoka Prefecture—have an Asian monsoon climate, with more precipitation in summer and less in winter (Fig. 6.2). Cold and dry winters and hot and wet summers are characteristic. The average temperature in Shizuoka (Shizuoka City) is 16.3 °C—typical of the temperatures in tea-growing areas worldwide. The average temperature in Kagoshima (Kagoshima City) is 18.4 °C; in this warmer area, the growing season is longer than it is in Shizuoka. Japanese climate data are available from the web site of the Japan Meteorological Agency (http://www.jma.go.jp/jma/indexe.html). The longitudinal trend in major tea-growing areas reveals a shift from central Japan, including Shizuoka Prefecture, to the warmer southwestern areas of Japan, including Kagoshima Prefecture on Kyushu. Figs. 6.3 and 6.4 are typical tea gardens in Shizuoka and Kagoshima Prefectures.

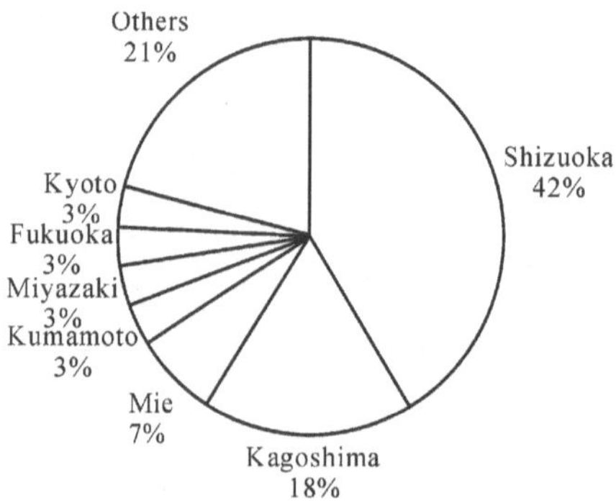

Fig. 6.1. Major tea producing prefectures and their share of the total tea-growing area of Japan

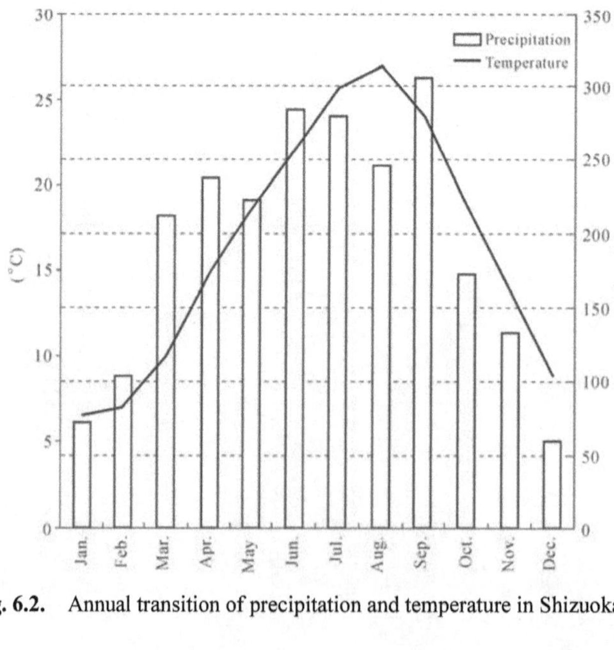

Fig. 6.2. Annual transition of precipitation and temperature in Shizuoka City

Fig. 6.3. Shimada tea garden in Shizuoka

Fig. 6.4. Makura tea garden in Kagoshima

6.1.6 Issues Facing the Japanese Tea Industry

The Japanese tea industry has recently faced many difficulties, including groundwater pollution caused by excessive fertilizer application. Many consumers are concerned about pesticide use. Large amounts of pesticides are being used on tea cultivations—especially against mulberry scale—and there is concern about their effects on the ecosystem. Moreover, in warm regions such as Kagoshima Prefecture, the synchronization of tea budding at first flush might be degenerating because of insufficient low-temperature exposure in winter. This budding synchronization is important to the quality of the first flush. Global warming may be behind this degeneration; breeding is therefore not the solution to every problem, but it is still very important.

6.2 The Challenge of Continuous Breeding Improvement

In this section, the Japanese tea breeding challenges and the continuous improvement of tea quality and yield ability are described.

6.2.1 Generational Acceleration and Appropriate Selection: Key Technologies in Tea Breeding

Viewed from a longitudinal perspective, breeding is the result of the integration of many beneficial genes. Integration of beneficial genes drives the generational change and the appropriate selection. This is a very simple but basic principle of breeding. Because tea is a tree crop, it has a long life cycle and takes a long time to reach the flowering and fruiting stage. In addition, it takes a long time to reach a harvestable age and quality.

The conventional process of Japanese tea breeding starts with the crossing and cultivation of a population of seedlings. The population is nurtured for several years, after which time suitable individuals are selected for their vigor and the quality of the tea manufactured from them. The selected individuals are propagated by cuttings and clonal strains are re-selected a few times. Finally, appropriate strains are selected as cultivars and planted in fields for crossing, which begins after several years. One generation of this process takes at least a few decades.

In terms of basic breeding principles, tea breeding has to surmount two laborious processes: (1) it takes many years to move from generation to generation and, (2) it also takes many years until the character of a cultivar can be evaluated. To overcome these problems in Japan, and especially at the Makurazaki Tea Research Station of the National Agriculture and Food Research Organization (NARO), breeding staff perform efficient crossings by using juvenile flowering (induced by restricting root growth in containers or pots) and juvenile selection by methods such as marker-assisted selection (MAS).

6.2.2 Marker-Assisted Selection

In Japanese tea breeding, MAS is used to select for mulberry scale resistance. Mulberry scale (*Pseudaulacaspis pentagona*) is typically a difficult insect to control and is the only pest that can kill Japan's tea plants. Development of resistant cultivars is the ultimate solution. An inoculation assay method has been developed (Mizuta, 2003), but it is laborious and time consuming. The green tea cultivar Sayamakaori has strong resistance to mulberry scale (Shige *et al.*, 1993). This cultivar has been crossed with the clonal strain Kana-Ck17 to construct a linkage map using RAPD and SSR marker systems to detect the resistant locus. Using this population, linkage maps of the parents have been constructed and the mulberry scale resistant gene *MSR-1* mapped (Fig. 6.5). Allele-specific markers are desirable for practical MAS; fortunately, the resistance locus was mapped in an area with a high density of markers and allele-specific RAPD and SSR screening were performed easily (Tanaka *et al.*, 2003; Tanaka, 2005). An easy DNA extraction method was developed to achieve practical MAS (Tanaka & Ikeda, 2002).

Many quantitative traits loci (QTLs) have been detected in tea genetic analyses in Japan. However, only mulberry scale resistance has been subjected to MAS. The reason why many of these QTLs are not highly valuable in breeding is that allele-specific markers have not been constructed. Hopefully, in the near future, MAS for anthracnose resistance will be put to practical use.

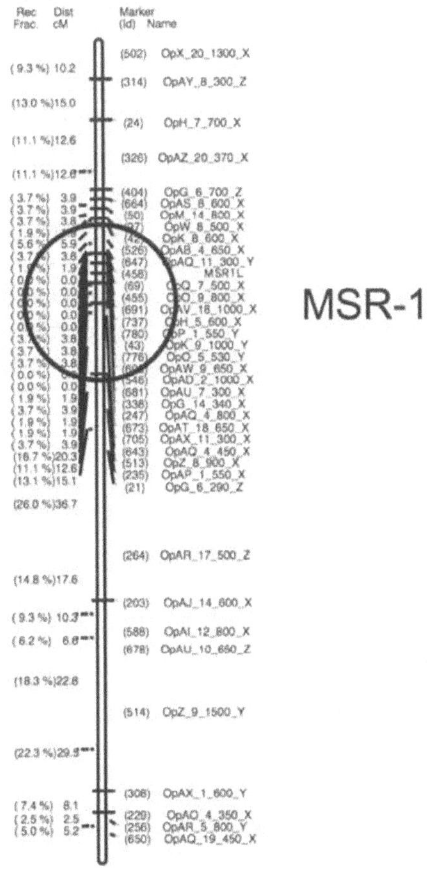

Fig. 6.5. Mulberry scale resistance QTL (*MSR-1*) on the linkage map of the resistant tea cultivar Sayamakaori

6.2.3 Importance of Germplasm for Breeding, Especially in Regard to MAS

The discovery of valuable alleles for breeding is the key to successful MAS. Sometimes there are valuable alleles in established cultivars. However, in advanced breeding programs, valuable alleles need to be mined from the germplasm. DNA markers for MAS can easily be established by using distant materials, since DNA markers stand out in the different DNA sequences in the comparison.

In Japan, NARO has found 4,000 collections of genetic resources as germplasm in tea and the wild relatives of tea from Japan and from countries such as China, India and Bangladesh. Some of the prefectural research institutes also have hundreds or thousands of genetic resources. These germplasms are conserved as clone bushes. This germplasm is being preserved and evaluated morphologically, chemically and using molecular biology, and subjecting it to allele mining should prove productive.

6.2.4 Construction of Integrated Genetic Linkage Maps

In Japan, genetic linkage maps have been constructed in accordance with pseudo-testcross theory (Grattapaglia & Sederoff, 1994) for QTL analysis in breeding programs. These maps have been constructed for individual cultivars or strains. Recently, at the Makurazaki Tea Research Station, progress has been made in the use of landmark markers, such as SSRs that have many alleles, to construct integrated linkage maps.

6.2.5 Toward an Era of Genomic Selection

Recently, the cost of SNP typing has been reduced dramatically by the use of microarray technology, and the cost of detection of SNPs should similarly be reduced dramatically by next-generation sequencer technology. This means that full genotype scans of DNA markers of many cultivars, strains and individuals will be very easy and inexpensive. Progress is being made in trials of genomic selection in other crops. Sharing of information on DNA markers, especially SNPs, may well be desirable.

References

Grattapaglia D, Sederoff R (1994) Genetic linkage maps of *Eucalyptus grandis* and *Eucalyptus urophylla* using a pseudo-testcross: Mapping strategy and RAPD markers. Genetics, 137: 1121-1137.

Matsumoto S, Takeuchi A, Hayatsu M, Kondo S (1994) Molecular cloning of phenylalanine ammonia-lyase cDNA and classification of varieties and cultivars of tea plants (*Camellia sinensis*) using the tea PAL cDNA probe. Theoretical and Applied Genetics, 86: 671-675.

Matsushita S (2002) Yamacha no kenkyu: nihoncha no kigen denrai o saguru. Iwata Shoin, Tokyo, p.242 (in Japanese).

Mizuta T (2003) Differences in development and reproduction of the mulberry scale, *Pseudaulacaspis pentagona* Targioni (Hemiptera: Diaspididae), on resistant and susceptible varieties of tea plant. Japanese Journal of Applied Entomology and Zoology, 47: 91-96.

Shige M, Nonaka T, Nagatomo S, Tanaka T (1993) The varietal difference and evaluation method of mulberry scale resistance in tea. Tea Research Journal, 78(S): 10-11 (in Japanese).

Tanaka J (2005) Practical marker-assisted selection for mulberry scale resistance of tea and its future. Research Journal of Food and Agriculture, 28(9): 39-44.

Tanaka J, Ikeda S (2002) Rapid and efficient DNA extraction method from various plant species using diatomaceous earth and a spin filter. Breeding Science, 52: 151-155.

Tanaka J, Shige M, Uezono Y, Taniguchi F, Mizuta T (2003) Marker-assisted selection in tea for the mulberry scale resistance derived from 'Sayamakaori'. Breeding Research, 5(S2): 105 (in Japanese).

Yamaguchi N, Tanaka J (1999) Classification of Japanese original tea by maternal inherited RAPDs. Breeding Science, 1(S2): 295.

7

Breeding of Tea Plant (*Camellia sinensis*) in Vietnam

Do-Van Ngoc

Northern Mountainous Agriculture and Forestry Science Institute—NOMAFSI, Phu Ho commune, Phu Tho town, Phu Tho province, Vietnam
Tel: (0084) 2103865073
E-mail: vienmnpb@vnn.vn

Abstract: The tea plant [*Camellia sinensis* (L.) O. Kuntze] has been cultivated in Vietnam for a long time. In past years, significant efforts were spent on tea breeding and correspondingly meaningful achievements were obtained. Some new tea cultivars of high yield and good quality, suitable for production of green, black and Oolong teas such as LDP 1, LDP 2, PH 8, PH 9, have been developed. Tea breeding, however, needs to be further strengthened, and to this end the following activities need to be promoted: collecting and conservation of tea genetic resources, hybridization of native and exotic cultivars of *C. sinensis* and *C. sinensis* var. *assamica*, and use of mutation for speeding up the breeding cycle.

7.1 General Introduction to the Tea Industry of Vietnam

Tea is a traditional plant in Vietnam and Vietnamese people have a daily habit of using tea (Quy & Oanh, 2008). Climate and soil conditions in Vietnam are suitable for tea growing. According to FAO statistics (2010), the total tea area in Vietnam was 131,200 ha, the average annual yield was 6.90 tonnes of fresh leaves/ha (equal to the world's average tea yield) and the total tea production was 178,100 tonnes in 2009. Currently, tea is planted in 35 provinces, mainly in the midlands and northern mountainous regions. In 2009, the tea growing area in the north was 104,200 ha, equaling 79.4% of the country's total tea acreage. The central highlands (Tay Nguyen) are the main tea area in the south with 27,000 ha, representing 20.6%.

7.1.1 Brief Tea Production History

Tea has been being planted in Vietnam for a very long time. However, only from the early 20th century was the tea industry developed. The history of Vietnamese tea development can be divided into main periods as below (Quy & Oanh, 2008).

7.1.1.1 Before 1882

During the Hung King Dynasty (about the 7th century B.C.), there were 2 tea regions (sources from Chinese literature and Le Qui Don): (1) In the Red River delta and the midlands, tea was planted for fresh tea leaves and flower-buds, and also for speciality dried tea of Hue and Bang. (2) In the mountainous areas Dao, Tay, Nung and H'mong, people used naturally growing tea for medicine and for traditional green tea products.

7.1.1.2 The Period from 1882 to 1945 (the Former French Colonial Period)

When occupying Vietnam, the French quickly developed tea production.

In 1890, the first large tea producing area of 60 ha was established in Tinh Cuong district, Phu Tho Province, for export to Europe, and this became the birthplace of the tea industry in Vietnam.

In 1918, the first Agricultural Research Station was set up in Phu Ho district, Phu Tho Province for conducting tea R&D activities applying techniques from Indonesia, Sri Lanka and using imported equipment from England. Later, two more tea research stations were established in the central highlands (Tay Nguyen), including the Tea Research Station in Bau Can district, Gia Lai Province, established in 1927 and the Tea Research Station in Bao Loc district, Lam Dong Province, formed in 1931.

In 1923, the first black tea production factory was built in Phu Ho district, Phu Tho Province, and this was the start of the new period of applying advanced technology and using modern equipment for tea processing in Vietnam.

In 1945, the total tea area of Vietnam was 13,505 ha, and the total tea production was 6,000 tonnes. Black tea produced in Vietnam was exported to North African markets (Algeria, Tunisia, Morocco, etc.). Vietnamese tea was recognized to have the same quality as Indian and Sri Lanka tea.

7.1.1.3 Period from 1945 to 1954

This period was considered to represent the degradation of the tea industry in Vietnam. Due to the war's impact, tea plantations were abandoned and both the tea area and production were reduced.

7.1.1.4 Period from 1954 to 1990

After 1954, when the French withdrew from Vietnam, tea was considered a plant of high economic value, important for economic development in the northern midlands and mountainous areas. Many state-run tea plantations were founded, such as Van Linh, Van Hung and Phu Son tea plantations in Phu Tho Province, Nghia Lo plantation in Yen Bai Province, Tan Trao plantation in Tuyen Quang Province and Song Cau and Quan Chu plantations in Thai Nguyen Province. Many black and green tea production factories were built and functioned using advanced equipment and technologies imported from the Soviet Union and China. In addition, tea growing cooperatives were established and the tea research station in Phu Ho (Phu Tho Province) was restored. This station's activities were then focused on the improvement of tea cultivars and cultivation techniques. A field germplasm collection was established and many advanced techniques were developed and applied. In the 1960s, experts from China and the Soviet Union came to assist Vietnam in tea research and production. Also, many Vietnamese students and staff went to study tea production techniques in China and the Soviet Union. At that time, the main tea products of Vietnam were green tea and black tea which were exported to countries of the former Soviet Union and Eastern Europe.

7.1.1.5 Period from 1990 to the present

Due to market fluctuations in the former Soviet Union and Eastern Europe in the late 1990s, tea production in Vietnam faced great difficulties. Traditional markets declined, new markets had not been opened and technological innovation was not enough to meet the requirements of the new markets of Western Europe. From 1995, along with the renewal of tea industry management, many joint ventures with Japan, Iraq, etc., using many advanced technologies, were started. The tea area, yield, production and value increased fast. This opened a new era of tea development in Vietnam. Tea production in Vietnam greatly increased in terms of both area and quality in the period 2000 – 2009.

7.1.2 Acreage, Productivity and Exports

Table 7.1 presents annual data about the tea areas, productivity and exports from 1995 to 2009. Tea production in 2009 increased 43,500 ha compared with 2000. Tea productivity increased 2.80 times. The average exported price was 1,108 US dollars per tonne in 2005, 1,049.3 US dollars per tonne in 2006, 1,144.5 US dollars per tonne in 2007 and 1,405.7 US dollars per tonne in 2008, respectively. It increased, but was still low compared with the average price in the world. Currently, tea production has fast developed and tea has been considered the strategic plant of the northern mountainous region and central highlands for solving unemployment problems and contributing to sustainable agricultural development.

Table 7.1 Tea acreage, productivity and export from Vietnam in1995 – 2009

Year	Total tea area (ha)	Yield (kilotonnes of dry tea)	Export (kilotonnes)
1995	66,700	40.2	—
1996	74,800	46.8	—
1997	78,600	48,2	—
1998	79,100	50.6	—
1999	84,800	52.5	36.0
2000	87,700	63.7	55.6
2001	95,600	76.8	52.1
2002	108,000	89.4	52.1
2003	116,000	107.0	62.2
2004	120,000	119.1	73.4
2005	123,742	133.4	85.4
2006	125,800	142.5	105.6
2007	128,402	157.5	126.0
2008	129,600	174.9	138.1
2009	131,200	178.1	136.0

7.1.3 Main Tea Products

Black tea is the main tea product of Vietnam, occupying 60% of production. Green tea production is 30%, the remaining 10% consists of Oolong tea, jasmine tea, yellow tea, etc.

Before 1994, traditional technologies were applied to black tea production. From 1994 up to the present, CTC technology has been applied in many factories. The cultivars appropriate for black tea production are var. *assamica* and var. *pubilimba* varieties, which are mainly cultivated in the midlands and low mountainous areas. Black tea production is often done on a large scale by big companies, including foreign companies (Khon & Phong, 1997).

Green tea is mainly produced for domestic consumption. Green tea production areas in Vietnam are concentrated in the mountainous regions such as in Lam Dong, Moc Chau (Son La) and Thai Nguyen. The green tea production equipment is often imported from China and the technology used is traditional, using heat rather than steam to fix the fresh tea leaves. Tea cultivars with small buds are appropriate for green tea production. Their growing areas are often small scale and scattered, mostly at household level.

Yellow tea is the traditional product of ethnic minorities in upland areas, produced from indigenous var. *pubilimba* tea cultivars of Ha Giang, Yen Bai. Yellow tea is traditionally produced by households for their own use, using traditional techniques. Currently, there are also some yellow tea production companies. These companies purchase raw materials from the Dao households and produce yellow tea mainly for export to the Mainland, Hong Kong and Taiwan of China.

Oolong tea is a new product which has been manufactured since the 1990s by Taiwanese companies in Lam Dong and Moc Chau (Son La). Tea cultivars Kim Tuyen, Thuy Ngoc and Oolong Thanh Tam are appropriate for Oolong tea production. Most Oolong production is exported to Taiwan. After 20 years, some new tea cultivars have been selected and cultivated for Oolong tea. Many production facilities have been established, not only for export but also for domestic consumption. The production scale of Oolong tea is often small but equipment and advanced technologies are imported from Taiwan. Appropriate soil and climatic conditions in some high areas, with good financial conditions, are important for producing Oolong tea. Oolong tea has the highest price in Vietnam.

On the other hand, many kinds of scented tea are produced from high quality green tea in Vietnam. Lotus green tea is a traditional product associated with the rich cultural history of the Ha Noi people. Jasmine teas are also produced around Ha Noi and Thai Nguyen. Some kinds of canned tea are also products of the Vietnam tea industry.

Vietnam exports tea to over 60 countries in the world. Black tea is mainly exported to the Middle East, South Asia, North East Asia and Russia. In addition, Vietnamese tea products have initially penetrated difficult markets such as the European Union, America and Japan. Green tea is mainly exported to Southeast Asia including mainland and Taiwan of China, etc. The price of export tea has fluctuated strongly. The price of the raw materials of high quality cultivars is 2 – 3 times higher than that of old cultivars, and this encourages growers to plant new tea cultivars.

7.1.4 Ecological Conditions of Some Main Tea Growing Regions

Temperature, rainfall and terrain influence both tea productivity and quality. In Vietnam, tea production areas can be divided into two main areas by geography and altitude: (1) a northern area in the northern mountainous areas and, (2) a southern area in Lam Dong highlands. The main ecological factors affecting tea production are listed in Table 7.2.

Table 7.2 Geographical and temperature characteristics of the main tea regions of Vietnam (2000)

No.	Regions	Average temperture (°C)	Rainfall (mm/year)	Humidity (%)	Total of temperture (°C)	Height above mean sea level (m)
1	North-western (Moc Chau)	18 – 23	1,500 – 2,000	85	2,800 – 4,200	800
2	Viet Bac (Tuyen Quang, Ha Giang, Yen Bai)	18 – 20	1,800 – 4,000	85 – 86	—	200 – 700
3	Northern midlands (Phu Tho, Thai Nguyen)	23.1	1,863	84	8,446	25 – 70
4	Highlands (Lam Dong)	20 – 23	1,800	70	7,500	850

Overall, average annual temperatures in the northern midlands are highest compared to those in all the other tea regions, with the annual total temperature of 8,446 °C . The annual average rainfall in the Viet Bac region is higher than in other regions. Climatic weather conditions in tea growing regions have a major influence on the quality of both raw materials and final products. We have conducted a study with two cultivars: Trung Du (TD) and Shan tea cultivars. The TD cultivar adapted to the lower regions and the Shan cultivar adapted to the uplands regions. Study results show that tea plants grow strongly with high rainfall and produce high quality raw materials at an elevation above 500 m. The tannin content of the two cultivars grown in different areas was analyzed and the results are listed in Table 7.3.

Table 7.3 Tannin component in 'two and a bud' (% dry matter, 2000)

Regions	Cultivars	Early harvest	Mid-harvest	Late harvest	Average
North-western: Moc Chau	Shan	25.27	33.00	27.30	28.66
	TD	21.20	31.00	21.80	26.66
Viet Bac: Ha Giang	Shan	26.68	28.82	30.10	28.53
	TD	28.46	29.64	27.22	28.44
Northern midlands: Phu Tho	TD	28.60	35.33	28.30	30.74
Northern midlands: Thai Nguyen	TD	30.10	34.00	31.60	31.90
Highlands: Lam Dong	TD	29.20	30.50	30.22	29.97

The soluble component is one of the important indicators in assessing the quality of tea products. Data from the analysis of soluble substances in 'two and a bud' plants cultivated in different areas are presented in Table 7.4.

Table 7.4 Soluble substances in 'two and a bud' (% dry matter, 2000)

Location	Cultivars	Early harvest	Mid-harvest	Late harvest	Average
North-Western: Moc Chau	Shan	40.05	42.80	40.05	41.21
	TD	39.90	41.81	40.00	40.56
Viet Bac: Ha Giang	Shan	33.41	41.76	38.10	37.76
	TD	39.67	41.28	39.62	40.19
Northern midlands: Phu Tho	TD	40.00	46.90	42.80	43.23
Northern midlands: Thai Nguyen	TD	40.80	48.10	41.10	43.33
Highlands: Lam Dong	TD	40.06	44.00	44.24	42.94

Table 7.4 indicates that the soluble component substances in 'two and a bud' in different ecological zones at mid-harvest are higher than at early harvest and late harvest. Soluble substances in the tea buds at mid-harvest in the northern midlands region is the highest compared with other tea regions. According to analysis, the annual average figure for soluble substances in the tea buds in the northern midlands is 43%, which is 2% – 3% higher than other tea regions. Data from the analysis of catechins component in 'two and a bud' plants cultivated in different areas are presented in Table 7.5.

Table 7.5 Catechins component in 'two and a bud'

Regions	Culti-vars	Percentage of catechins (%)					Total of catechins (mg/g dry matter)
		EGC	GC	EC+C	EGCG	ECG	
North-western: Moc Chau	Shan	16.78	10.58	16.01	41.68	15.95	162.37
Viet Bac: Yen Bai	Shan	12.84	9.53	13.67	41.29	22.67	113.35
	TD	16.00	10.38	13.80	38.19	21.63	110.60
Northern midlands: Phu Tho	TD	18.19	8.26	9.06	48.10	18.39	131.03
Northern midlands: Thai Nguyen	TD	18.77	9.19	8.83	48.23	11.93	118.25
Highlands	Shan	15.10	11.90	15.20	34.70	23.10	92.30
	TD	19.40	11.90	12.90	36.40	19.40	82.40

Overall, the total of catechins in tea shoots harvested from different regions varies from 82.40 to 162.37 mg/g dry matter. Tea plants grown in Phu Tho had the highest catechins level compared to that in other tea regions.

7.2 Tea Genetic Resources in Vietnam

The purpose of collecting, conservation, evaluation and use of tea germplasm accessions is to enrich the genetic material source for tea breeding.

7.2.1 Tea Germplasm Plays Crucial Role in Tea Breeding

To meet the requirements of tea production, the tea industry should strengthen germplasm collecting activities in order to diversify and enrich genetic material stock for varietal improvement (Guo, 2005).

The first collection activities were under taken by a French scholar named Baux in 1885 in the north of Vietnam, and later by Pavie in 1890 in the Red River and Mekong River deltas. Exotic germplasm accessions were also introduced. Twenty-seven cultivars were collected and maintained in Phu Ho station from 1918 to 1935 (La, 1998). The number was 99 cultivars for the period from 1960 to 1963, and 54 cultivars for the period from 2000 to 2008. The present total number of tea germplasm accessions is 180 cultivars.

7.2.2 Situation of Collecting, Conservation and Evaluation

Most of the collected tea cultivars are in the form of seeds (124 accessions, occupying 68.8%), and the rest are in the form of cuttings (occupying 31.2%) that were mainly made from 2000 to the present. All cultivars are now preserved in Phu Ho (5 plants for each cultivar), except Shan teas are preserved in Ha Giang (800 m above sea level). Germplasm accessions have originated from 10 countries in the world. Introduced cultivars account for 65.0%, native ones 35.0%. This is confirmation of collection and conservation in order to improve diversity in the cultivars of tea in Vietnam. Table 7.6 gives details of the sources of Vietnam tea germplasms.

Table 7.6 The number of tea accessions from different regions

Region	No. of cultivars	*C. sinensis* (L.) O. Kuntze var. *sinensis*	var. *assamica* (Masters) Kitamura	var. *pubilimba* Chang
Vietnam	63	21	18	24
China	54	54		
Japan	23	23		
India	10	2	8	
Georgia	10	10		
Sri Lanka	10		8	2
Laos	3		1	2
Korea	3	3		
Indonesia	3	2	2	1
Myanmar	1	1	1	
Total	180	113	38	29

Germplasm accessions are characterized in terms of stem, branch, leaf, flower and fruit morphology. According to morphological characteristics, tea accessions are divided into 3 types as below (Chen *et al.*, 2006).

(1) Small-leaf tea (*C. sinensis* (L.) O. Kuntze var. *sinensis*): 113 cultivars occupy 62.8% of the tea germplasms collection: a shrub, short growing branches, small and thick and dark green leaves, 4 – 15 cm length, 5 – 7 cm width with 6 – 9 pairs of veins/leaf, numerous flowers, light shoots, low yield but good quality. These cultivars can tolerate up to –12 °C and are grown in the northern high mountainous regions and the northern midlands regions of Vietnam. These cultivars mainly introduced from China, Georgia and Japan.

(2) *assamica* tea (*C. sinensis* var. *assamica* (Masters) Kitamura): 38 cultivars occupy 21.1% of the tea germplasms collection: a tall plant (10 m in nature), tall growing and occasional branches, large, green, thin leaves of elliptic shape, 20 – 30 cm length with 12 – 15 pairs of veins/leaf, big shoot weight, high yield and quality, few flowers and fruits, poor tolerance to draughts and low temperature. They are cultivated mainly in the midlands, such as at Phu Tho, PlayKu and Lam Dong.

(3) Shan tea (*C. sinensis* var. *pubilimba* Chang): 29 cultivars occupy 16.1% of the tea germplasms collection: a tall plant (reaching 6 – 15 m in nature), 20 –

60 cm stem diameter, tall growing and occasional branches, large, light green and thin leaves of elliptic shape, 15 – 18 cm length with 10 pairs of veins/leaf, big shoot weight with lot of bright hair, high yield, good quality, few flowers and fruits, good tolerance to low temperature. The Shan type cultivars are very special and occupy a large area in high mountains such as Cao Bo (Ha Giang), Suoi Giang (Yen Bai), Cho Long (Son La), Tam Duong (Lai Chau) and Bao Loc (Lam Dong).

The 'two and a bud' weight of small-leaf tea (*C. sinensis* var. *sinensis*) is 0.38 – 0.55 g; of *assamica* type (*C. sinensis* var. *assamica*) and Shan type (*C. sinensis* var. *pubilimba*) varies from 0.85 – 0.88 g (Thu & Tien, 2001). Results of analysis of the mechanical components in the tea bud of selected cultivars are presented in Table 7.7.

Table 7.7 Mechanical and chemical components in 'two and a bud'

	Cultivars	Shoot weight (g)	Yield (tonnes fresh tea/(ha·year)	Tannin (%)	Dissolved compound (%)	Production
C. sinensis var. *assamica*	PH1	0.88	25.6	35.44	44.12	Black tea
	1A	0.85	23.4	34.90	44.10	Black tea
C. sinensis. var. *pubilimba*	TRI777	0.89	18.7	33.10	41.52	Green and black tea
	Shan Chat Tien	0.88	23.3	33.00	41.10	Green and black tea
C. sinensis var. *sinensis*	Trung Du	0.55	9.2	30.24	42.16	Green and black tea
	Keo Am Tich	0.38	7.2	26.10	41.54	Green and Oolong tea

C. sinensis var. *assamica* (cultivar PH 1, 1 A): High tannin, dissolved compound and yield (25.6 tonnes/ha). This one is suitable for black tea.

C. sinensis var. *pubilimba* (cultivar TRI 777, Shan Chat Tien): Medium tannin (33.0% – 33.1%) and low dissolved compounds, good yield (18.7 – 23.3 tonnes/ha). This is suitable for black and green tea.

C. sinensis var. *sinensis* (cultivar Trung Du, Keo Am Tich): Low tannin (26.10% – 30.24%), dissolved compounds (41.54% – 42.16%), low yield (7.2 – 9.2 tonnes/ha). This one is suitable for green tea and Oolong tea.

Evaluation results of tea germplasm accessions in Vietnam are shown in Table 7.8.

Table 7.8 Results of tea germplasm assessment and use

No.	Evaluation traits	Number of accessions	Note
1	Descriptors for tea	180	Character notes
2	Yield	81	
3	Quality	81	
4	Genetic assessment	33	RAPD method
5	Using for cross hybrid	81	
6	New recognized cultivars	6	Selection 2, hybrid 2, imported 2
7	New clones recognized provisionally	12	Selection 3, hybrid 2, imported 7

7.3 Breeding and Selection Techniques

Recently, Vietnam has selected and bred some new tea cultivars to meet the needs and improve the quality of tea products.

7.3.1 Procedures of Tea Breeding

It takes about 10 to 15 years to breed a new cultivar according to the below procedures (Fig. 7.1).

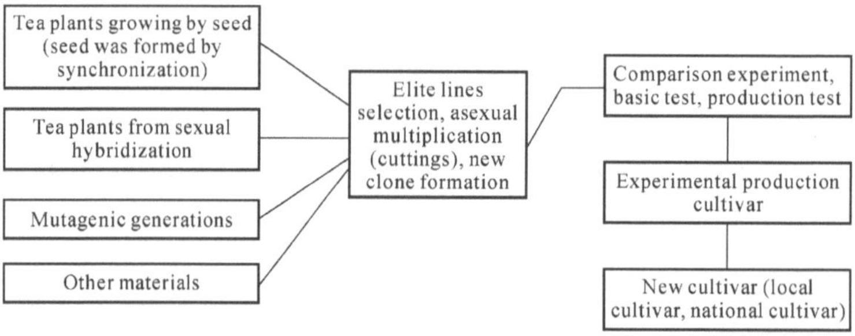

Fig. 7.1. The procedures of tea breeding

7.3.2 Methods of Tea Breeding and Selection

Tea breeding materials include natural materials, i.e., tea cultivars in the genetic nursery garden, local cultivars and imported cultivars, artificial materials from sexual hybridization and mutagenic offspring. Currently, there are three main tea breeding methods in Vietnam, individual selection, hybridization and mutagenic method. Selection from importing cultivars is also a supplementary method.

7.3.2.1 Individual Selection

From cultivars in the nursery garden and existing cultivars in production, after assessing the economic characteristics of the starting materials, elite individuals were selected and propagated asexually (cuttings) to form new tea clones (Niem, 1998; Toan & Loan, 1994). Then tests of these clones were conducted over 4 to 6 years to select the best ones in terms of productivity, quality and resistance to pests. A production test was conducted after 4 to 6 years of monitoring and evaluation to select good cultivars. The Ministry of Agriculture and Rural Development (MARD) recognizes them as experimental production cultivars, and then continues to test the complete production over 2 to 3 years. MARD then proposes to recognize them as new cultivars (national cultivars). The evaluating criteria during the breeding process include: growth duration, yield and yield components, quality and pest resistance. By the application of an individual selection method, 5 tea cultivars have been developed, including PH 1, 1 A, PH 11, PH 12, PH 14. Also, 13 elite plants of Shan tea were selected (Ngoc *et al.*, 2006, 2007, 2008).

7.3.2.2 Hybridization

There are many good features of crossing. In particular it is suitable for selecting appropriate characteristics from parental plants (Toan *et al.*, 1998; Li *et al.*, 2005; Toan & Phuong, 2006). It is necessary that the flowering times for both mother and father plants are similar. The mother plant must be able to produce fruit and the pollen of the father plant has a long duration. Parental plants do not have a close relationship in term of genetics and origin.

Hybridization is the most popular and effective tea breeding method in Vietnam. The following are some successful examples:

(1) ♀ Fuding Dabaicha with high quality ×♂ PH 1 with high yield to produce F_1 generation with high yield and good quality.

The father cultivar PH 1 belongs to the *C. sinensis* var. *assamica* with high yield, medium quality. The mother cultivar Fuding Dabaicha (from Fujian Province, China) belongs to *C. sinensis* var. *sinensis*, with low yield, but high quality, aroma and light taste. Four elite individuals, namely LDP 1 (Fig. 7.2), LDP 2 (Fig. 7.3), TDP and CLT, had been selected until 1988 with high yield and quality. LDP 1 and LDP 2 have been selected after 10 years of experiment (LDP 1 was recognized in 2001 and LDP 2 in 2006, respectively). The planting area of these cultivars occupies 32% of the total tea area, mainly at altitudes below 500 m. They are planted for both black and green tea production.

Fig. 7.2. LDP 1 cultivar

Fig. 7.3. LDP 2 cultivar

(2) ♀ TRI 777 with special flavor and suitable in Vietnamese conditions ×♂ Kim Tuyen with high quality to produce F_1 generation with high yield and good quality.

Father plant cultivar Kim Tuyen: originated from Taiwan, imported into Vietnam in 1994 and planted in Lam Dong, Ha Tay, Yen Bai, Son La provinces. Kim Tuyen has some characteristics such as early shoot emergence, very strong growth, high yield of 17 tonnes/ha (8 years old), good resistance to pests, medium drought resistance, shrub stem, slightly spread branches, thick branch density, glossy green oval leaves, horizontal leaves, leaf length of 7.2 cm, width of 3.1 cm. Shoot is light green, with a purple felt when young. Kim Tuyen was approved for large scale production in 2007 and now is planted in an area of over 1,000 ha mainly in Lam Dong, Lang Son, Yen Bai, Phu Tho and Thai Nguyen.

Mother plant cultivar TRI 777: originated from Cho Long commune, Moc Chau district, Son La Province, Vietnam. Seeds of TRI 777 were sent to Sri Lanka in 1937, and TRI 777 has been recognized as a national cultivar and planted in the highlands of Sri Lanka at altitudes of 800 – 1,200 m. It was imported back to Vietnam in 1977. TRI 777 cultivar belongs to the *C. sinensis* var. *pubilimba*. It has good growth, furry and snowy shoots, low shoot density, narrow branch angle, relatively wide spread. Its gives an approximate yield of 17 tonnes/ha (in 10 years old plants in Phu Ho). This is a good quality cultivar, appropriate for high quality green and black tea. The survival rate of cuttings is high, with healthy plantlets and high survival rates after planting. TRI 777 can be heavily damaged by mosquito bugs and aphids. A suitable growing area is 500 m above mean sea level with good conditions for intensive farming. TRI 777 was recognized as a new cultivar in 1997 and is now planted in over 200 ha, mainly in Thai Nguyen, Son La and Lang Son.

7.3.2.3 Mutagenic Method

This method is not popular in Vietnam and is less applied. There was research work by Menh *et al.* (1999), aimed at using gamma rays on tea seeds of PH1 and TRI 777 with a dosage from 0.8 Kr (kilo X-ray) to 5.0 Kr. This work was conducted over 10 years (1990 to 1999).

The results showed that a small dosage (0.8 Kr) of gamma rays acted as stimulation for seeding. With a dosage of 5.0 Kr, the gamma rays affected the emergence of seeds of PH 1 and TRI 777 cultivars 2 months after sowing. PH1 tea seeds are more sensitive than TRI 777. The dosage for killing half (LD_{50}) of the gamma radiation on PH 1 is 4.5 Kr and on TRI 777 is 5.0 Kr. Dissociated morphological traits of PH 1 and TRI 777 are not in the rules. Four plants including 4.0 and 5.0 from TRI 777, P 20 and P 52 from PH 1, have been selected. These selected plants were grown, multiplied, tested and evaluated. The data is shown in Table 7.9.

7.3.2.4 Selection from Importing Cultivars

When importing cultivars, the geography, temperature, hours of lighting, soil conditions, humidity and rainfall are all factors that are related to successfully imported tea plants.

Depending on objectives, in choosing the requirements for new cultivars, territories should note the following: (1) The relationship between environmental factors and other regional and imported raw product areas. (2) The adaptation of imported cultivars: cultivars are well able to cope with the environmental conditions when in a new location despite a big difference in environmental conditions. The success rate should be higher than when compared with regions with poor adaptability.

Therefore, attention should be paid to facilitate the use of imported cultivars adapted to conditions where the import of seeds and the testing and evaluation are good, compared to the breeding of new indigenous materials. Imported cultivars should be tested and cultivated on a small scale as well as being checked strictly.

Five cultivars including TRI 2023, TRI 2025, TRI 2043, TRI 777 and DT 1 were imported from Sri Lanka to Vietnam in 1977, of which TRI 777 was approved for pilot production in 1992 and for large scale production in 1997.

In 1994, many cultivars were imported from Chinese Taiwan such as Kim Tuyen, Thuy Ngoc, Oolong Thanh Tam, Tu Quy Xuan for planting in Lam Dong Province. These cultivars were later experimented with in Lang Son, Son La, Yen Bai, Phu Tho by the Tea Research and Development Center (now belonging to NOMAFSI). In 2003, Kim Tuyen and Thuy Ngoc were approved for pilot production and in 2007 for large scale production. They are now planted in Lam Dong, Lang Son, Son La for Oolong tea.

Table 7.9 Yield of selected clones in Phu Ho (tonnes/ha)

Clones	May	June	July	Total	% compared to the control
TRI 777 (Control)	1,577	0,830	0,913	3,320	100.00
4.0	1,940	0,885	1,937	4,762	143.43
5.0	5,810	3,600	2,158	11,568	348.43
PH 1 (Control)	1,382	1,830	1,162	4,374	100.00
P 20	4,150	1,548	1,058	6,756	154.46
P 52	2,670	2,666	1,121	6,457	147.62
LSD 0.05	2.05	1.27	1.29	—	—

To assess the shoot quality of selected clones compared to PH 1 and TRI 777, the data is shown in Table 7.10.

Table 7.10 Quality of shoot material of selected clones in Phu Ho

Clones	Tannin (%)	Soluble component (%)	Testing mark
TRI 777 (Control)	30.10	42.30	Aromatic, light taste
4.0	30.28	41.60	Less aromatic, strong taste
5.0	33.43	43.17	Aromatic, light taste
PH 1 (Control)	33.38	44.17	Less aromatic, bitter
P 20	31.57	42.47	Aromatic, light taste
P 52	32.19	42.53	Less aromatic, bitter

Two tea clones have been selected with different morphology compared to the mother plant and with higher yield, and they are promising cultivars for production.

7.3.3 Some Newly Developed Tea Cultivars

In the past few years, Vietnam has developed some new tea cultivars with high yield, good quality and suitable for green tea, black tea and Oolong tea. Currently, there are five nationally released tea cultivars as follows. In Vietnam, only MARD is the agency allowed to recognize a crop cultivar): (1) PH 1 cultivar: released in 1980, selected from individuals in tea gardens originating from India, high yield with 20 – 25 tonnes/ha, good for black tea. (2) TRI 777 cultivar: released in 1980, selected from individuals in tea gardens imported from Sri Lanka in 1975, high yield with 10 tonnes/ha, good for green tea. (3) LDP 1 cultivar: released in 2003, selected from cross between Fuding Dabaicha and PH 1, high yield with 15 tonnes/ha, good for green tea and black tea. (4) LDP 2 cultivar: released in 2007, selected from cross between Fuding Dabaicha and PH 1, high yield with 20 – 27 tonnes/ha, good for black tea. (5) Phuc Van Tien cultivar: released in 2008, selected from individuals in tea gardens imported from China in 2000, high yield with 10 – 12 tonnes/ha, good for green tea and black tea. In addition, there are some other provisional tea cultivars including PH 8 (Fig. 7.4), PH 9 (Fig. 7.5), PH 10, PH 11, Shan Chat Tien and Shan Tham Ve (Niem 1988; Ngoc *et al.*, 2006, 2007, 2008; Toan *et al.*, 1998; La, 1998; Toan & Phuongh, 2006). Characteristics of some new tea cultivars are presented in Table 7.11.

Fig. 7.4. PH 8 cultivar

Fig. 7.5. PH 9 cultivar

There are more and more new cultivars in production in Vietnam. Before 2000, the ratio of new tea cultivars was only 12% and the average tea yield was only 3.5 tonnes/ha (600 kg of dried tea/ha); in 2008, the ratio of new tea cultivars reached 48% and the average tea yield was 6.9 tonnes/ha (over 1.3 tonnes/ha of dried tea). New cultivars with high quality produce tea that commands a 2 – 3 times higher price; high quality materials results in high quality tea products which can meet market demands and thus the final tea products have increased by 20% – 30%. Planting new cultivars increases the quantity, quality and the value of tea.

Table 7.11 Some new tea cultivars in Vietnam

Cultivars	Source	Yields kg/(ha· year)	Year of release	Variety type	Criteria for release	Resistance to pests/diseases	Adaptability/Stability
LDP 1	Fuding Dabaicha×PH 1	15,000	2003	Hybrid between var. *sinensis* and var. *assamica*	Good growth, wide spread, early secondary branches, thick shoot density, give high yield early. Good for black tea and in some areas for green tea	Good resistance to drought and pests	Wide adaptability. Suitable for planting in the low hills below 500 m above sea level
LDP 2	Fuding Dabaicha×PH 1	20,000 – 27,000	2007	Hybrid between var. *sinensis* and var. *assamica*	Light green leaves, long, time enabled early buds. Good growth, wide spread, early secondary branches. Good for black tea	Good resistance to drought and pests	Wide adaptability. Suitable for planting in the low hills below 500 m above sea level. Area in production reached 18%
PH 8	TRI 777×Kim Tuyen	17,240	2009	Hybrid between var. *pubilimba* and var. *sinensis*	Thick, dark green, the horizontal leaves. High yield early, is an intensive farming responsive cultivar. High quality Oolong tea	Moderate pest resistance	Wide adaptability, drought and cold resistance. Easily propagated by cuttings
PH 9	TRI 777×Kim Tuyen	15,800	2009	Hybrid between var. *pubilimba* and var. *sinensis*	Thick, dark green leaf. Good growth, early bud emergence, snowy buds. High yield early. High quality green tea	Moderate pest resistance	Wide adaptability, drought and cold resistance. Easily propagated by stem cuttings. Adapts well in north Vietnam
PH 10	Selected from individuals in tea garden that imported from China in 1994	7,400	2010	*C. sinensis*	Dark green, small leaf. Buds are green and purple, much hairy snow. Medium growth, medium yield. High quality Oolong tea	Resistance to pests pretty good	Easily propagated by cuttings, strong growth in the nursery, high survival rates after planting

(To be continued)

(Table 7.11)

Cultivars	Source	Yields kg/(ha· year)	Year of release	Variety type	Criteria for release	Resistance to pests/diseases	Adaptability/Stability
PH 11	Selected from collected tea garden that originating from Indian since 1993	13,500	2010	*C. sinensis* var. *assamica*	Leaves are bright yellow green. Fat buds, bright yellow green, hairy little snow. Strong growth, high bud density. High yield early. Good for black tea	Good pest resistance, such as greenhoppers and bauxite, mosquitoes	Easy for cuttings, high survival rates after transplanting
Shan Chat Tien	Individually selected from Shan Ha Giang and collected in tea genetic garden in Phu Ho	18,000	2006	*C. sinensis* var. *pubilimba*	Big and strong branches. Yellow green leaves, big shoots. Good for black tea	Good pest resistance	Suitable for planting in Phu Tho, Yen Bai, Ha Giang, Nghệ An provinces
Shan Tham Ve	Individually selected from Shan Ha Giang and collected in tea genetic garden in Phu Ho	8,000	2006	*C. sinensis* var. *pubilimba*	Branches in the lower position, big and strong branches. Big and snowy shoots. Good for green tea	Good pest resistance	Suitable for planting in Phu Tho, Yen Bai, Ha Giang provinces occupying 1% total tea area in Vietnam

7.4 Propagation and Extension System of New Tea Cultivars

Before the year 1972, the propagation method in Vietnam was mainly from tea seeds (Niem, 1998). Tea fruits were collected from the tea plots that produced tea leaves. This caused poor quality, low yield and a low coefficient of propagation. There were not enough tea fruits for fast expansion of the production area of preferred cultivars. After 1972, when the PH 1 cultivar was released and the cutting technique was improved, propagation by tea cuttings was gradually expanded. From 2000 up to the present, this has been used for all propagation purposes. In addition, grafting is also used for some cultivars (Toan & Phuong, 2006).

7.4.1 Cutting Technique

Cutting is the most popular vegetative propagative method for new tea cultivars.

7.4.1.1 Technique of Planting and Cultivating Tea Gardens for Production of Tea Cuttings

If we want to get good tea cuttings we need to have a good mother plant and the technique for rearing tea cuttings needs to be done in accordance with the correct process. Mother plants are planted for cuttings for propagation in a nursery garden. This garden must be planted from tea cultivars asexually propagated from cuttings of collected-purebred tea cultivars.

7.4.1.2 Technique for Rearing Tea Cuttings

There are two seasons for rearing tea cuttings in Vietnam: the summer-autumn season and the winter-spring season. If we want to collect cuttings in July-August (the summer season) then we need to start choosing the main litter picking in April to May for rearing cuttings (not plucking). If we want to collect cuttings in November then we need to start choosing the main litter picking in August for rearing cuttings. The time to start rearing the cuttings is when the twigs on the cuttings have 5 – 6 mature leaves (90 – 105 days).

7.4.1.3 Nursery Technique

There are two best cutting seasons in the north of Vietnam: the summer-autumn season and the winter-spring season. In the winter-spring season, we can start doing cutting from mid-August to mid-December. In the summer-autumn season, we can start doing cutting from mid-June to mid-August. In the south of Vietnam (Tay Nguyen and Bao Loc), the cutting season is from April to August.

7.4.1.4 The Standard for Plantlets

For planting in the field, plantlets need to meet the following requirements (Table 7.12).

Table 7.12 The standard for planting and time for planting

No.	Targets	Requirement
1	The age of cultivars (month)	8 – 12
2	The height of the cultivar (cm)	22 – 30
3	The diameter of the stem (mm)	≥ 2.5
4	Browning level of stem (%)	≥ 50
5	The number of tea plant's leaves (leaf)	≥ 8

7.4.2 Grafting Technique

Nowadays, there are many tea propagation methods. Grafting is one of the new propagation methods. This method overcomes the weakness of the propagation method by seeding and by cutting. The basic of this propagation method is that of using a mother tea plant that has good characteristics such as strong growth, high quality and good adaptability so as to produce good cultivars.

7.4.2.1 Planting of Stocks

Seeds of selected tea plants with good growth and adaptability are sown in polyethylene (PE) bags or directly in the field. Irrigation, fertilization and pest control are realized as required.

7.4.2.2 Tea grafting Technique

Choose young tea plants that are planted by seeding or by cutting. Tea plants that are 4 months old, having 5 – 6 mature leaves and a stem diameter of over 0.15 cm can be used as stock. The best grafting season is in February-May and July-August.

There are three grafting methods as follows:

-Graft the tea plant's terminal shoots with leaves onto stocks.

-Graft the tea plant's twigs without leaves.

-Graft the tea plant's twigs without leaves and cover with a nylon bag (use this method only in the nursery).

Remove all plastic bags and nylon wires after 25 – 30 days. When the young tea plants have 6 – 8 mature leaves, are above 20 cm in height with a browning level about one third up the stem and no pestilent insect infection, then they can be planted.

7.4.3 Institutions for the Extension of New Tea Cultivars

New tea cultivars are propagated by science research institutes, provincial agricultural extension centers, district extension stations, plant seed companies etc.

Newly developed tea cultivars are only multiplied and planted when they are approved by the Government's appropriate authorities for pilot or large scale production. Research institutes will provide those cultivars' cuttings to plant seed companies, extension centers, etc. for planting and multiplication.

There is one tea nursery garden in every tea producing area. The tea nursery gardens provide cuttings to tea seedling producers in the area for further propagation (by cutting) to provide seedlings for tea producers. The owners of tea nursery gardens are only allowed to provide quality cuttings which meet standards approved by the provincial Department of Agriculture and Rural Development.

7.4.4 The Targets of New Cultivar Development

In order to increase tea yields and quality, the tea industry needs to speed up planting new cultivars. By 2015, the share of new cultivars will be 60% of the total tea area. By 2020, the figure will be 70%. In addition, the tea industry needs to improve the value of Vietnamese tea products as well as diversifying the types of products.

Breeding and multiplication are necessary so as to create new cultivars with high quality, to adjust to changing climatic conditions and to resist insect pestilence. There is a need to focus on hybridization to collect new cultivars, using the mother plant from native cultivars and the father plant from imported cultivars that belong to *C. sinensis* and *C. sinensis* var. *assamica.* It is necessary to pay attention to tea breeding methods by treating mutation to obtain new cultivars with salient features. In addition, one must also pay attention to overall solutions, from developing new tea cultivars to appropriate cultivation techniques and processing technology and equipment that meet the requirements for diversifying tea products to increase the economic benefits, and promote the potential and advantages of Vietnam's tea plants.

In order to speed up the transfer of new cultivars and new technologies to producers, supportive policies are required.

7.5 Development of Tea Research Institutions

The tea research in Vietnam was initiated since 1918. It became the Vietnam Tea Research Institute in 1988. The Center for Tea Science and Development of the Northern Mountainous Agriculture and Forestry Research Institute was founded in 2005. Tea breeding has got bigger budgets, more equipment and staff training and achievements in tea breeding have contributed an increase of new cultivars to 48% so far.

7.5.1 History of Formation and Development

The French established the Agricultural and Forestry Experimental Station in Phu Ho, Phu Tho in 1918 (Quy & Oanh, 2008). This station had the task of doing research in agriculture and forestry with the main focus on tea. After 1945, when the Democratic Republic of Vietnam was founded, this station belonged to the Ministry of Agriculture and Rural Development.

In 1960, the station belonged to the Agricultural and Forestry Academy. In 1968, the station belonged to the Industrial Crops, Fruit Tree and Medicinal Plant Institute. In 1988, the station was renamed as the Tea Research Institute and belonged to the Vietnam National Tea Corporation. In 2005, the Center for Tea Research and Development was founded based on the Tea Research Institute, belonging to the Northern Mountainous Agriculture and Forestry Science Institute. The headquarters are at Phu Ho commune, Phu Tho town, Phu Tho Province. This center is the only tea research institution in Vietnam.

The center has 24 ha of experimental land and 3 research departments: the Tea Breeding Department, the Tea Cultivation Techniques Department and the Tea Quality Control Department. In addition, the center has technology transfer and administration units. The center's tasks are research, consulting, transferring techniques, exporting and importing tea products and processing equipment. The center's functional budget is provided by the Government through projects (by tender) or from provincial projects (by tender or order). The center's income comes from consultancy contracts, the transfer of advance techniques, the selling of tea products or tea seedlings. The center's new cultivars and new advanced techniques are protected by licenses granted by the Government.

7.5.2 Scientific Research Fields

From 1995 up to the present, tea breeding has enjoyed help from the Government in the form of bigger budgets, more equipment and the training of staff. There are now 8 PhD's, 12 MSc's working at the center. And there are also 6 postgraduates and 8 masters-trainees in the center. In the past 10 years, achievements in tea breeding have contributed to an increase in the area of new tea cultivars which now cover 48% of Vietnam's total tea area, raising tea production, improving tea quality and the value of tea products. However, there are problems facing tea breeding in Vietnam. Normally, the users of newly developed cultivars do not pay for these new cultivars, and thus it is difficult to encourage the researchers to do their work.

7.6 Conclusions

Vietnam has great potential and has advantages in the development of tea production (Quy & Oanh, 2008). In order to improve yield, quality and the value of tea, there is a need to apply comprehensive solutions. Among them, breeding and multiplication of new tea cultivars with high yield and good quality play a key role. During recent years, owing to the Government's assistance to researchers, great achievements have been realised in tea breeding.

For the further development of tea production in Vietnam, tea breeding and propagation need to be promoted. We need to collect and introduce genetic materials, promote crossing between native cultivars and imported cultivars among *C. sinensis* and *C. sinensis* var. *assamica.* Cultivation techniques also need to be improved. We need to train our researchers and develop cooperation with international institutions in various ways, especially in tea breeding.

References

Chen L, Yu FL, Yang YJ (2006) Germplasm and Genetic Improvement of Tea Plant. Beijing: China Agricultural Science and Technology Press, pp.20-40 (in Chinese).

Food and Agriculture Organization (FAO) (2010) http://faostat.fao.org/.

Guo JC (2005) Varietal characters and genetic variations of Oolong tea germplasms. In: Proceedings of 2005 International Symposium on Innovation in Tea Science and Sustainable Development in Tea Industry. 11-15 November, Hangzhou, China, pp.381-388.

Khon TK, Phong TT (1997) 100 years of the world tea industry. Translation document, Vietnam Tea Corporation, Ha Noi.

La NH (1998) Research on morphological characteristics of tea germplasm garden in basic construction period in Phu Ho in order to provide starting materials for breeding. Results of scientific research and technology development of tea 1988-1997. Ha Noi: Agricultural Publisher, pp.407-408.

Li XH, Ye TM, Huang QW, Fu DH, Zhang CZ, Zeng L (2005) Study on distant hybridization for commercial tea production. In: Proceedings of 2005 International Symposium on Innovation in Tea Science and Sustainable Development in Tea Industry. 11-15 November, Hangzhou, China, pp.389-395.

Menh L, Ngoc DV, Hoang NH (1999) Research results of tea breeding by mutation method, Annual reports of Northern Mountainous Agriculture and Forestry Science Institute.

Niem NV (1988) Study proceedings of industrial crops and fruit trees. Ha Noi: Ha Noi Agricultural Publisher, pp.13-24.

Niem NV (1998) Procedures of PH1 selection for production. Results of scientific research and technology development of tea 1988-1997. Ha Noi: Ha Noi Agricultural Publisher, pp.152-156.

Ngoc DV, Phuong NTM, Toan NV (2006, 2007, 2008) Results of tea selection by multiplication method. Annual Reports of Northern Mountainous Agriculture and Forestry Science Institute.

Quy DN, Oanh DTN (2008) Science on tea culture of the world and Vietnam. Ha Noi : Ha Noi Agricultural Publisher, p.82.

Thu VT, Tien DH (2001) Chemical compounds in tea and some popular analysis methods in tea production in Vietnam. Ha Noi: Ha Noi Agricultural Publisher.

Toan NV, Loan TV (1994) Some characteristics of tea leaf and its meaning in selection. Results of scientific research and technology development of tea 1989-1993. Ha Noi: Ha Noi Agricultural Publisher, pp.33-46.

Toan NV, Phuong NTM (2006) Multiplication methods in tea in Vietnam. Results of scientific research and technology transfer in the period of 2001-2005. Northern Mountainous Agriculture and Forestry Science Institute (NOMAFSI), Ha Noi: Ha Noi Agricultural Publisher, pp.65-73.

Toan NV, Lu TT, Niem NV (1998) Tea selection methods. Results of scientific research and technology development of tea 1988-1997. Ha Noi: Ha Noi Agricultural Publisher, pp.309-325.

8

Tea Plant (*Camellia sinensis*) Breeding in Korea

Byeong-Choon Jeong[1], Young-Goo Park[2]

[1] Mokpo Experiment Station, National Institute of Crop Science, RDA, Muan 534-830, Korea
E-mail: bcjsro77@hanmail.net
[2] Department. of Forestry, Kyungpook National University, Daegu 702-701, Korea
E-mail: ygpark@knu.ac.kr

Abstract: Tea plant was introduced into Korea from China more than 2,000 years ago in the Kaya Dynasty. During the Goryeo Dynasty (918 – 1392), the tea culture flowered along with Buddhism. However, it gradually declined until the 1980s, when a period of economic growth in Korea combined with a rapid recovery of the tea culture and industry. Scientific tea research in Korea began at Boseong in 1992. The results of these studies are not concluded regarding tea breeding, cultivation and manufacture. In 2004, a tea research laboratory was opened at Mokpo Experiment Station, National Institute of Crop Science, Rural Development Administration (RDA). Over 80% of the total Korean tea gardens are seedlings on mountain slopes consisting of individuals with various sprouting times and growth types causing low yield, inconsistent product quality, as well as difficult to use machinery. In order to solve this problem, it is necessary to develop and supply clonal cultivars with disaster-resistant, high quality and high yield features. Several clonal tea cultivars are already registered at Boseong Tea Experiment Station. A new strategy for shortening the process of clonal tea breeding, characteristic examination, clonal tests, local adaptability tests and propagation of promising lines has been carried out simultaneously at 6 local adaptability test stations. The highly adaptable clones will be selected for use throughout Korea.

8.1 Tea Culture and Production

Most of the tea producing area in Korea is situated in the southern region of the Korean Peninsula to avoid frost damage in the winter. The northern limit of tea cultivation is expected to move upward with global warming. The soil of the tea farms is mostly sandy loam or clay loam with granite as its base material. The acidity of the soil is adequate for cultivating tea plants and it is rather sterile. Conditions such as temperature and rainfall are not ideal. The main tea product is green tea, but lately the production of fermented tea has been increasing. The tea gardens are small and located on sloping land and are mostly developed with native tea seeds. The productivity is low and due to their location labor is scarce and mechanization is difficult. Accordingly, the cost of production is high, which raises the consumer price which in turn increases tea imports. In this situation, the rate of increase in the consumption of home-produced tea is less than the rate of increase in production. As a result inventory is piling up and some farms are giving up tea production.

8.1.1 Introduction to the History of Tea in Korea

The tea from the three East Asian nations, Korea, China and Japan, has developed into an integrated culture that has led to exchanges of local and temporal customs and traditions. Tea has not only spread as a kind of popular beverage, but has also formed a unique culture in which each nation's spirit is fused.

According to tradition, the tea plant was introduced from China into Korea during the Kaya Dynasty by Empress Heo in 48 A.D. but there is not sufficient historical evidence to support this. According to Korean historical records, Daryum, who was an envoy of the Silla Dynasty, introduced tea seeds from China to Sankuksaki in 828 A.D. The king of Hungduk ordered the cultivation of the seeds on Mt. Jiri. However, historical records indicate the use of tea in Silla long before 828 A.D. Tea culture saw prosperity in the early era of the Silla (57 B.C. – 935 A.D.) and the Goryeo (918 – 1392) Dynasties. Entering the Joseon Dynasty (1392 – 1910), it was oppressed by the policy of anti-Buddhism and pro-Confucianism. However, it has survived even though it has encountered many difficulties such as excessive tea taxation, lack of interest from the court or the government, severe cold, inferior black tea goods scandals and competition from the alcohol and coffee industries. The Korean tea culture also survived during the Japanese imperialist rule (1910 – 1945) and the National Liberation and the events thereafter. All these unconstructive circumstances resulted in the beginning of scientific and systematic breeding research on tea plants being delayed by a hundred years in Korea as compared to Japan. In the 1970s, a new generation of tea entrepreneurs provided the foundations for the resurrection of the dormant tea culture and industry in Korea.

8.1.2 Location of Natural Growth Tea

Korea is situated between latitude 33° and 38° N and longitude 124° and 132° E. It is estimated that there are around 300 locations where tea plant (*Camellia sinensis* (L.) O. Kuntze) grows 'wild' in Korea. Estimates of the coordinates of the east, west, north and south limits of the range where tea plant can be cultivated, based on the results of a survey on native habitats during the period from 2004 to 2006, are shown in Fig. 8.1 and Table 8.1. They show that the western limit is Goeup Village at Sinjeong-ri, Haeje-myeon, Muan-gun, Jeollanam-do situated at longitude 126°17′09″ E and the eastern limit is Daun-ong, Jung-gu, Ulsan-si, Kyungsangnam-do at longitude 129°24′03″ E. On the mainland, excluding Jeju Island, the southern limit is Gohyeon-ri, Hyeonsan-myeon, Haenam-gun, Jeollanam-do at latitude 34°27′48″ N and the northern limit is Mt. Hamra at Ungpo-myeon, Iksan-si, Jeollabuk-do situated at latitude 36°02′50″ N. The point of lowest altitude is Gohyeon-ri, Hyeonsan-myeon, Haenam-gun, Jeollanam-do (29 m above mean sea level (amsl)) and the highest altitude is the mountain at the back of the Agricultural Cooperative Federation in Daap-myeon (582 m amsl), Gwangyang-gun, Jeollanam-do. These results are likely to change if further precise surveys are made in the future (Jeong *et al.*, 2007).

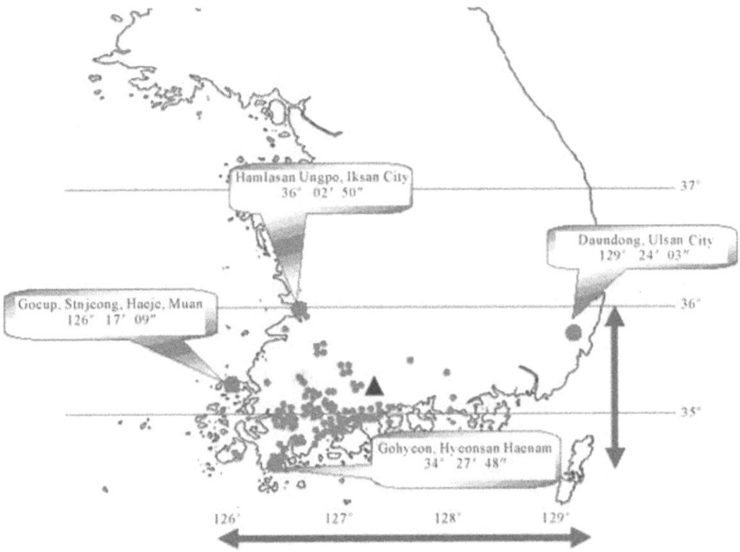

Fig. 8.1. Distribution of tea plant natural habitats in Korea

Table 8.1 Addresses and coordinates of the endmost areas of tea plant natural habitats

Endmost area	Address of natural habitats	Coordinates
East	Daundong, Ulsan City	Longitude 129°24′03″
West	Goeup, Sinjeong, Haeje, Muan County	Longitude 126°17′09″
South	Gohyeon, Hyeonsan Haenam County	Latitude 34°27′48″
North	Hamlasan Ungpo, Iksan City	Latitude 36°02′50″

8.1.3 Current State of the Green Tea Industry

Tea, a preference food, is a well-being food with many beneficial health components. The cultivating area, products and consumption of tea have increased since 1985 in Korea. The Korean green tea industry, which prospered from 2000 to 2004, began to accumulate its stock in 2005, owing to the increasing consumption of fermented teas, with lower import duties, from China. This made tea producing farmers anxious. The agrochemical detection scandal in 2007 which resulted in the return of tea goods from displays in markets, led to an even greater decrease in the consumption of locally produced tea. This increased the number of tea producing farms halting production.

Tea producing countries across the globe, including Korea, are making enormous efforts to improve the competitiveness of their domestic tea industries. On the side of cultivation management, they seek to scale up their operations and achieve mechanization, resulting in cost-saving production due to labor reduction and efficient production systems. By introducing environmentally friendly farming methods, they try to preserve the environment, strengthen the agricultural infrastructure and reduce the damage caused by natural disasters. In addition, they take much interest in creating plant-specific tea gardens by breeding excellent cultivars. On the side of distribution, they seek to secure price competition by supplying quality products at lower prices to meet consumer demands. To meet diverse consumer needs, they focus on product diversification and differentiation. They try to enhance functionality, safety and convenience in the process of product development, while developing their tea cultures with an eye towards boosting the domestic consumption of Korean green tea products.

8.1.3.1 Increasing Trends in the Cultivation Area, Production and Per Capita Consumption

The domestic green tea industry had enjoyed relatively favorable market conditions until 2004. Since then the supply surpassed the decreasing demand and at the end of 2006 a surplus of 900 tonnes of tea had accumulated (Korea Tea Production Association, 2006) (Table 8.2). Excessive supply came earlier than expected due to the increase in tea imports with the opening of the agricultural market because of the World Trade Organization/Doha Development Agenda, agricultural negotiations and the conclusion of the inter-country Free Trade Agreement. Brokers and importers import low-tariff products such as tea bags and tea Pu'er. The increasing import of fermented tea is dividing tea consumption, which has been mainly concentrated on green tea, into green tea and fermented tea, resulting in a decrease in consumption of locally produced green tea.

There are various ways to overcome this difficult situation. Here we will explain the current situation and major problems in the Korean tea industry and the development of new strategies and directions for future research.

Table 8.2 The total stock of home-grown green tea (tonnes)

Year	Total	Boseong	Hadong	Gurye	Suncheon	Gwangyang	Others
2005	130	50	30	10	20	10	10
2006	900	350	300	30	50	20	150
2007	480	200	150	20	30	10	70
2008	136	60	40	7	10	6	13

Data: Estimated volume, Korean Tea Production Association (2005 – 2008)

Table 8.3 shows that tea production in Korea has increased remarkably in the last 20 years (1985 – 2006). This is due to increased cultivation areas, production and the per capita consumption of tea. The cultivation area increased 8.2 times, production 8.6 times and per capita consumption 7.2 times. Likewise, the number of tea farm households increased 57 times, from 92 in 1985 to 5,244 in 2006. Most of these are very small scale farms, covering an area of about 0.7 ha. As a result, production of green tea per individual farm household has decreased from 5.2 tonnes in 1985 to 0.8 tonnes in 2006 even though overall productivity has increased. From 2005, the upward trend in cultivation areas and production of tea began to decline due to an increase in the total surplus stock of green tea. Consequently, if the tariff barrier is lifted and tea is imported, small domestic tea farms and manufacturers will be at risk. Therefore, it is important that the tea industries exert all efforts to enhance their competitiveness in all aspects.

Table 8.3 Tea cultivation area, production and consumption in Korea from 1985 to 2006

Items	1985	1996	1999	2003	2006	Ratio of 2006 to 1985
Planting area (ha)	449	829	1,400	2,225	3,692	8.2
Total amount of production (tonne)	476	947	1,193	2,053	4,080	8.6
Amount of production per ha (tonne)	1.06	1.14	0.90	1.32	1.38	1.3
Number of farmhouses	92	1,171	1,979	3,281	5,244	57.0
Planting area per farmhouse (ha)	4.9	0.71	0.707	0.703	0.704	0.14
Production amount per farmhouse (tonne)	5.2	0.8	0.6	0.6	0.8	0.15
Consumption per capita/year (g)	11.6	20.8	25.5	48.5	84.5	4.2

Consumption per capita/year (g) excluding imported tea.
Data source: Korean Informational Marketing Research (KIMR, 2005), The Ministry of Agriculture and Forestry (MAF, 2007)

Table 8.4 shows the import and export of three types of tea from 2000 to 2010 in Korea. Annually imported green tea ranged from 25 to 175 tonnes, while imported black tea ranged from 322 to 717 tonnes. Annually imported mate tea ranged from 0 to 198 tonnes. In 2006, the amount of imported black tea increased by more than 276 tonnes compared to the amount imported in 2004. This situation led to the imbalance between supply and demand, increasing the amount of stock in the domestic green tea storehouses of the tea farmers as a poor-selling item. Korea's tea export is still very small and only amounts to below 0.1% of the world tea market. The annually exported green tea ranged from 27 to 767 tonnes, while black tea ranged from 281 to 1,629 tonnes. Almost all exported teas have been shipped to China by Korean Tea Companies.

Table 8.4 The import and export of three types of tea in Korea from 2000 to 2010 [tonnes (1,000 US$)]

Year	Import			Export		
	Green tea	Black tea	Crude tea	Green tea	Black tea	Crude tea
2000	88(126)	322(1,620)	—	27(143)	344(844)	—
2002	59(172)	372(1,592)	0.2(1)	319(7,951)	1,061(2,641)	—
2004	175(296)	460(1,745)	5(20)	417(1,578)	1,629(3,746)	—
2006	174(300)	717(6,168)	6(20)	767(2,131)	774(2,283)	—
2008	30(53)	421(3,082)	34(96)	313(990)	324(922)	—
2010	25(124)	561(4,037)	198(661)	427(2,300)	281(2,098)	0.06(4)

Crude tea with mate: Yerba mate has been used as a tonic and a central nervous system stimulant.
Data source: Korea Agro-Fisheries Trade Corporation (www.kita.net)

8.1.3.2 High Consumer Price Due to High Production Cost

With the price of Korean green tea products increasing due to low productivity and high production costs, consumers avoid buying such expensive green tea products, resulting in shrinking consumption and a surplus of green tea. Abroad, the market price of Korean tea is lower than that of Japanese tea, but the domestic market price of Korean green tea is higher than imported Japanese green tea. Recently, this surplus issue has become a serious problem. Higher production costs due to higher wages and fuel costs combined with a larger supply of domestic green tea may push the Korean green tea market's competitiveness to its limit. Table 8.5 shows comparison of consumer prices between Japan, China and Korea. Table 8.6 shows comparison of yield and production costs between Japan, China and Korea in 2003.

Table 8.5 Comparison of consumer prices between Japan, China and Korea (US $/100 g)

Country	Japan (A)	China (B)	Korea (C)	C/A	C/B
High quality	12.0	5.0	45.0	3.7	9.2
Middle quality	6.0	3.0	30.0	4.4	10.0

Source: 2004 Korea Information Marketing Research Institute

Table 8.6 Comparison of yield and production costs between Japan, China and Korea (2003)

Country	Japan (A)	China (B)	Korea (C)	C/A	C/B
Yield (kg/ha)	8,810	6,160	5,030	0.6	0.8
Cost (US $/kg)	2,765	1,125	3,520	1.3	3.1

Source: 2004 Korea Information Marketing Research Institute

8.1.3.3 Small Scale Farming: Limit Enhancing Competitiveness

Most tea farms in Korea are small in size. In 1985, the average area per tea farm was 4.9 ha due to the small number of tea farms at that time. Due to the promising future of the tea industry, the city and county governments started giving financial support to expand tea farm areas. Consequently, the number of small-sized tea farms increased rapidly, from 1,979 in 1999 to 5,244 in 2006. The average cultivation area per tea farm decreased to as little as 0.44 ha in the Hadong area and 0.88 ha in the Boseong area in 2006. When the other areas are included, the average cultivation area per tea farm in Korea is 0.7 ha. Problems such as lack of distribution of high-quality cultivars, insufficient manufacturing facilities and tea farms located on mountain slopes resulted in the tea industry reaching its limit in enhancing its competitiveness in response to the opening of the global tea market. Table 8.7 shows comparison between tea number of farmhouses in 2006.

Table 8.7 Comparison between tea planting areas and number of farmhouses (2006)

Districts	Number of tea farms	Planting area (ha)	Cultivating area per farmhouse (ha)
Hadong	1,917	853	0.44
Boseong	1,165	1,023	0.88
Others	2,162	1,816	0.84
Total	5,244	3,692	0.70

Source: 2006 Korea Information Marketing Research Institute

8.1.3.4 Labor-Intensive Operation with Low Productivity and High Production Cost

All tea farms are located on mountain slopes except those in Gangjin, Haenam and Jeju-do, so it is difficult to mechanize fertilization and plucking. Also, 80% of tea farms are planted with native cultivars which are not uniform in sprouting and growth, so mechanical harvesting is not possible and the yield of fresh leaves is low since hand plucking has to be done several times. This increases the production costs. Furthermore, management facilities for irrigation and protection against frost and wind, farming roads and harvesting machines are in short supply (Kim, 2000).

8.1.4 Type of Tea Produced and Technologies Used

Korea produces green tea. The most popular Korean method for manufacturing green tea is to bake and roll the tea leaves. The tea leaves are put into a large metal cauldron and gently tossed and turned. From there they are placed on a mat and rolled by hand. This process is repeated approximately 3 times before the leaves are ready to be sorted and packaged. The whole process is performed by hand:

picking, rolling, sorting and sifting.

Once the tea leaves are picked, they are processed right away; either baked or steamed. In terms of manufacturing green tea, there are two types of manual methods in Korea: Bucho-cha and Jeung-cha. When using the Bucho-cha method, the order is as follows: Pick leaves → Roast in cauldron at 350 °C, roll and rub on a rough straw mat → Separate individual leaves → Repeat several times: roasting at a lower temperature, roll and separate → Final drying. When using the Jeung-cha method, the leaves are steam-roasted at 100 °C for 45 s in a steam pipe. The rolling, rubbing, separation and drying processes are done mechanically.

8.1.5 The Climate in Korea for Tea Cultivation

Korea has a temperate climate with four distinct seasons. The season changes gradually, but each season has distinctive seasonal characteristics. Spring and autumn are relatively short while summer and winter are rather long. Summer begins in June with high temperatures and humidity and lasts until August with monsoon rains and typhoons that account for 60% of the annual precipitation. Heavy showers accompanied by thunder and lightning occur during this season causing occasional floods. Winter is dry and cold due to the northwesterly wind sweeping down from Siberia. Korea has hotter summers and colder winters compared to the other countries located in the same latitude zone of the continent. The annual average temperature is 12 to 14 °C in the central region and 3 to 10 °C in the northern region.

Annual rainfall ranges from 600 to 1,600 mm with an uneven seasonal distribution. The rainy season, which begins in late June, lasts approximately 30 days. Two to three typhoons affect Korea between June and October every year. As mentioned earlier, 50% to 60% of the annual rainfall is concentrated during summer causing an unstable water supply.

8.1.5.1 Temperature

The areas of cultivation of tea plant in Korea show an annual average temperature of 12.8 to 15.5 °C during the winter (December to February), with the lowest temperatures ranging between –0.4 and 4.3 °C (excluding Jeju Island). The highest temperature during the summer ranges between 25.3 and 29.6 °C (Table 8.8). The extreme minimum temperature ranges from –3.3 to –15.6 °C and the extreme maximum temperature ranges from 23.9 to 29.7 °C (Table 8.9). The altitude of the main producing areas is between 200 m and 800 m amsl, favorable for the production of high leaf quality for green tea due to the huge temperature differences between day and night.

Table 8.8 Average temperatures of the main tea producing areas (°C) (2000 to 2006)

Season	Spring				Summer				Autumn				Winter				Average
Month	3	4	5	Average	6	7	8	Average	9	10	11	Average	12	1	2	Average	Average
Gwangju	1.7	7.8	24.3	11.3	27.7	29.5	30.3	29.2	26.8	21.8	4.5	17.7	–1.0	–2.8	–2.0	–1.9	13.5
Boseong	0.4	5.4	23.2	9.7	26.3	29.0	30.3	28.5	26.9	22.0	2.5	17.1	–2.7	–4.3	–3.2	–3.4	12.8
Mokpo	2.3	8.2	22.3	11.0	26.0	28.2	29.9	28.0	26.2	21.3	6.2	17.9	0.4	–1.5	–1.1	–0.7	13.8
Jinju	0.0	5.7	24.6	10.1	28.1	30.2	30.6	29.6	27.1	22.1	1.6	16.9	–3.6	–5.5	–3.9	–4.3	13.1
Namhae	2.5	8.0	23.1	11.2	26.9	29.2	30.4	28.8	26.6	21.9	5.5	18.0	–0.4	–2.2	–1.4	–1.3	14.0
Pohang	3.9	9.8	22.7	12.1	26.5	28.9	29.5	28.3	25.3	21.5	6.5	17.8	0.3	–1.5	–0.1	–0.4	13.8
Jeongeup	0.4	6.5	24.2	10.4	27.8	29.9	30.5	29.4	26.7	21.4	3.4	17.2	–2.5	–4.2	–3.2	–3.3	12.8
Jeju	6.2	10.7	18.5	11.8	22.4	26.3	27.2	25.3	23.0	17.6	10.3	17.0	5.6	3.7	3.6	4.3	15.5
Gangneung	2.4	8.4	21.9	10.9	26.0	27.9	28.3	27.4	24.0	20.3	5.5	16.6	–0.7	–3.0	–1.4	–1.7	12.9
Average	2.2	7.8	22.7	10.9	26.4	28.8	29.7	28.3	25.8	21.1	5.1	17.4	–0.5	–2.4	–1.4	–1.41	13.6

Source: 2008 Korea Meteorological Administration.

Table 8.9 Extreme minimum and maximum temperatures of the southern and coastal tea producing areas (°C) (2000 to 2007)

	Haenam	Mokpo	Jeongeup	Jinju	Ulsan	Gangneung	Seogwipo
Maximum	26.8	26.2	25.7	26.2	25.9	23.9	29.7
Minimum	–10	–10.2	–15.1	–15.4	–10.6	–15.6	–3.3

Source: 2008 Korea Meteorological Administration

8.1.5.2 Rainfall

Because a large volume of water is required during the growth period from spring to autumn, tea plant is liable to suffer from drought during this period. If the moisture content of the soil decreases to 30% of the maximum water holding capacity, the plant withers. In general, high soil moisture content is good for plant growth of the above-ground part, but for the roots excessive moisture results in an insufficient supply of oxygen, which retards growth. The optimal soil moisture content for the growth of tea plant is 80%. The annual rainfall in Korea is 1,200 to 1,800 mm, but in many cases rainfall is not sufficient during the sprouting period in March to April. This delays sprouting and therefore irrigation is necessary. Table 8.10 shows the annual precipitation of tea cultivating areas in 2003.

Table 8.10 The annual precipitation of tea cultivating areas (2003)

Month	1	2	3	4	5	6	7	8	9	10	11	12	Total
Gwangju	31.5	34.4	69.1	82.1	92.0	168.8	222.6	201.2	189.5	51.9	42.9	56.8	1,222
Boseong	13.5	58.2	70.0	184.0	138.3	273.1	218.7	254.9	125.9	45.4	42.7	35.0	1,459
Gangjin	34.0	32.8	71.2	164.2	101.2	254.2	204.8	232.1	135.4	47.0	45.4	36.2	1,227
Gurye	30.5	37.7	62.2	97.5	96.7	205.2	320.4	264.0	134.8	50.7	48.3	25.8	1,374
Hadong	49.2	55.6	73.1	158.4	108.1	188.0	201.8	226.4	120.0	87.5	48.6	42.0	1,358
Seogwipo	57.8	77.2	108.0	188.7	205.9	282.3	282.6	216.5	160.9	71.3	78.4	41.8	1,776

Source: Boseong Tea Research Experiment Station, Sancheong: 1,480 mm, Hamyang: 1,169 mm

8.1.5.3 Topography

The major tea producing areas in Korea are situated in Jeollanam-do, Jeollabuk-do, Gyeongsangnam-do and Jeju-do where the temperature and rainfall are relatively high. However, these areas are mountainous regions and the farms are located on steep slopes of 5° to 40°. In the future, productivity should be improved by labor saving through cooperative cultivation in plain zones (Table 8.11, Fig. 8.2).

Table 8.11 The conditions and locations of the tea gardens in the tea producing areas

Region	Gradient	Soil texture	Planting direction	Remarks
Gochang	5° – 15°	Loam	South, East	Around the Seonunsa, altitude 36 – 67 m
Gwangju	40°	Loam	East, West	Around the Jeungsimsa, altitude 234 m
Jangseong, Nammyeon	10° – 15°	Loam	East, West	Altitude 100 m
Boseong, Hoicheon	10° – 40°	Loam	South, East	Seashore, warm, foggy
Gangjin, Weolnam	10°	Loam	South, West	Under Weolchulsan, shuts out the north wind
Gwangyang, Da-ap	10° – 40°	Loam	South, East	The end of Bakunsan, near Seomjin river
Gurye, Piagol	10° – 40°	Loam	East, South, West	Piagol valley in Jirisan, around the Seomjin river
Yeongam, Deokjin	5°	Loam	South, West	Plain zone, the bottom of Weolchul mountain
Hadong	10° – 40°	Loam	South, West	The bottom of Jiri mountain, riverside of Seomjingang
Sancheong	30° – 50°	Loam	South, East	The bottom of Jiri mountain, foggy
Jeju, Seogwipo	5°	Volcanic ash soils	South	Seashore, altitude 250 m

Fig. 8.2. Topographies of tea gardens in Korea

8.1.5.4 Soil

The Korean Peninsula is located in the heart of the North Western Pacific region. It lies contiguous to China and Russia and extends southward into the Pacific Ocean. It is surrounded by the Sea of Japan to the east, the East China Sea to the south and the Yellow Sea to the west. The Korean Peninsula encompasses 221,000 km^2 of which 45% (99,600 km^2) constitutes the Korea.

The soil quality of tea farms has a considerable effect on the growth, production and quality of tea. Tea grows well in acidic soil with pH ranging from 4.5 to 5.5, but it can also grow at lower pH values ranging from 3 to 4. In addition, well-drained soil with good water penetration (water permeability), air permeability, water retention and nutrient retention capacity and a plow layer deeper than 90 cm are required. Korean soil is generally sandy, low in organic matter content and fertility and is susceptible to erosion. It is formed from acidic rocks of coarse, low base granite and granitic gneiss on mountainous topography. Table 8.12 shows soil chemicals found in the Korean tea gardens.

Table 8.12 Soil chemicals found in the Korean tea gardens

Soil series name	Soil layer	pH (1:5 H_2O)	O. M. (g/kg)	Av. P_2O_5 (mg/kg)	Ex. K ($cmol^+$/kg)	Ex.Ca ($cmol^+$/kg)	Ex. Mg ($cmol^+$/kg)
Asan	Top	4.7	16.6	52	0.13	0.17	0.15
	10 – 25 cm	4.9	8.3	46	0.11	0.11	0.09
	25 – 55 cm	4.7	5.9	50	0.15	0.10	0.19
Jangsan	Top	4.7	30.1	11	0.25	0.38	0.31
	13 – 35 cm	5.2	19.1	11	0.09	0.13	0.48
	35 – 55 cm	4.9	9.8	16	0.11	0.09	0.14

O.M.: Organic matter; Av.: Available; Ex.: Exchange.
Source: 2009, Korean Soil Information Systems, Rural Development Administration.

8.1.5.5 Environmental Characteristics of Major Domestic Production Areas

The annual average temperatures of the major tea production areas in Korea range from 12.8 to 15.5 °C, but temperatures can drop to –10 to –17 °C in Jeollanam-do and Gyeongsangnam-do and down to –4 °C in Jeju-do (Seogwipo). This presents the possibility of frost damage. Optimal rainfall for the growth of tea plant is over 1,500 mm per year and over 1,000 mm during the growth period from March to October. Rainfall in Korea is adequate during this period except for the dry spring season. The northern limit of tea plant cultivation is latitude 45° N, and it is 38° N in China, 42° N in Japan and 36° N in Korea (the site of the Imhae Temple at Ungpo-myeon, Iksan-si). Low temperatures in winter, late frost, a dry spring and low rainfall are responsible for the low northern cultivation (survival rate of the tea plant) limit of tea plant in Korea.

8.2 Tea Germplasm Collection, Conservation and Appraisal

The Boseong Tea Experiment Station selected around 2,300 superior tea plants and investigated their characteristics. These results were placed in a database that is utilized in breeding programs (Kim, 2008). The Mokpo Experiment Station (MES), National Institute of Crop Science (NICS), also investigated the characteristics of 700 individual tea plants collected in 1988. The superior individuals were selected and their characteristics were published. Currently, the MES is examining around 3,000 individuals selected from a group of 15,000 plants collected from native habitats throughout the country. The Korean Forest Research Institute and Kyungpook National University have been studying gene conservation, protection of genetic diversity, utilization of genetic resources and cryopreservation of tea as a forest tree.

8.2.1 Collection and Conservation

Korea has a long history of tea cultivation and culture, but little research has been done on breeding specific high quality tea cultivars compared to Japan and China. For a private farm to breed a reliable cultivar is almost impossible. Therefore, almost all tea gardens consist of seedling tea plants originating locally and wildly with great morphological variations. At the Boseong Tea Experiment Station 2,300 native tea plants were collected as genetic resources from 1994 to 1998. Their characteristics were investigated and the data were stored in a central database.

At MES in 1988, 330 individual tea plants were conserved out of tea plants collected from Boseong, Moodeung Mt. and Wolchul Mt. and their morphological characteristics were investigated (Song *et al*., 2005a). Within the limits of this investigation of the morphological characteristics, many variations were observed in the number of stems, stem length, leaf area, fresh weight and leaf color. The leaf shape index was 2.5±0.3 and the chlorophyll contents of the tea leaf showed a significantly negative correlation with the value of the colorimeter as –0.98. Negative correlation was observed in the number of stems and stem length, the number of leaves. No correlation was observed in leaf width, number of serrations, number of veins and petiole length. The number of shoots in the group collected from Boseong was more abundant than those from Moodeung Mt. and Wolchul Mt. However, the group collected from Wolchul Mt. showed the longest length of shoot among all the collected groups. In addition, the number of leaves per shoot, leaf area, leaf length and fresh weight were highest in the group collected from Wolchul Mt. The contents of major components related to tea quality were different among the different native sites. In particular, the biggest difference was observed in tannin and chlorophyll content. Table 8.13 shows morphologial characteristics and Table 8.14 shows leaf characteristics of tea plants collected in the Jeonnam area.

Table 8.13 Morphological characteristics of wild tea plants collected in the Jeonnam area

Region	No. of stem	Stem length (cm)	No. of leaf	Leaf length (cm)	Leaf width (cm)	Index (length/ width)	Leaf thickness (mm)	Petiole length (mm)	No. of veins	No. of serrations
Wolchul	40.7	21	8.3	7.9	3.3	2.39	0.556	3.7	7.3	27.9
STD	11	4.7	1.3	1.3	0.5		0.12	0.6	1.6	5.2
CV (%)	27.15	22.60	15.45	16.13	14.34		23.15	17.19	21.40	18.63
SV	−0.62	0.84	0.78	0.14	−3.91		−0.31	−0.42	−0.44	−0.10
Boseong	73.5	8	5.8	6.8	3.2	2.50	0.711	3.8	6.9	23.4
STD	20	4.1	1.1	0.9	0.4		0.05	0.6	0.7	2.8
CV (%)	27.21	49.19	18.37	12.79	11.18		23.15	16.70	10.81	11.77
SV	0.85	−1.23	−0.74	−0.84	−0.03		1.01	−0.31	−0.73	−1.05
Moodeung	64.6	11	6.1	7.7	3.1	2.15	0.615	4.3	8.5	29.1
STD	23	4.5	1.2	1.1	0.5		0.09	0.7	1.1	4.4
CV (%)	35.51	13.89	19.81	13.89	15.91		6.52	15.91	13.02	14.98
SV	0.45	−0.61	−0.58	−0.09	−0.21		0.20	0.35	0.38	0.13
Mean	**59.6**	**13.3**	**6.7**	**7.5**	**3.2**	**2.34**	**0.627**	**3.9**	**7.6**	**26.8**

Table 8.14 Leaf characteristics of tea plants collected in the Jeonnam area

Region	Leaf area (cm^2/leaf)	Leaf area (cm^2/plot)	Flesh weight (g/plot)	Leaf color***		
				L	a	b
Wolchul	12.1	3,875	127	26.4	−17.2	25.1
STD	4	1,192	33	7	6.4	11.8
CV (%)*	32.80	30.78	25.78	24.85	−37.02	16.90
SV**	0.30	0.12	0.30	−0.74	−0.69	−0.37
Boseong	6.5	2,578	80	32.5	−10.9	14.8
STD	2	551	15	2	1.3	2.5
CV (%)	32.18	21.40	18.78	4.59	−11.67	16.90
SV	−1.01	−0.83	−0.74	0.28	0.47	−0.28
Moodeung	10.1	3,629	103	34.3	−10.7	10.1
STD	4	1,466	51	2	1.5	4.2
CV (%)	41.86	40.41	49.30	6.24	−14.03	87.81
SV	−0.19	−0.06	−0.22	0.57	0.52	−0.28
Mean	**9.6**	**3.361**	**103**	**31.0**	**−12.9**	**16.7**

*Coefficient of variation; **Standard variation; ***Measured by Minolta CM-508d

The conservation of the forest tree gene is necessary, due to the increase in endangered or extinct species, protection of genetic diversity and utilization of genetic resources.

A genetic conservation program for forest trees might include studies on exploration, collection, evaluation, conservation and utilization, as shown in Fig. 8.3. Exploration and collection are not easy procedures because en of high polymorphism and genetic variation within a forest species (Park & Son, 1996). Isoenzyme and DNA variation are evaluated in the analysis of forest genetic variation.

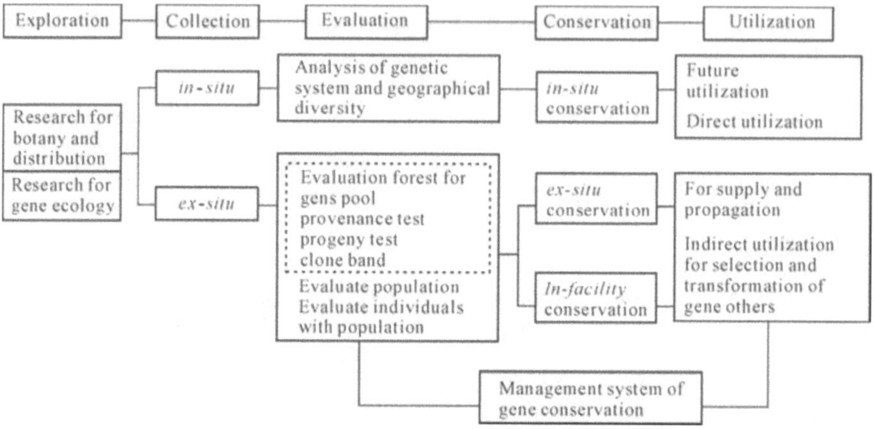

Fig. 8.3 The processes of forest gene conservation from exploration to utilization

There are three methods of gene conservation: *in-situ, ex-situ* and *in-facility*. *In-situ* conservation provides the opportunity to preserve the broadest range of species, whereas *ex-situ* collections may be more appropriate when access to a specific or endangered population is desired. *In-facility* conservation, germplasm is preserved in a controlled environment using tissue culture techniques and/or cooling to subzero temperatures (cryopreservation).

8.2.2 *The Genetic Diversity of the Korean Tea Population*

The morphological and genetic variation of the Korean tea population was investigated. The variation of leaf size was distributed as follows: 1.1% were small (<5.5 cm), 54% were middle-sized (5.5 to 8.4 cm) and 54% were large (>8.4 cm). When the leaf shape (leaf length/leaf width) was investigated it was found that 8.8% were long-oval (<2.1), 48.2% were oval (2.1 – 2.5) and 42.9% were egg shaped (>2.5).These characteristics are similar to the morphological characteristics of Japanese tea. In RAPD analysis using 20 populations, the frequency of polymorphism was 30%. It was proposed that each heterogeneous group should be conserved for gene conservation (Kaundun *et al*., 2000).

The evaluation of genetic diversity was conducted by RAPD-PCR in 6 Korean tea populations. Highly significant differences were found among these populations, upon analysis of molecular variance. A noticeably higher proportion of diversity was observed within populations 84%, compared to the 16% between populations (Park *et al*., 2002).

Nonetheless, in the Korean tea population the genetic variation is smaller than the genetic variation of the Chinese or Japanese wild tea populations. Because the Korean tea population size is small, gene conservation is very important (Park *et al*., 2002; Kaundun *et al*., 2000).

8.2.3 Cryopreservation of Tea

Although seed storage methods have been adopted as part of the genetic conservation program for numerous species, for a relatively high number of species they are inappropriate or even impossible because of limited seed viability and a lack of knowledge of cryophysiology.

Cryopreservation is the process where cells or tissues are preserved by cooling them to subzero temperatures, typically −196 °C. Procedures of cryopreservation and freezing strategies are illustrated in Fig. 8.4. There are 5 possible freezing strategies: simple freezing, slow freezing, encapsulation-dehydration, vitrification (using cryoprotectants—antifreeze-like liquids) and encapsulation-vitrification. When preserving tea embryos by cryopreservation, the highest survival rate is obtained when tea embryos are incubated in Plant Vitrification Solution 2 (PVS2) for 60 min. Cryopreservation is a reliable and applicable method for tea gene conservation without genetic variation (Park & Kim, 2007).

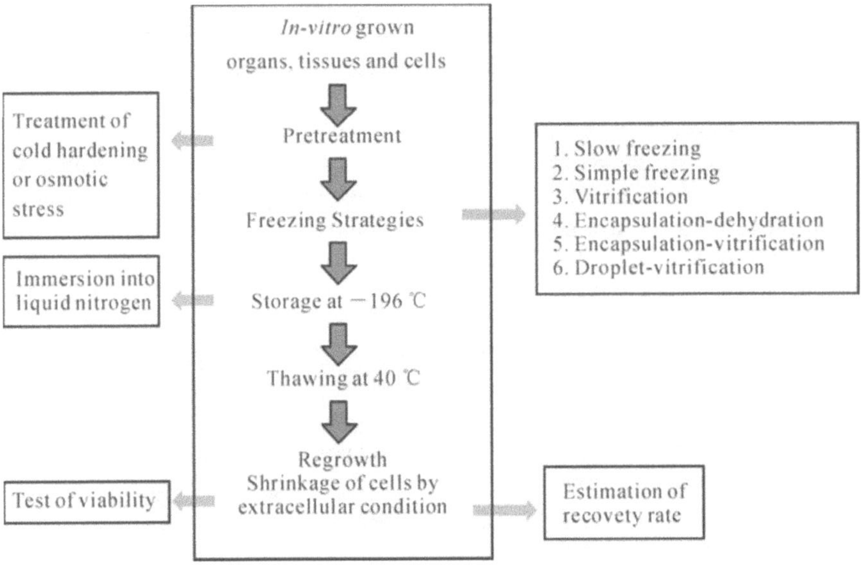

Fig. 8.4 Cryopreservation procedures and freezing strategies

8.2.4 Appraisal and Utilization

A native species is endemic to an area, with its origins sometimes going back several hundreds of thousands of years. It is a mixture of many types and is well adapted to the environment. The native Korean tea plants observed in several regions have characteristic small leaves, 6 to 9 petals and a few stamen longer than the pistil (Choi, 2000). The contents of major components related to tea quality are similar among different sites. Native cultivars have higher total amino acids, vitamin C, total nitrogen and chlorophyll, indicating that they could be good crossing parents for the breeding of high quality cultivars (Song *et al.*, 2005b; Han *et al.*, 2005).

The Boseong Tea Experiment Station investigated the crossing rate and pollen germination ability of the native tea plants and found the following:

(1) The crossing rates between the different cultivars are 13% to 15% and 3% to 5 % within the same cultivar.

(2) The fertilization rates for inter-specific crossing are between 2.9% and 4.1%: (a) Yabukita: 3.2% is strictly selfing and 3.1% is selfing intra-plant, (b) Boseong native tea: 2.9% is strictly selfing and 4.1% is selfing intra-plant.

(3) The pollen germination ability (%) of Daewonsa native and Yabukita tea collected at different flowering stages is as follows: (a) One day after bloom: 75% to 86%, (b) Bloom: 84% to 97%, (c) One day before bloom: 87% to 90%, (c) Two days before bloom: 61% to 94%, (d) Five days before bloom: 67% to 85%.

(4) Change in the pollen germination ability (%) of Daewonsa native and Yabukita tea after a short storage period (storage temperature 0 to –5 °C): (a) One day: 82% to 92%; (b) Three days: 83% to 89%; (c) Five days: 78% to 83%; (d) Fifteen days: 66% to 80%; (e) Twenty-five days: 64% to 73%.

(5) Change in the pollen germination ability (%) of Daewonsa native tea pollen collected at different flowering stages after a longer storage period (storage temperature 0 °C, storage period 3 to 12 months): (a) One day after bloom: 8%; (b) Bloom: 60% to 66%; (c) One day before bloom: 69% to 86%; (d) Two days before bloom: 48% to 74%; (e) Five days before bloom: 25% to 35% (Choi, 2000). Table 8.15 shows characteristics of the leaf and flower of native tea plants. Table 8.16 shows contents of tannin, caffeine, vitamin C, chlorophyll and total amino acids in native tea leaves.

Table 8.15 Characteristics of the leaf and flower of Korean tea plants

Location (12 counties, 25 sites)	Leaf length (cm)	Leaf width (cm)	No. of stomata (ea.)	No. of petals (ea.)	No. of stamen (ea.)	Stamen length (mm)	Pistil length (mm)
Mean	9.8	4.2	121	7.1	218	10.5	12.2
Range	7.7 – 12.5	3.0 – 4.9	91 – 152	5.7 – 8.6	169 – 268	8.4 – 12.7	9.6 – 14.9

Source: 2000 Boseong Tea Experiment Station

Table 8.16 Contents of tannin, caffeine, vitamin C, chlorophyll and total amino acids in Korean tea plants

	Tannin (%)	Caffeine (%)	Vitamin C (mg/100 g)	Chlorophyll (mg/100 g)			Total amino acids (mg/100 g)	T-N (%)
				a	b	Total		
Mean	14.5	2.53	224.3	180	62.0	242.0	2,222	4.29
Range	12.5 – 18.3	2.21 – 3.11	167.9 – 271.8	140.5 – 247.4	46.9 – 85.3	187.4 – 332.7	1,888 – 2,500	3.59 – 4.89

Number of locations: 29; T-N: total nitrogen

In 2004, the researchers of MES began a project with the aim of selecting around 3,000 tea plants as genetic resources. Over 3 years, 143 kg of tea plant seeds and 16,700 cuttings were collected from 120 locations in Korea. These collected seeds and cuttings, along with 9 Japanese cultivars, were planted in the same experimental field. The results of the evaluation were reviewed after planting of the collected germplasm; the leaf shape index was 2.5. Leaf color showed the same level after planting without the collected regions, though they differed from each other in their original region.

The quality of the Korean tea accessions was found to be higher than the Japanese cultivars when the functional components such as caffeine, catechins and the amino acid were analyzed. The Korean tea accessions showed higher caffeine, EGC and ECGC contents compared to the Japanese tea cultivars. Specifically, the content of caffeine in Wolchul 3 – 13 was 6 times higher than the Japanese tea cultivar Saemidori. Using this information, 30 superior tea plants were selected and, since 2006, regional adaptation tests have been conducted at 6 different locations. Using 13 selected primers from a total of 65 primers that were tested, 5 bands (500 to 2,000 bp) were observed on average. The tea plants could be classified into 2 groups: Japanese cultivars and Korean tea plants. Table 8.17 shows chemical components of the soil from the areas from where the tea plants were collected in 2005.

Table 8.17 Chemical components of the soil from the areas from where the native tea plants were collected in 2005

	pH (1:5 H_2O)	O. M. (g/kg)	T-N (%)	Av. P_2O_5 (mg/kg)	Ex. ($cmol^+$/kg)			C. E. C. ($cmol^+$/kg)	EC (ds/m)
					K	Ca	Mg		
Max	6.6	112.6	0.46	280	0.93	15.04	2.60	1152	0.58
Min	4.28	17.0	0.14	8	0.21	0.40	0.20	6.5	0.115
M	5.02	49.38	0.298	60.75	0.428	3.778	0.89	41.59	0.25
SD	0.42	24.42	0.082	64.08	0.16	3.43	0.56	180.12	0.12
CV (%)	8.46	49.45	27.39	105.48	37.21	90.77	62.06	433.91	46.73

C.E.C: cation exchange capacity; EC: specific electrical conductance.
Source: 2006 Report of collecting of wild tea plants project of RDA

8.3 Tea Breeding and Selection Techniques

In tea production, clonal cultivars are preferred for the advantages of high production, high quality and easier management due to the good and uniform characteristics. Currently, 80% of the total tea cultivation area is still planted with native tea plants, 17% with clonal tea plants of a Japanese green tea cultivar 'Yabukita' and the remaining 3% with cultivars from other countries.

In response to the tea farmers' demand to shorten the time period, it takes to distribute clonal cultivars, new strategies must be developed to reduce the breeding period. One strategy is the selective breeding method that breeds cultivars within 10 to 13 years. This method reduces the breeding period by conducting 3 types of experiment required in the breeding process in parallel. Local adaptation tests have been conducted for selected superior strains since 2006. Seven cultivars, Bohyang, Myungseon, Chamnok, Seonhyang, Mihyang, Jinhyang and Oseon have already been developed by the selection breeding method and registered by the Boseong Tea Experiment Station (Kim, 2008). Several universities are also experimenting with tea breeding using new biotechnology techniques such as transformation of superior individuals with cold tolerance genes, etc.

8.3.1 Breeding and Registration System for New Cultivars

To meet the urgent need for clonal cultivars of industrial areas such as tea plant farms and tea manufacturers, local self-governing bodies and the tea culture circle, it is pre-requisite to select fine ones among collected individuals and develop them into new cultivars. A major breeding program consists of the selection of promising plants from the field of genetic resources as the source for the production of improved clonal cultivars or hybrids.

In order to create an efficient breeding system, we are preparing to establish guidelines about breeding and how to execute evaluation in the selection of high-quality cultivars.

At present, in addition to high yield and high quality, the objectives of tea breeding are: (1) high tolerance to cold, diseases and pests, (2) high content of useful constituents and, (3) vigorous growth under low nitrogen conditions in the soil.

The cross breeding system, a popular tea breeding system used in countries such as Japan and China, requires a breeding period of 18 to 24 years (Takeda, 2000; Gong, 2000). Korean tea breeders are also using the cross breeding method, as shown in Table 8.18. With regards to tea breeding organizations in Korea, there is one national experimental station, one provincial experiment station and one company where tea breeding is done (Fig. 8.5). Adaptability testing is carried out by 4 provincial tea research teams and two stations.

Table 8.18 The comparison of conventional breeding and shortening procedure

Tea cross breeding procedure and duration		Conventional breeding (years)	Shortening procedure (years)
Step 1	Genetic resources collection testing period		
Step 2	↓ Selection of crossing parents		
Step 3	↓ Artificial pollination and seed production	1	1
Step 4	↓ Raising hybrid seedlings (in PE house)	1	1
Step 5	↓ Individual selection	5 – 6	3 – 4
Step 6	↓ Individual cutting, nursery screening	1	1
Step 7	↓ Clonal test	5 – 7	steps 7 and 8 are combined 4 – 7
Step 8	↓ Local adaptability test	5 – 8	
Step 9	↓ National level registration		
Total		18 – 24	10 – 13

Tea breeding organizations
↓
Adaptability testing stations
↓
National level registration
↓
National tea plant propagation stations
↓
Nationwide growing plan

Fig. 8.5. Tea breeding and popularization system in Korea

8.3.2 *Strategy of Shortening Tea Breeding Period*

The main objective of tea breeding is to improve the yield and quality of the shoot tips ('two and a bud') or end product. Only about 10% to 18% of the total biomass is harvested and used to make tea products (Magambo & Cannell, 1981; Banerjee, 1991). As the tea plant is woody crop, the development of new cultivars requires long breeding periods. In order to achieve the above-mentioned objective, the first goal is to shorten the breeding term. By 2015, the Korean tea breeding researchers hope to have successfully shortened the breeding period by the process depicted in Table 8.18 (Jeong *et al.*, 2005, 2006). The key factors of this strategy compared to the longer processes shown in Table 8.18, are a shorter individual selection period and the initiation of clonal and local adaptability tests in parallel at local adaptability test stations. Another important factor is the shorter propagating period achieved by starting propagation at an earlier experimental phase.

In advanced countries, cross-breeding is now the most popular method for developing new cultivars. Parents with desired characteristics, such as high quality, high yield and/or disease resistance, are chosen from among the preserved genetic resources. The plants those meet the requirements in each cross combination are selected in the individual selection fields. Another method is direct selection from seedling nurseries which have great genetic variation. In both cases, the procedures after individual selection are similar. In Korea, the method that shortens the tea breeding period, as shown in Table 8.18, will be the method of choice until 15 tea cultivars are registered.

8.3.3 *Biotechnology of Tea Breeding*

When using a biotechnology approach in tea breeding, mass propagation is achieved by establishing multi-shoots and somatic embryos. Somatic embryos are induced from the embryogenic calli from buds and zygotic tea embryos, suitable for conservation and genetic transformation. The tissue culture procedures are as follows (Park, 2007): growth induction → cell suspension → bioreactor culture → maturation → somatic embryos → dormancy induction → encapsulation → direct seeding → gel suspension → fluid drilling → seed tape technology → transplants → transplanting (Fig. 8.6).

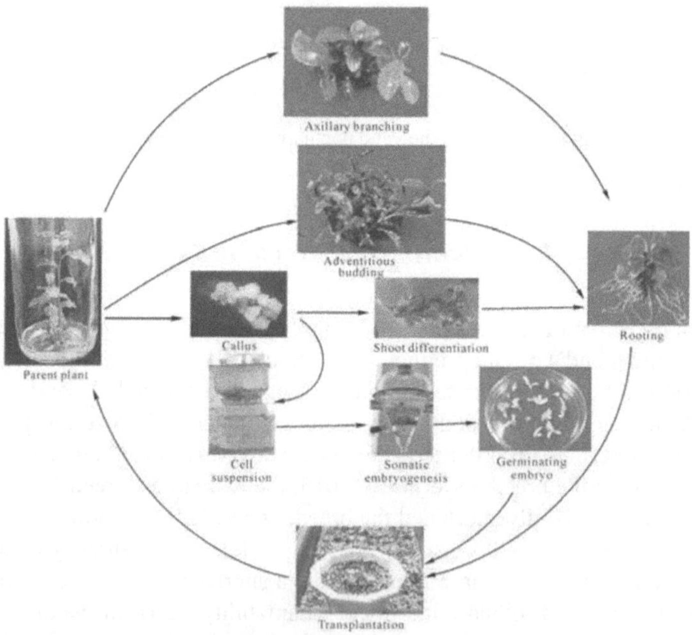

Fig. 8.6. Clonal propagation by plant tissue culture

8.3.3.1 Induced Multi-Shoots and *Somatic Embryogenesis by Tissue Culture*

For mass propagation of tea plants *in vitro*, the growth of calli and multi-shoots was investigated on 6 different growth mediums containing various phytohormones incubated at (25 ± 0.5) °C for 12 weeks. The WPM (Woody plant medium) and MS (Murashige and Skoog) medium with 5 μM of BA (Benzyl aminopurin), 2ip (isopentenyl adenine) and TDZ (Thidiazuron) showed good propagation of multi-shoots. Multi-shoot growth was the best on the WPM with 5 μM BA and 5 μM GA_3 (Gibberellic acid). Multi-shoots grown on the MS medium containing 30 μM IBA (Indolebutric acid) were found to have a balanced growth of roots and shoots. Multi-shoots grown on the MS mediums containing an IBA concentration higher than 30 μM showed unbalanced root growth. Multi-shoots were acclimatized to artificial soil after rooting on the MS medium containing 30 μM IBA. The multi-shoots were well rooted after 3 to 4 weeks after inoculation in the artificial soil (Park *et al.*, 1997; Son *et al.*, 2007) (Fig. 8.7).

Somatic embryos are very easy and convenient to use in plant breeding systems. Tea somatic embryos were established using zygotic embryos on the MS medium containing 5 to 20 μM cytokinin. The MS medium containing cytokinin higher than 20 μM resulted in a decreased number of somatic embryos. The best growth was obtained for somatic embryos grown on 1/2 MS and 1/4 MS containing 10 μM IBA. The somatic embryonic callus was established by culturing pre-embryos on the MS medium with high concentrations of auxin or cytokinin. The somatic embryos were divided from the somatic embryonic calli. Different concentrations of auxin and cytokinin were required to stimulate root or shoot growth (Lim *et al.*, 2008).

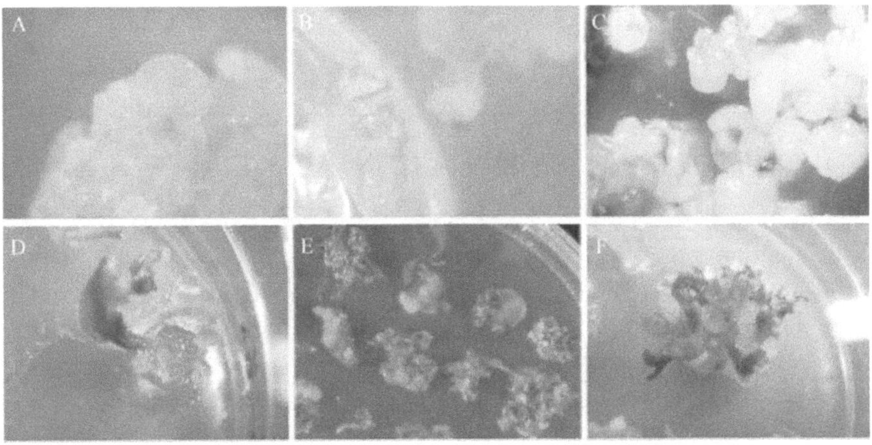

Fig. 8.7 Regeneration callus induction from zygotic embryos, somatic embryogenesis and micropropagation of the tea plant

A: callus induction, B: embryogenic callus, C: somatic embryos, D: germination of somatic embryos, E and F: micro-propagation

8.3.3.2 Gene Transformation of Tea Plants with Cold Tolerance Genes

Generally, tea breeding programs are focused on quality improvement rather than traits such as quantity, efficiency, functional ingredients and tolerance to stress (Fig. 8.8). A breeding program aimed at the improvement in quality can be complemented by adopting molecular technologies. By reviewing breeding history, recent advances in micropropagation and molecular assisted genetic transformations for feasible future tea breeding strategies can be proposed. Putative transgenic plants were obtained by introducing CLP (chitinase-like protein, an anti-freeze protein) genes into somatic embryos and/or axillary buds. The target gene was recombined into the pGA748 plasmid and then transferred into *Agrobacterium* (LBA 4404). Co-cultivation of the microorganism and explant ensured transformation of the somatic embryos. Using a selection medium containing 50 mg/L kanamycin, successfully transformed transgenic plants could be selected. Further experiments, such as PCR analysis using NPT- II (Neomycin phosphotransferase II) and CLP specific primers, hybridization in southern blots, and/or western blots, confirmed successful transformation of the target protein. Field test results showed that 6 of the transgenic plants were cold resistant. This study suggests that further research techniques are warranted for developing reliable systems/methods to breed cold tolerant plants molecularly (Lim *et al.*, 2008). The results obtained in this study demonstrated the potential of molecular breeding in tea.

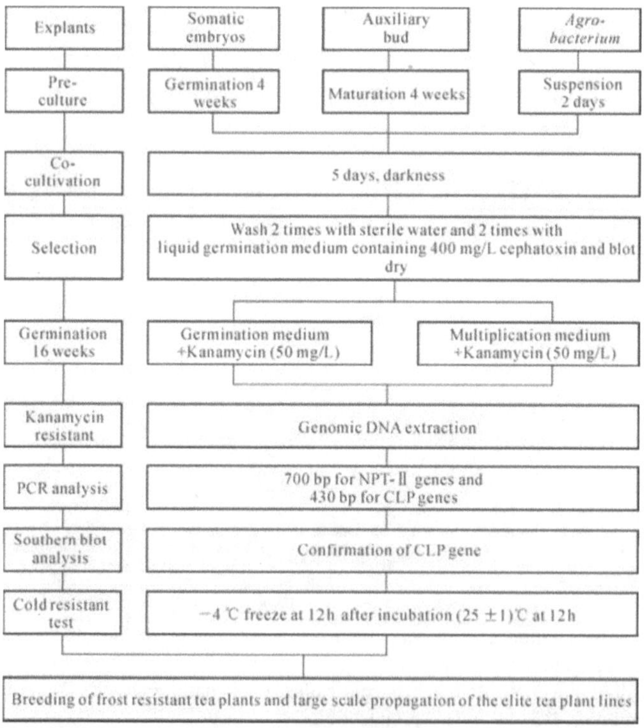

Fig. 8.8. A breeding scheme for tea plants using molecular approaches

8.4 Propagation Techniques of New Cultivars

The effective tunnel type cutting method developed in Japan was not available in Korea due to high temperature injury occurring in summer and cold injury occurring in winter during the nursery sapling stage. To address this problem, the "double shading and tunnel type-non watering two node cutting method" was developed and is now widely used in Korea (Moon *et al.*, 2008). High and low temperature injury is overcome by using a vinyl tunnel in winter and double shading over the vinyl tunnel in summer (Fig. 8.9). Using vinyl tunnels instead of vinyl house nurseries reduces the costs involved by half. For this method the suitable cutting time is the middle and the last 10 days of June. Rooting duration is 45 to 60 days.

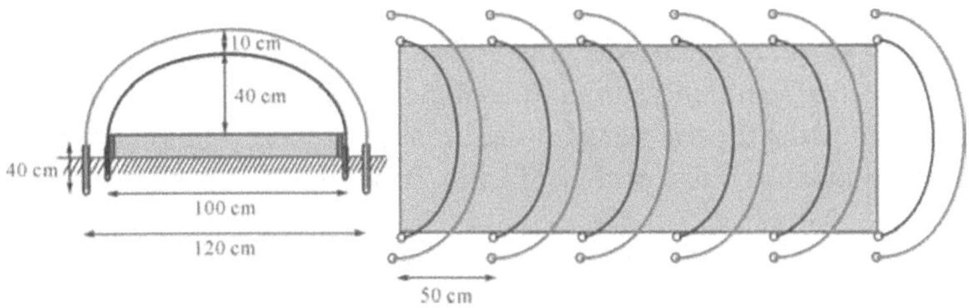

Fig. 8.9. Establishment of an iron pile and prop-stick for a sail in a tea plant cutting tunnel

The temperature of the double shaded cutting nursery during the rooting period remains between 23 to 32 °C (optimum for rooting) whereas the temperature of a single shaded cutting nursery reaches up to 40 °C. In winter time, no frost damage was observed in nursery plants growing in a double plastic tunnel in the Jeju Island region, whereas in the Muan region an intermediate degree of frost damage was observed in nursery plants growing in a single plastic tunnel (Fig. 8.10).

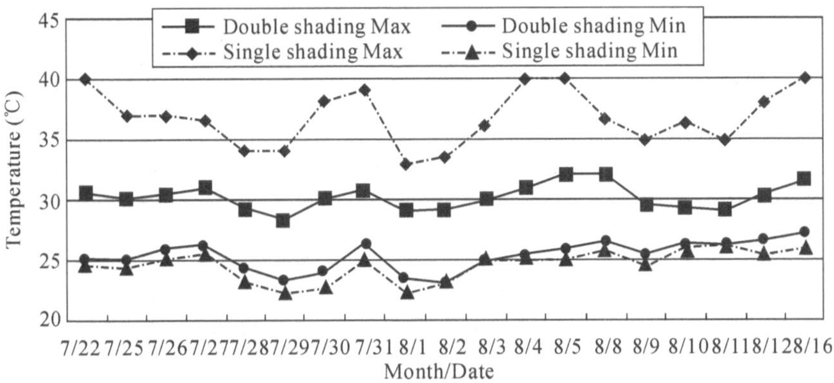

Fig. 8.10. Changes of temperature in sealing nursery by shading methods during the rooting period in tea plant cutting

This experiment was conducted to establish suitable index among the physiological properties of rooting media for tea plant cutting. In this experiment, the rooting media for tea plant cutting were clay (red soil), river sand, river sand and clay mixture (2:8, 5:5, 8:2, *v/v*), vermiculite + perite mixture (5:5, *v/v*) and sand loam. The separated sand ratio of clay which was measured with separating method was 51.1% and it increased in proportion to amount of river sand. With secondary regressive analysis, a suitable separated sand ratio of rooting media which can get more than 80% of rooting ratio were estimated to 63% – 74%, 64% – 78% and 63% – 79% in nursery, container and paper pot cutting, respectively. In secondary regressive analysis between separated sand ratio and osmotic coefficient, the osmotic coefficient were algebraically increased in proportion to separated sand ratio ($R^2 = 0.91$). The osmotic coefficient is a time to permeate media of water by osmotic pressure, and it increase according to the proportion to particle size of rooting media.

Although this R^2 value was lower than that of separated sand ratio and permeability coefficient, measuring of osmotic coefficient was suitable method to select rooting media for tea plant cutting because of their simple instruments and short measuring time (Moon *et al.*, 2007, Fig. 8.11).

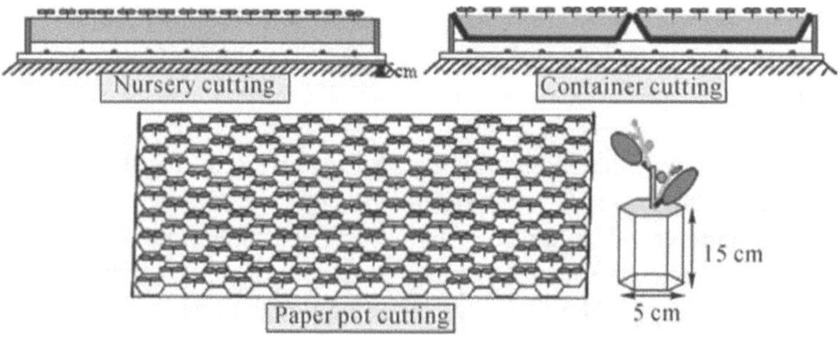

Fig. 8.11. Used cutting methods in tea plant cutting trial

8.5 Conclusions

Following conventional methods, tea breeding takes 24 years after fertilization. Since the Boseong Tea Experiment Station at was built in 1992, 7 clonal cultivars selected by the segregation breeding method have been registered. The Mokpo Experiment Station has established a strategy to shorten the breeding period of clonal cultivars from 24 years down to 10 to 13 years. As part of this strategy, regional adaptation tests are already being conducted successfully at 6 different locations across the country since 2006.

In order to improve the tea quality and yield, several farmers' old tea gardens cultivated with seedling populations have been replaced by imported clonal tea cultivars. Twenty percent of Korea's tea gardens are cultivated with clonal cultivars imported from Japan and China. The "airtight tunnel non-watering, two-node cutting" method was especially developed for the Korean climate by Moon *et al.* (2007). The suitable cutting time for this method is the middle and the last 10 days of June. The rooting duration is 45 to 60 days. Almost all Korean tea farmers have been trained to use this cutting method.

Some cultivars, germplasm and advanced techniques used by other countries should be introduced and adapted locally to solve some domestic problems. Through joint research with other countries more developed in tea plant breeding technology, we want to introduce cultivars for green tea or fermented tea and use them to develop new cultivars. In addition, we plan to introduce technologies for selecting cold tolerant and high functionality cultivars, optimal cultivars by tea type and optimal cultivars by use. According to an adage "a crisis is a chance" it is inevitable that we learn about the tea breeding technologies of other countries to strengthen the competitiveness of the Korean tea industry.

References

Banerjee B (1991) Tea: Production and Processing Dynamics. New Delhi: Oxford and IHB Publishing Co. Pvt. Ltd.

Choi HK (2000) Tea breeding and cultivation in Korea. Journal of Korean Tea Society, 6(2): 121-138.

Gong SY (2000) Processing and multiple application of tea in China. Journal of Korean Tea Society, 6(2): 79-96.

Han SK, Kim KS, Song YS, Moon YH, Jeong BC, Bang JK (2005) Difference of biochemical contents in conserve Korean tea and introduced tea cultivars. Journal of Korean Tea Society, 11(3): 56-68.

Jeong BC, Song YS, Moon YH, Han SK, Bang JK, Kim JW, Kim JH, Park YG (2005) Tea plant breeding plans for the tea industry in Korea. In: Proceedings of 2005 International Symposium on Innovation in Tea Science and Sustainable Development in Tea Industry. 11-15 November, Hangzhou, China, pp.322-332.

Jeong BC, Song YS, Moon YH, Han SK, Bang JK (2006) Shortening period of clonal tea breeding in Korea. Journal of Korean Tea Society, 12(3): 93-99.

Jeong BC, Song YS, Moon YH, Han SK, Bang JK (2007) Collection and multiplication of superior germplasm for development of the new tea tree cultivar. Joint Research Report for Tea Germplasm in Korea, pp.1-46.

Kaundun SS, Zhyvoloup A, Park YG (2000) Evaluation of the genetic diversity among elite tea (*Camellia sinensis* var. *sinensis*) accessions using RAPD markers. Euphytica, 115: 7-16

Kim JW (2008) Characteristics of Korean tea plant (*Camellia sinensis* (L.) O. Kuntze) and breeding of superior lines. PhD Thesis, Mokpo University, Korea. pp.1-65.

Kim YG (2000) The status and prospect of Korean green tea industry. Journal of

Korean Tea Society, 6(2): 41-64.

Korean Tea Production Association (KTPA) (2005 – 2008) Cultivating method of tea plant for future. Boseong County of Jeollanam-do.

Lim CS, Choi CH, Park YG (2008) Transformation of tea plant (*Camellia sinensis* (L.)) using cold tolerance related gene. Journal of Korean Tea Society, 14(3): 109-127.

Magambo MJS, Cannell MGR (1981) Dry matter production and partition in relation to yield of tea. Experimental Agriculture, 17: 33-38.

Moon YH, Song YS, Han SK, Jeong BC, Bang JK (2007) Selection of suitable rooting media by osmotic coefficient measurement for tea plant (*Camellia sinensis* (L.)) propagation. Journal of Korean Tea Society, 13(3): 149-158.

Moon YH, Song YS, Han SK, Jeong BC, Bang JK (2008) Optimum depth of rooting media and wintering method for double shading and tunnel type cutting nursery of tea plant (*Camellia sinensis* (L.)). Journal of Korean Tea Society, 14(2): 59-68.

Park YG (2007) Strategy of gene conservation of *Camellia sinensis* in Korea. Journal of Korean Tea Society, 13: 125-140.

Park YG, Son SS (1996) Forest genetic conservation based on *in vitro* culture systems. In: Park YG, Sakamoto S (eds.) Biodiversity and conservation of plant Gentic Resources in Asia. Tokyo: Japan Science Press, pp.157-174.

Park YG, Kim KW (2007) Cryopreservation of forest tree species in Korea. The Fifth International Conference "Propagation of Ornamental Plants". 5-8 September, 2007, Sofia, Bulgaria, p.25.

Park YG, Ahn IS, Bozhkov P (1997) Effect of exogenous plant growth regulators on morphogenetic response *in vitro* by embryo and leaf cultures of *Camellia sinensis* (L.) O. Kuntze. Korean Journal of Plant Tissue Culture, 24: 129-135.

Park YG, Kaundun SS, and Zhyvoloup A (2002) Use of the bulked genomic DNA-based RAPD methodology to assess the genetic diversity among abandoned Korean tea population. Genetic Resources and Crop Evolution, 49(2): 159-165.

Son YJ, Lim CS, Yang BH, Park YG (2007) Development of mass propagation techniques of *Camellia sinensis* through *in vitro* culture. Journal of Korean Tea Society, 13: 79-92.

Song YS, Moon YH, Han SK, Jeong BC, Bang JK (2005a) Morphological characteristics of progeny population in collected wild tea (*Camellia sinensis*). Journal of Korean Tea Society, 11(2): 93-105.

Song YS, Han SK, Moon YH, Jeong BC, Bang JK (2005b) Variation of total nitrogen and tannin contents in collected wild tea trees. Journal of Korean Tea Society, 11(3): 45-55.

Takeda Y (2000) History and development in Japanese tea breeding. Journal of Korean Tea Society, 6(2): 139-158.

9

Tea Germplasm and Improvement in Bangladesh

Abdul Qayyum Khan
Bangladesh Tea Research Institute, Srimangal-3210, Moulvibazar, Bangladesh
E-mail: khanbtri003@yahoo.com; scientist08@yahoo.com

Abstract: Tea was first introduced to the Halda Valley in Chittagong and Surma Valley of greater Sylhet in British India between 1840 and 1857. Tea management in Bangladesh can be distinguished in two broad categories: the sterling companies, the inland companies and proprietors. The sterling companies cover 39% of tea cultivated areas and their production share is 48% of the total. A recent survey reveals that the situation has been improved. The good middle, lower middle estate category managed 58.7% of total tea area, 21.0% being absolute clones plus 3.6% biclonal and polyclonal seedling teas, the rest being seedling teas of various sources. Bangladesh contributes only 1.2% in production and 0.5% in exports to the world tea trade. We should not be downcast, considering the production in the next two decades will include a small increment of 50 kilotonnes.

9.1 General Introduction

Tea is a widely consumed non-alcoholic beverage worldwide, consumed as a health drink. This will lead to wide popularity as the proceeds and consumption are likely to go around the world. Tea is grown in more than 50 countries. Bangladesh, located in the south of Asia, ranked as the 11th tea producer in 2008. Its tea acreage was 58,005 ha and produced 59,000 tonnes of made tea. Of this, 8,393 tonnes were exported, accounting for approximately 2.1% in tea acreage, 1.2% in production and 0.5% in exports to the world, respectively (FAO, 2010;

ITC, 2009). Although, our first increase in domestic consumption coincided with a small decrease in the price worldwide, we should not be downcast considering our long term objective is to double our production in another 2 decades with a small increment of 50 kilotonnes. It seems unlikely that this will lead to a surplus in production. Increasing productivity and producing more is essential for us.

The Bangladesh tea plantations predominantly consist of 3 basic seedling plant types, *Camellia sinensis* (L.) O. Kuntze var. *sinensis*, var. *assamica*, var. *assamica* ssp. *lasiocalyx* and their hybrids. In general, seedlings are inherently poor in yield and quality. A seedling plantation is arbitrarily classified as light-leaved var. *assamica*, dark-leaved var. *assamica* ssp. *lasiocalyx*, and var. *sinensis* or hybrid *jats*. It is generally believed that the light-leaved var. *assamica* type plant produces the best quality but is normally less hardy than the dark-leaved type. The var. *sinensis* type is considered to produce good flavors but there are too many variations within a *jat* and even within individual bushes regarding the yield, quality and resistance to pest and disease. A major portion of the seedling field consists of a relatively smaller number of tea bushes. In tea production, 10% of the bushes produce only 2% of the total crop and about 0.5% of the bushes produce high quality tea as a cash crop because of their superior genetic ability. And the crop yield of an individual bush in a field may vary up to 500% between the lowest and the highest yielding bushes (Wright, 1963). On average, one out of 40,000 bushes in a seedling population will be outstanding in yield and quality (Wight, 1958). All these facts led to the breeders concentrating on the selection and propagation of elite mother bushes having good yield and quality potential. Tea improvement in Bangladesh started with the establishment of the Bangladesh Tea Research Institute (BTRI) in 1957.

9.2 Germplasm Collection and Evaluation

Germplasm is the basis of plant improvement. The wide genetic reserves of tea provide the opportunities for evolving a wide range of superior cultivars with greater diversity. The tea germplasm is important not only because it supplies genes to modify and improve cultivars and hybrids, but also because in the germplasm the mutant genes that originated naturally have accumulated over hundreds of years. These natural resources have to be saved in order to develop better tea cultivars with high yield and quality potential. The genetic resources of tea consist of cultivated species/varieties, different plant types, old seed *jats*, local cultirars, improved clones and seeds, materials introduced from other countries. These genetic resources of tea are being lost and the base of genetic perversity of tea is being narrowed down due to uprooting of old seed plantations and clonal predominance. The available genetic diversity of tea needs to be conserved before becoming extinct. With this consideration, initiatives have been taken in the collection and preservation of diverse genetic materials of tea at BTRI, from home and abroad.

9.2.1 Local and Introduced Tea Genetic Materials at BTRI

The BTRI totally preserved 386 accessions of tea germplasm (Fig. 9.1), including 317 clones, 69 seed stocks; 328 local collections and 58 introduced accessions (Table 9.1).

Fig. 9.1. Tea germplasm garden in Bangladesh

9.2.2 Present Status of Seedling and Clonal Cultivars

Tea management in Bangladesh can be distinguished in two broad categories: the sterling companies, the inland companies and proprietors. Though the sterling companies covered 39% of tea cultivated areas, they shared about 48% of the total production in an earlier assessment (Alam, 1994). It was calculated that out of the total plantation, at least 90% were still seedlings, the remaining 10% was clonal garden largely planted with BTRI clones.

But a recent survey reveals that the situation has been improved. The good mid, lower mid-estate category accounted for 58.7% of the total tea areas, 21.0% being absolute clones plus 3.6% biclonal and polyclonal seedlings, the rest being seedling teas of various sources. The clonal ratio will increase to 35% – 40% by 2020.

Table 9.1 Tea germplasms preserved in the BTRI

Type		Origin	Accessions
Clones	Native	1. BTRI released clones	17
		2. BTRI non-released clones	107
		3. Collected from tea estates	96
		4. Local garden clones	5
		5. Seed *barie* collection	60
		Subtotal	285
	Introduced	1. Tocklai released clones	15
		2. North East Indian garden clones	9
		3. UPASI clones	4
		4. Sri Lankan clones	3
		5. Kenyan clones	1
		Subtotal	32
Seed stocks	Native	1. BTRI released biclonal stocks	4
		2. BTRI non released biclonal stocks	16
		3. Local garden polyclonal stocks	2
		4. Local garden seed stocks	17
		5. Other seed stocks	4
		Subtotal	43
	Introduced	1. Tocklai biclonal stock	4
		2. Sri Lankan polyclonal stock	1
		3. Other seed stocks	21
		Subtotal	26

Source: Botany Division, BTRI

9.2.3 Prominent Tea Cultivars in Bangladesh

The BTRI has released 17 clones, i.e., BT-1 to BT-17, and 4 biclonal seed stocks. The main characteristics can be found in Table 9.2.

Table 9.2 Yield and quality performance of Bangladesh released clones and seed stocks

Clone/Seedling	Source	Released year	Yield (kg/ha) Immature	Mature	Higher	Quality*	Category
BT-1 Clone	Baraoora	1966	1,614	3,298(14)	4,683	AA	Standard
BT-2 Clone	Rajghat	1975	1,820	3,627(14)	4,876	AAF	Standard
BT-3 Clone	Rajghat (Udnacherra)	1975	1,476	3,431(14)	4,504	AA	Standard
BT-4 Clone	Baraoora	1981	1,418	2,581(14)	3,757	E	Quality
BT-5 Clone	BT1×TV1	1987	2,083	2,811(7)	4,313	AA	Standard
BT-6 Clone	BT1×TV1	1987	2,189	2,916(7)	4,102	E	Quality
BT-7 Clone	Rajghat(Burmacherra)	1991	1,646	2,790(7)	4,004	AA	Standard
BT-8 Clone	Baraoora	1992	2,140	3,316(12)	5,410	E	Standard
BT-9 Clone	Daragaon	1994	2,773	3,784(7)	4,763	AA	Standard
BT-10 Clone	Doloi	1995	3,730	4,600(7)	5,136	E	Standard
BT-11 Clone	B207/39×TV1	1999	2,515	3,713(5)	5,179	AA	Standard
BT-12 Clone	Hatimara	2000	1,917	4,018(5)	5,209	AA	Standard
BT-13 Clone	Shamsernugger	2000	1,502	3,203(4)	3,870	AA	Standard
BT-14 Clone	BT2×TV9	2002	1,683	3,450(4)	4,051	AA	Standard
BT-15 Clone	BT1×TV1	2002	1,938	3,735(4)	4,830	AA	Quality
BT-16 Clone	Shamsernugger	2005	1,288	3,604(7)	4,231	AA	Standard
BT-17 Clone	BT2×TV1	2006	1,873	3,897(8)	5,238	AA	Standard
BT S1Seed	BT1×TV1	1985	1,586	3,217(2)	3,744	AA	—
BT S2 Seed	B207/39×TV1	1985	1,672	3,381(10)	3,860	AA	
BT S3 Seed	BT1×TV19	2001	1,811	3,381(2)	3,956	E	
BT S4 Seed	B207/39×TV19	2001	1,687	3,303(2)	3,890	AA	

Source: Botany Division, BTRI. The figure in parenthesis indicates number of cropping years taken for average calculation. *AA-Average, E-Excellent, F- Flavour

9.2.4 Stock of Tea in Cultivation in Bangladesh

Tea was first introduced in the Halda Valley in Chittagong and the Surma Valley of greater Sylhet in British India between 1840 and 1857, about the same time as in North East India, which includes Assam, Derjeeling, Terrai and Dooars. Since then, until the first quarter of the 20th century, there was collaboration in the tea plantations with those in North East India, because of intimate interaction and exchange of materials among the estates in an undivided tea industry. The tea stock was thus enriched with various types from different sources. The initial introduction from China of the var. *sinensis* hybrid type was soon discouraged and the indigenous var. *assamica* variety was introduced into the new plantations. Due to local adaptation and selection pressure, only dark green and medium to large leaf plants became predominant, light green and large leaf types were less preferred. Many seed orchards were established for seeds.

The var. *assamica* variety with intermediate leaf size and early flushing habit was the speciality of many seed *jats*. In Monipore Tea Estate, the first such standard seed source was established in Bangladesh by Stiefalhhgen in 1860 with seeds originally collected from Burma. Such Monipore stocks, e.g. Monipore,

Amo, Balisera, Mirtinga and Luskerpore *jats* were known in the tea world as early as 1918.

The evolution of the genetic stock of Bangladesh tea may be attributed to four major arrhythmic phases of stagnancy and the development of the industry.

(1) The 1950s decade of post partition stagnancy and slow growth.

(2) The 1960s decade of mandatory extension and development.

(3) The 1970s decade of Bangladesh post liberation, a period of planning and rebuilding activities.

(4) The 1980s and 1990s decades of utilization of research innovations in the development of the tea industry.

All situations affecting preservation prospects and the prosperity of genetically improved stocks were vitally important for the growth of the tea industry. In the above perspective, a contribution to the productivity of Bangladesh tea came from genetic materials using mostly inherited seeds and clones by way of garden selection, exotic introduction or development by BTRI.

9.3 Breeding and Selection Techniques

There is much scope for non-conventional approaches to tea improvement. But the potential of non-conventional methods is limited by lack of knowledge of some aspects of tea genetics. Nevertheless, there have been several advances in tea breeding and selection.

9.3.1 Selection Criteria for Seed Cultivars

The general approach in the selection process is evaluation of progenies raised from seeds produced by crossing or hybridizing between pairs of morphologically different bushes for vigor and uniformity. The highly heterogeneous F_1 seeds are evaluated for vigor, and the more vigorous progenies are re-evaluated at different stages of growth and productivity.

The selection and breeding of seed cultivars was a time consuming and laborious process as it often took about 25 years to develop improved cultivars.

In comparison with clones, the seed plants, due to their hybrid vigor and good root system, are in general more adaptable to divergent ecological and growing conditions. In seed grown tea, the existence of wide variations is well known to the seed breeders and this probably made them aware of the advantage of clonal propagation for establishing a large and uniform population. Clones are plants genetically identical to the parent, originating by vegetative propagation such as by cuttings or grafts taken from the tea bush. Clonal selection also involves considerable time and is a laborious process but a widely adapted method of plant improvement in tea. It usually takes 7 – 10 years before a clone can be released for commercial plantation (Bezbaruah, 1968).

9.3.2 Selection Criteria for Yield

In a bush in tea field, we look for certain yield contributing characteristics. The yield contributing components are plant type, branching habit, shape, size and pose of leaf density and distribution of plucking points, and number of buds per shoot. It was found that the plants produce shoots rapidly yielded about a 30% greater crop than those produce new growth slowly (Wellensick, 1933). The numbers of buds per pruning stick, evenness of flush and fresh weight of pruning are also considered to be important criteria contributing to the yield of a tea bush.

9.3.3 Selection Criteria for Quality

Quality is an inherent characteristics and an important criterion in selecting potential clones. Tea quality varies geographically, seasonally and between clones/seed *jats*. Pubescence has a significant correlation to the quality of orthodox tea (Wight, 1963). Quality is also associated with the greenness' of the leaf. Leaf color was shown to be correlated to the quality of processed tea (Wellensick, 1947). He observed that a better tea quality was found in light green leaves of the var. *assamica* type as compared to the dark green leaf type. Kanthamani (1969) further discovered that the leaf hairs contained large number of polyphenols and amino acids. A good number of organic acids present in the leaf portion were also found in the leaf hairs. Caffeine was found to be present to an appreciable extent. Besides this, a large number of carbonyl compounds important for the development of aroma were detected in the hairs and very few of these compounds were traced in the non-hairy portion. The leaf color was shown to be correlated to the quality of processed tea (Wellensick, 1947). The frequency of calcium oxalate crystals in the parenchymatous cells of the leaf petiole designated phloem index was a useful parameter indicating quality. Plants with a low phloem index are usually of lower quality value in comparison with those with a higher phloem index.

9.3.4 Mutation

Attempts at inducing mutation in tea seeds, cuttings, pollen grains with physical

(X-rays and γ-rays) and chemical mutagens (ethyl methane sulphonate) have not yielded the desired success in producing superior mutants. Mutants produced by these treatments have stunted growth, reduced vigor and less foliage and branches (Singh, 1984). The apparent genetic variation in the response of mutagens suggests that the technique has potential in tea breeding.

9.3.5 *Polyploidy*

Most of the cultivated tea is diploid ($2n = 30$). Natural triploid ($2n = 3x = 45$) resulting from open pollination has been reported from Japan and other countries. Triploid and tetraploid had a bigger leaf and heavier shoots but were poorer in cup quality than diploid. The density and morphology of sclereids and stomata proved to be useful indices in determining the ploidy level. The tetraploid shows higher density and size of celeries and has the largest stomata. Polyploids with favorable characteristics could be cloned and used in the improvement of the genetic stock.

9.3.6 *Tissue Culture*

The potential of tissue culture in various aspects of plant improvement has already been recognized and has attracted the attention of scientists. Rapid multiplication of propagation materials through the tissue culture, particularly in the initial selection, assumes an importance in plant improvement programs. Tissue culture studies in tea have been carried out at a number of centers in China, Japan, Sri Lanka, Thailand and the former USSR. Tissue culture study at BTRI was undertaken to develop suitable protocols for the development of homozygous diploid plants through anther culture. The method can also make a major impact on the industry in the area of rapid multiplication of a large number of clonal plants, regeneration of pest and disease free plants, production of pure breeding lines, storage and exchange of germplasm, inter-specific and inter-generic hybridization, development of useful polyploids and mutants and the improvement of yield and quality (Mohan & Newton, 1990).

Recently, BTRI has initiated research on tissue culture. An initial success in callus formation has been achieved, but full organogenesis remains to be achieved.

9.4 Conclusions

The development of 17 clones and 4 seed stocks has been the result of the exploitation of natural variability to a limited extent and through conventional breeding at the institute, yet little progress has been made in this direction due to the scarcity of a wide genetic base. It is necessary to build up a broad genetic base for tea through international collaboration and exchange. Moreover, some unexplored research areas need to be addressed to increase the relatively low harvest index of tea. We need to increase the genetic diversity of tea by inducing, recognizing and regenerating chromosomal changes through mutation, polyploidy, tissue culture and genetic engineering. Thus, a coordinated approach combining the conventional methods of selection and hybridization with non-conventional methods is of prime necessity for tea improvement.

Acknowledgements

This paper is supported by the Botany Division, BTRI.

References

Alam AFMB (1994) Influence of improved genetic material towards higherroductivity of tea in Bangladesh. Proceedings of the International Seminar on Integrated Crop Management in Tea towards Higher Productivity. 26-27 April, Colombo, Sri Lanka, pp.33-49.

Bezbaruah HP (1968) Genetic improvement of tea in North East India—Its problems and possibilities. Indian Journal of Genetics, 28A: 126-134.

Food and Agriculture Organization (FAO) (2010) http://faostat.fao.org.

International Tea Committee (ITC) (2009) Annual Bulletin of Statistics, London.

Mohan JS, Newton RJ(1990) Prospects of biotechnology for tea improvement. Proceedings of Indian National Science Academy, 56(5&6): 441-448.

Kanthamani S (1969) Tea clones and quality. UPASI Scientific Department Bulletin, 27: 35-37.

Singh ID (1984) Advances in tea breeding in North-East India. In: Iyer RD (ed.) Proceedings of PLACROSYM-V. CPCRI, Kasargod. 15-18 December, 1982, Indian Society of Plantation Crops, pp.88-106.

Wellensick SJ (1933) Floral biology and technique of crossing with tea. Arch Thicket, 12: 27-40.

Wellensick SJ (1947) Foundation of general plant breeding. Breeding of Tea. Tjeekwillink and znHarlum (2nd Edition), pp.305-347.

Wight W (1958) Theagrotype concept in tea taxonomy. Nature, 181: 893-895.

Wight W (1963) Improved method & clonal selection. Two and A Bud, 8(2): 3-5.

10

The Development of High Yielding Tea Clones to Increase Indonesian Tea Production

Bambang Sriyadi, Rohayati Suprihatini, Heri Syahrian Khomaeni

Indonesia Research Institute for Tea and Cinchona (RITC), Gambung Village, P.O. Box 1013, Bandung City 40010, West Java, Indonesia
E-mail: sriyadi_b@yahoo.co.id; rohayatis2000@yahoo.com; hskbreeder@gmail.com

Abstract: Generally, tea productivity in Indonesia is low and difficult to improve due to the fact that the majority of its plants are seedlings and of older age. Efficient plantation management requires replanting of older plants using clones with high yield potential. 3,500 to 4,200 kg/(ha·year) potentially yielding GMB 1-5 and, 4,000 to 5,500 kg/(ha·year) potentially yielding GMB 6-11 have been released as leading tea clones as substitutes for the seed-based older plants. However, the development of these clones has lagged behind due to badly-performing replanting programs implemented in almost all the plantations in Indonesia. Using these GMB series tea clones should potentially increase Indonesian tea plantations productivity to 3,000 kg/(ha·year).

10.1 General Introduction

Indonesia is an important tea producing country, ranking as No. 7 tea producer in the world (FAO, 2010). Its tea acreage was 127.4 kilohectares, its tea production was 136.5 kilotonnes, and the export was 92.3 kilotonnes in 2009, respectively (ITC, 2010).

10.1.1 The Development of the Indonesian Tea Industry

Tea is not an original Indonesian plant. It was introduced to Indonesia from Japan by the German, Andreas Cleyer, in 1684, and planted as an ornamental plant in Jakarta. In 1827, tea was succesfully planted on a bigger scale at the Cisurupan Experimental Garden, West Java, and from that moment tea plantations started to be developed in Java.

Rudolf Edward Kerkhoven brought a new *Camellia sinensis* var. *assamica* (Masters) Chang type of tea plant in 1877 to Indonesia (Java) from Sri Lanka (Ceylon) and planted it at the Gambung Estate, West Java. It nowadays is the Indonesian Research Institute for Tea and Cinchona Head Office. Since then, tea planting has developed rapidly in Indonesia. Now, tea plantation are not only found in West Java Province, but also in Central Java, East Java, Sumatera, and a small part of Sulawesi (Celebes) provinces (Table 10.1).

Table 10.1 Area and production by provinces and category of producers in Indonesia in 2008

Province	Smallholder		Government estate		Private estate	
	Area (ha)	Production (tonnes)	Area (ha)	Production (tonnes)	Area (ha)	Production (tonnes)
Nangroe Aceh D	0	0	4,781	4,333	0	0
North Sumatera	0	0	0	0	205	302
West Sumatera	3,035	2,333	577	1,388	40	77
Jambi	0	0	2,625	5,858	0	0
South Sumatera	0	0	1,470	2,377	0	0
Bengkulu	0	0	1,418	1,177	0	0
West Java	53,128	33,905	28,976	48,999	21,781	21,828
Banten	24	5	0	0	0	0
Central Java	5,213	4,444	1,550	2,323	3,605	3,585
Yogyakarta	192	255	0	0	0	0
East Java	56	30	1,345	2,388	419	645
Central Sulawesi	0	0	0	0	1,870	1,015
South Sulawesi	0	0	0	0	129	132

Most tea plantations in Indonesia are at a high elevation (Fig. 10.1), from 700 m above mean sea level until more than 2,000 m. The temperature range is from 13 – 25 °C with the relative humidity higher than 70%. The soil type consists mostly of Andisols and some areas are latosols and podzolik.

10.1.2 Tea Germplasm in Indonesia

Tea genetic resources in Indonesia were divided into seed propagation plants and clonal plants. Seed propagation plants consist of two groups, i.e., illegitimate plants (plant with vague parentage) and plants from seed garden with certain parentage. Clonal plants consist of the first generation of clones, the second generation of clones, and the expected excellent clones. The first generation of clones was bred by planters and some clones were introduced from other countries, such as TRI 777, TRI 2023, TRI 2024, and TRI 2025 from the Tea Research Institute of Sri Lanka, and some clones had been advised as "recomended clones" for planters on account of their qualities (Van der Knaap, 1955). A total of 600 clones of the first generation are collected and preserved *ex-situ* at Gambung, Pasir Sarongge, and Simalungun experimental garden.

Fig. 10.1. Tea field in Indonesia

10.1.3 The Development of Productivity

The average productivity of an Indonesian tea plantation is still low, 1,084 kg/(ha· year) in 2009. Smallholder tea plantations' productivity was the lowest, averaging 898 kg/(ha· year), while the State owned tea plantation companies production was 2,170 kg/(ha· year) and private owned tea plantation companies' was 1,212 kg/(ha· year), respectively.

One of the reasons for this low productivity is that the majority of these tea plantations are seed-based plants with generally lower yield potential and older age (Sukasman, 1996). According to Hadfield (1971), in the future the seed-based, older tea plantations would not be profitable and should be rehabilitated immediately through rejuvenation accompanied by infilling or replanting with related environmental and soil improvement to increase their productivity.

The major problem with plantation rehabilitation is the loss in production during the growth periods. The magnitude of these losses would depend heavily on the rehabilitation methods selected, soil fertility, the quality of plant materials used (Sukasman, 1996). Production losses inherent in juvenility based rehabilitation are generally the result of lagging plant growth in the first year (Khandiah & Wimeladharma, 1980), weeds between plants, and high mortality rates (Tolhurst, 1956). Replanting is one of the rehabilitation methods requiring expensive investment, but is generally better than the rejuvenation method (Mwaka, 1983), owing to the fact that the latter, on average, would not achieve the intended production increase. For example, the tea plantations rejuvenation program in India was only able to increase production from 1,000 kg/(ha· year) in 1950 to 1,380 kg/(ha·year) in 1970 and reached the highest increase of up to 1,926 kg/(ha· year) in 1979 (Copanna, 1984), while replanting using tea clone in less fertile soils resulted in higher yields than the previous plants (Winddows, 1953; Rashid, 1987).

The contemporary tea industry in Indonesia is facing unfavorable circumstances. Tea prices tend to be decreasing and the production costs tend to be increasing from year to year. The price of made tea is 1.00 – 1.70 US$/kg. The production costs tend to be increasing with the increase in the regional minimum wages (RMW), fossil based fuels (FBF), electricity's base rate (EBR), and production facilities such as fertilizers, and so on. On the other hand, the Indonesian tea industry could not determine the product's selling price so that the only effort it could pursue was to cut production costs. Production cost cutting could be achieved through the efficiency of the production processes and the improvement in plantation productivity. To improve productivity, one of the efforts that could be pursued was the improvement of cultivation techniques and the substituting of superior plants with greater genetic potential than the previously inferior ones, through a replanting program. Through such a program, the plants' genetic potential would not only improve, but the population per area unit would be enhanced as well so that plant maintenance would be easier. Kartawijaya (1995) reported that clonal plants' productivity increased significantly, and found that the higher the clonal plants percentage, the better the plants' productivity improvement would be (Bezbaruah, 1984; Sriyadi & Astika, 1993; Wachira & Njuguna, 1994). The replanting program aims at high potential yield with superior tea clones, disease-resistant and with good material qualities.

The Indonesian Research Institute for Tea and Cinchona has discovered superior tea clones, GMB (Gambung) 1, GMB 2, GMB 3, GMB 4, and GMB 5, which had been formally released by the Minister of Agriculture of the Republic of Indonesia under decrees Nos. 260, 264, 265, 266, 267/KPTS/KB.230/4/1988 on 21 April 1988, respectively. These clones have 4,000 kg/(ha· year) potential yields, with disease resistance, and better initial growth in medium and high grown areas (Astika *et al.*, 1990). Furthermore, the Minister of Agriculture of the Republic of Indonesia further released 6 superior tea clones on 9 October 1998 under his decrees Nos. 684, 684a, 684b, 684c, 684d and 684e, labeled as GMB (Gambung) 6, GMB 7, GMB 8, GMB 9, GMB 10 and GMB 11, respectively (Astika *et al.*, 1999). These clones have almost similar characteristics to the GMB 1 to 5 clones, except that the latter clones have a higher potential yield. Some of them could achieve 5,800 kg/(ha· year) potential yield. In fact, the development of these GMB series were lagging behind owing to the nonexistence of a continuous replanting program in major private and state-owned tea plantations. To attract attention and give assurance to tea farmers so that they will be willing to implement replanting programs, dissemination of information on the potential of these GMB series clones' for improving tea productivity should be carried out.

10.2 Improving Tea Estates Productivity

The conventional method of plant breeding is still used. The selection of mother bushes for the seed propagation plant population is still ongoing. However, now the breeder is doing more intensive controlled pollination between clones that have good characteristic to achieve the objective with positive mass selection and multilocation testing. Generally, the selection objectives involve obtaining new elite tea clones with high productivity, good quality, good resistance to disease and pests, and good adaptability in extreme growing conditions. Biotechnology is still rarely applied in tea breeding in Indonesia; however some breeders have applied this for some specific purposes.

After the use of vegetative-based tea cloning in 1968, Indonesian tea plantations productivity has improved significantly among smallholder tea plantations, major private owned plantations and state owned tea plantations over a 10-year period (Table 10.2). These productivity improvements were supported by research results from tea cultivation systems such as proper fertilization, pruning, plucking and controlling of pests and disease and recommendations for the use of high potential yield tea clones.

Table 10.2 Productivity of types of tea plantation in Indonesia (kg made tea/ (ha· year))

Year	Smallholder	Private estate	Government estate	National average
1970	397	379	860	556
1980	495	673	1,685	972
1990	612	1,029	1,926	1,176
2000	588	921	1,900	1,138
2009	898	1,212	2,170	1,084

Plantation productivity in 1970 was still low, i.e., 556 kg/(ha· year) of made tea was the national average. At that time, the Indonesia Research Institute for Tea and Cinchona issued its recommended tea clones with their high potential yields (Cup *et al.*, 1971; Danimihardja & Suprapto, 1974; Astika & Muchtar, 1976, 1978) so that in developing these recommended tea clones, replanting activities substituted seed-based tea plants that in turn led to a two-fold increase in plantation productivity by 1978. Tea clones recommended during that time resulted from selection of mother bushes at several plantations. The clones such as PS 1, Mal 2, Cin 143, SA 35, Kiara 8, PG 18, Skm 118, KP 4, and introduced clones TRI 2024 and TRI 2025 with their approximately 2,500 kg/(ha· year) potential yield in large areas and at various elevation.

In the 1990s, the productivity of smallholder's and privately owned tea plantation companies' was bolstered two-fold, while the state-owned tea plantation companies improved by 50% from their 1980s productivity. These improvements were strongly associated with replanting activities performed in several plantations and the release of superior tea clones, GMB 1, GMB 2, GMB 3, GMB 4, and GMB 5 with their related development in some estates.

While tea plants are long-lived perennial plants, their productivity will general be at a maximum between 20 – 30 years of age and will gradually deteriorate until they are 50 years old (Gazi, 1978). This implies 50 years of useful economic life for tea plants when they should be replaced by new tea clones with higher potential yields. The actual practice in tea plantations involves replanting 2% of the total area every year so that the production potential will improve through the gradual introduction of superior tea clones.

10.3 GMB Series Tea Clones

Since the first release of GMB 1 to 5 clones in 1988, there are now 11 GBM series clones released in total. They are high yield, high quality and resistant to blister blight.

10.3.1 The Breeding of GMB Clones

GMB series tea clones are the second generation of tea clones; because they resulted from the selection of F_1 plants generated from hybridization of the oldest, first generation clones, i.e., Cin 143, GP 3, GP 8, KP 4, Mal 2, Mal 4, Mal 15, PS 1, Kiara 8, and PS 324. Their hand pollination had been performed in 1972 (Astika *et al.*, 1978). By 1974, F_1 plants from 47 cross-breeding combinations had been planted on the actual fields. After plucking-area formation, observation of tea leaf yields were performed in 1977 to select tea lines with higher potential yields, resulting in 20 tea lines that have subsequently been cloned vegetatively. In 1985, multi-location tests were done in 12 Indonesian plantations. Observations on potential yields, potential qualities, adaptability, and blister blight resistance resulted in GMB 1 to 11. The GMB 1 to 5 were released in 1988 due to their higher potential yields, and started to be harvested when 18 months old (Astika *et al.*, 1996). The GMB 6 to 11 clones were released in 1998 due to their higher potential yields, better qualities, and blister blight resistance. Observations on potential yields of GMB 1 to 11 clones, relative to TRI 2025 clones, in 3 tested locations, their blister blight resistance, and their catechins content are listed in Table 10.3. Potential qualities are indicated by their catechins content. Table 10.4 shows the potential qualities of sensor evaluation results over the superior tea clones of GMBs (Sriyadi *et al.*, 1993). The observations regarding potential qualities used 2 professional tea testers to evaluate color, taste, aroma, and infused leaf parameters. The scoring level for color, aroma and infused leaf are on a 1 – 5 scale, with the meaning 1 = worst, 2 = bad, 3 = average, 4 = good, and 5 = best. For taste, the scoring level is 20 – 50, with 20 – 29 = weak and not fresh, 30 –

39 = strong but not so fresh, and 40 – 50 = stronger and fresh.

Table 10.3 Potential yields of GMB 1 to 11 clone series (their blister blight resistance and catechins content)

Clones	Crossing parentage	Potential yield (kg/(ha· year))	Blister blight resistance	Catechins content (%)
GMB 1	KP 4×PS 1	4,021	Resistant	16.7
GMB 2	PS 1×KP 4	4,023	Resistant	16.0
GMB 3	Cin 143×PS 1	4,247	Resistant	14.6
GMB 4	Mal 2×PS 1	3,464	Resistant	17.1
GMB 5	Mal 2×PS 1	3,527	Resistant	15.5
GMB 6	PS 324×PS 1	4,400	Resistant	16.0
GMB 7	Mal 2×PS 1	5,800	Resistant	15.9
GMB 8	PS 324×PS 1	4,200	Resistant	14.9
GMB 9	GP 3×PS 1	4,700	Resistant	17.0
GMB 10	Mal 2×PS 1	4,800	Resistant	16.8
GMB 11	Mal 2×PS 1	5,500	Resistant	13.9
TRI 2025	Introduced from Sri Lanka	2,800	Resistant	15.7

Table 10.4 Potential qualities of GMB 6 to 11 (their color, taste, aroma and infused leaf)

Clones	Parameters				Total score
	Color	Taste	Aroma	Infused leaf	
GMB 6	4.66	41.00	3.33	4.00	52.99
GMB 7	4.66	41.00	3.33	4.00	52.99
GMB 8	5.00	40.33	3.33	4.00	52.66
GMB 9	5.00	45.00	4.33	4.00	58.33
GMB 10	5.00	42.33	3.00	4.00	54.33
GMB 11	5.00	42.33	3.33	4.00	54.66
TRI 2025	4.33	41.66	3.00	3.66	52.65

Notes: Color, aroma and infused leaf scores are on a 1 – 5 scale. The range of taste scores is 20 – 50

10.3.2 The Discriminatory Characteristics of GMB Clones

The GMB series clones have a common ancestor in PS 1. Five of the 11 GMB series clones, i.e., GMB 4, GMB 5, GMB 7, GMB 10, and GMB 11 have the closest relationships due to hand pollination of Mal 2×PS 1, so that they have many similarities that could lead to difficulty in identifying the exact clones. Field experiences led to the conclusion that there are some morphological characteristics that could be used in identifying the exact GMBs, namely:

GMB 1 clone is selected from F_1 of hand pollination between KP 4×PS 1, with typical characteristics of yellowish green, large and wide leaves, clear wavy leaf surfaces containing wax, long internodes, a high percentage of buds, good branching, high rate after pruning growth, and blister blight disease resistance.

GMB 2 clone is selected from F_1 of PS 1×KP 4 having characteristics that are almost similar to the GMB 1 clone except that it has round, large leaf shapes and a very convex leaf surface.

GMB 3 clone is selected from F_1 vegetative reproduction from Cin 143×PS 1 crossbreeding, with typical characteristics of light green leaves, slightly convex leaf surfaces, longitudinal leaf shapes with conical tip, an erect leaf position, medium internodes, disease resistant, but with a hard trunk so that it is more difficult to prune and less resistant to orange insects.

GMB 4 clone is selected from F_1 of Mal 2×PS 1 hand pollination, with typical characteristics of light green buds, blurred older leaves, slightly plain leaf surfaces, semi-erect leaf position, good branching but less resistant to orange mite.

GMB 5 clone is selected from the population similar to GMB 4 clone's, with typical characteristics more convex than GMB 4's, yellowish green leaves, light yellow on the middle part of its leaves. This clone has the most defects among other GMB clones such as having, high *banji* (dormant shoot) percentage, less disease resistance, high rate of dieback, it is non-responsive to pruning, and difficult to prune owing to its hard branches.

GMB 6 clone is selected from F_1 of PS 324×PS 1 hand pollination, with typical characteristics of light green leaves, longitudinal and very conical leaf shapes, medium internodes, and with high bud growth. It has high rate after pruning growth, is easy to pluck but has less disease resistance.

GMB 7 clone is selected from F_1 of Mal 2×PS 1 hand pollination, like GMB 4 and GMB 5. This is the best GMB clone among the other GMB clones owing to high potential yields (5,800 kg/(ha· year)) (Fig. 10.2). The typical characteristics of the GMB 7 clone are light green leaves, thick wax on leaf surfaces that make them sparkling, slightly convex leaf shapes, medium internodes, a semi-erect leaf position and very good branching, in addition to a high percentage of buds. It is easy to prune, and there is rapid younger leaf growth after pruning. It has disease resistance and draught resistance (Fig. 10.2).

Fig. 10.2. Tea clone GMB 7

GMB 8 clone is selected from F_1 of the same hand pollination of GMB 6 so that it has similar leaves and leaf tips, i.e., light green and conical, except that it has wavy leaves, a semi-erect leaf position, and bottom to top branching. It has a high percentage of bud growth, good branching that enhances pruning, and is easy to pluck.

GMB 9 clone is selected from F_1 of GP 3×PS 1 hand pollination, so that it is the hybrid approaching a var. *sinensis* type with very good quality, like its GP 3 parent. It has slightly small buds, in large quantities and a high growth rate. It has slightly small, oval leaves, a violet leaf tip, a semi-erect leaf position, and small branches so that it would be more difficult to prune.

GMB 10 clone has typical characteristics of light green leaves with very convex leaf surfaces, floppy leaves, and long internodes. It has very good branching so that it is easy to prune and has rapid after pruning leaf growth.

GMB 11 clone has typical characteristics of dark green leaves with a high percentage of bud growth, and long internodes so that it is easy to pluck. It has wavy leaves with wax layers. It has black nodes on its older leaves with a semi-erect leaf position. It has a defect, in that it is less resistant to orange mite pests during dry seasons.

10.4 Concludsions

The GMB clone series consists of superior new clones resulting from plant preservation that requires a long period of time, labor and high costs. It has been felt that the development of these clones was too slow due to the fact that replanting programs have not been declared compulsory activities among tea plantations in Indonesia. Consequently, plantation productivity will be difficult to enhance. Introducing continuous replanting programs every year through the use of these GMB series clones would bolster the productivity of the tea plantations in Indonesia to 3,000 kg/(ha· year).

References

Astika W, Muchtar D (1976) Anjuran bahan tanaman teh tahun 1976. Warta BPTK, 2(3/4): 297-306.

Astika W, Danimihardja S, Muchtar D (1978) Persilangan buatan pada tanaman teh. Warta BPTK, 2(3/4): 273-288.

Astika W, Muchtar D (1978) Anjuran bahan tanaman teh tahun 1978. Warta BPTK, 4(3/4): 296-297.

Astika W, Muchtar D, Sutrisno (1990) Pelepasan klon unggul teh. Warta TEH DAN KINA, 1(1): 20-22.

Astika W, Muchtar D, Sutrisno (1996) Klon-klon teh baru yang dilepas oleh Balai

Penelitian Teh dan Kina Gambung. Warta TEH DAN KINA, 7(1/2): 6-15.

Astika W, Muchtar D, Danimihardja S, Sriyadi B, Sutrisno (1999) Pelepasan klon teh seri PPS 1, PPS 2, MPS 5, MPS 6, MPS 7, dan GPPS 1. Prosiding Pertemuan Teknis Teh Nasional 1999, Bandung, 8-9 November, 1999, pp.34-42.

Bezbaruah HP (1984) A revised method for selection of vegetative clones. Two and A Bud, 3(1): 13-16.

Copanna MA (1984) Factors responsible for increased productivity in high range. RITC Weekly Seminar, Gambung, 24 September, 1984, p.9.

Cup GA, Madjid A, Schoorel AF (1971) Anjuran bibit tanaman teh tahun 1971. Sidang Komisi Teknis Perkebunan III.

Danimihardja S, Suprapto AM (1974) Anjuran bahan tanaman teh tahun 1974. Menara Perk, 42(2/3): 93-97.

Food and Agriaclture Organization (FAO) http://faostat.fao.org/.

Gazi MS (1978) Distribution pattern of yield and vacancy of tea in Bangladesh. Tea Journal of Bangladesh, 14(2): 19-22.

Hadfield D (1971) The challenge of old age. Two and A Bud, 18(2): 22-28.

International Tea Committee (ITC) (2010) Annual Bulletin of Statistics, London.

Kartawijaya W (1995) Peranan tanaman klonal dalam peningkatan produktivitas kebun teh. Warta TEH DAN KINA, 6(3/4): 74-80.

Khandiah S, Wimeladharma S (1980) Studies on the physiology of pruning 3: The implication of removing ageing tea field by rejuvenation pruning and infilling. Tea Quarterly, 42(2): 13-19.

Mwaka E (1983) Rehabilitation of moribund tea plants. Tea, 10(2): 124-133.

Rashid A (1978) Improvement of tea in Bangladesh. Tea Journal of Bangladesh, 14(1): 18-22.

Sriyadi B, Astika W (1993) Perbandingan hasil bahan tanaman teh asal biji, klon assamica, dan klon sinensis. Jurnal Penelitian Teh dan Kina, 2(1-3): 41-45.

Sriyadi B, Astika W, Sutrisno (1993) Potensi kualitas-dalam beberapa klon teh anjuran. Jurnal Penelitian Teh dan Kina, 2(1-3): 29-36.

Sukasman (1996) Rehabilitasi kebun teh tua dan permasalahannya. Warta TEH DAN KINA, 7(1/2): 29-39.

Tolhurst JAH (1956) Ideas on the experimental replanting of tea. Tea Quarterly, 27: 60-66.

Van der Knaap WP (1955) Results of clones Tea Experiments at Pasir Sarongge. Archives Teecultivated, 19(2): 47-98.

Wachira FN, Njuguna CK (1994) Clonal yield performance of some cambod teas *Camellia sinensis* var. *assamica* subsp. *lasiocalyx*. Tea, 15(2): 70-73.

Widdows JO (1953) Vegetative propagation in relation to the replanting of poor yielding areas. Tea Quarterly, 24: 43-45.

11

The Tea Industry and Improvements in Turkey

Sezai Ercisli

Department of Horticulture, Faculty of Agriculture, Ataturk University 25240 Erzurum, Turkey
E-mail: sercisli@gmail.com

Abstract: The aim of this chapter is to give detailed information about the introduction, history, culture, production, achievements, etc. of tea in Turkey. The first attempt to introduce tea to Turkey at the turn of the 19th century from China was unsuccessful because the wrong location was selected. However, later attempts from 1924 to 1937 with imported seeds from Georgia within the former USSR succeeded. A tea law was completed by the government in 1940 and until today tea farmers have been supported by the government. In 1947, the first tea factory with 60 tonnes/day capacity was built in Rize, capital of the tea industry in Turkey. The tea industry and growing areas developed and expanded very fast between 1950 and 1960, and in 1965 the tea production in Turkey reached a self-sufficient level. The country is now one of the most important tea producers in the world in terms of the production. Currently, all tea plantations in Turkey are established by seedlings and show huge heterogeneity. Around 1970s, some selection studies were carried out on these seedlings to obtain promising clones and some clones were created and released but not widely commercialized. The tea industry in Turkey solely consists of black tea production but more recently product diversification (green tea, mixed with some fruits) has been initiated. In addition, tea improvement activities which concentrate on clonal selection to decrease seedling populations have also been initiated again.

11.1 General Introduction to the Turkish Tea Industry

Tea belongs to the family Theaceace, genus *Camellia* L., section *Thea* (L.) Dyer and usually involves one species, including two or three botanical varieties, i.e., *C. sinensis* (L.) O. Kuntze, *C. sinensis* var. *assamica* (Masters) Kitamura, *C. sinensis* var. *pubilimba* Chang (Chen *et al.*, 2005). Individuals in most countries around the world drink tea; tea-drinking has attained ceremonial status in many places as both a social and medicinal beverage.

11.1.1 Location Map of Main Tea Growing Areas

Tea is grown in very restricted areas in Turkey. The main tea growing areas are Rize, Artvin and Trabzon, as shown in Fig. 11.1.

Fig. 11.1. Main tea growing areas (Trabzon, Rize, Artvin) in Turkey

11.1.2 Historical Introduction to the Tea Industry

The first introduction of tea to Turkey was made from China in 1888 during the time of the Ottoman Empire. Both seeds and nursery materials of tea plants had been brought from China to the Marmara region in Turkey and the first tea plantation was established near Bursa city. This first attempt was unsuccessful and after 4 years a second introduction was made from China again. The second attempt was also unsuccessful because of unfavorable climatic conditions in the Marmara region for tea production. In 1917, a team from Halkali Agriculture School from Istanbul visited the tea growing area of Batumi and the Caucasian region in Georgia (formerly in the USSR) and prepared a report for the Turkish government implying that the Black Sea region in Turkey might possess favorable conditions for tea growing because it was close to the Batumi region of Georgia which had favorable conditions. This report was seriously considered by the

Turkish government because at that time there was a serious economic crisis in Turkey, particularly in the Black Sea region, just after the First World War. During that time, most of the people in the region were unemployed and were searching for work. The number of people migrating from the region to Istanbul was high. Thus, the Turkish government decided to establish experimental tea plantations with seed materials introduced from Georgia in the Eastern Black Sea regions in 1924, in the early years of the Republic, as an alternative income source for unemployed people. From 1924 to 1937 the first adaptation study was successful and 30 tonnes of tea seeds in 1937 and 40 tonnes in 1940 were imported from Georgia to establish big scale tea plantations in the Eastern Black Sea region. The government enacted tea legislation in 1960 and from that date the tea farmers have been supported by the government mainly through a support price provided by the tea factories owned by the State. Support prices were determined by the government annually taking into consideration the supply and demand balance in the international tea market, the inflation rate and production costs. In 1947, the first tea factory with 60 tonnes/day capacity was built in Rize. The tea industry and growing areas in Turkey were developed and expanded very quickly between 1950 and 1960. Tea production in Turkey reached a self-sufficient level in 1965 and the Turkish government stopped importing tea from the other tea producing countries. The tasks of buying, processing and selling tea were conducted by the Tekel (Monopoly of the State) General Directorate until 1971. These responsibilities were transferred to the state-owned tea corporation (Caykur) and in 1984 the monopoly on tea was lifted and the private sector was given the right to operate in the tea industry. Currently, Turkey is one of the important tea producing countries. Furthermore, Turkey is one of the few tea producing countries that does not use pesticides in its tea plantations (FAO, 2005; Mendilcioglu, 2000) (Figs. 11.2 and 11.3).

Fig. 11.2. A typical tea plantation in Turkey

Fig. 11.3. Tea plantation in Turkey

11.1.3 Global Standing of the Turkish Tea Industry

The total tea plantation areas of Turkey are 76,000 ha and 65% of these plantations are located in the Rize region, 21% in the Trabzon region, 11% in the Artvin region and the remaining 3% in other areas (Ordu, Giresun) in the Black Sea region in Turkey. The size of tea plantations in Turkey is as follows:

80% is between	0.05 – 0.5 ha
17% is between	0.6 – 1.0 ha
2% is between	1.1 – 1.5 ha
1% is	>1.6 ha

Therefore, it can be concluded that 97% of tea plantations in Turkey are family businesses. Tea is an economically valuable plant for the Eastern Black Sea region in Turkey and a means of subsistence for more than 200,000 farmers. Currently, Turkey is one of the important tea producing countries, ranking 5th in the world after China, India, Kenya and Sri Lanka with an annual 209,000 tonnes of dry tea production in 2009 (FAO, 2010). As mentioned before, Turkey does not import tea from other countries. The export level is 25,000 – 30,000 tonnes per year. The age of the tea industry in Turkey is about 50 – 60 years.

The production trends for dry tea of Caykur and the private sector are shown in Table 11.1. Currently the governmental organization (Caykur) has a higher amount of dry tea production than the private sector. In fact, Caykur has fewer fresh tea processing factories (50) in the Black Sea regions compared to the private sector (230). The fresh tea processing capacity of Caykur is 6,720 tonnes per day and the private sector has 8,746 tonnes per day. It can be concluded that the private sector cannot compete with the governmental organization, namely Caykur. There are several reasons which explain this situation but the most important reason is familiarity with Caykur among tea farmers for so long a time.

Table 11.1 Government organization (Caykur) and private sector dry tea production in Turkey (tonnes)

Years	1985	1990	1995	2000	2005	2006	2007	2008*	2009
Caykur	133,000	96,000	83,000	91,000	109,000	115,000	122,000	124,000	112,000
Private	5,000	38,000	83,000	54,000	106,000	89,000	86,000	89,000	97,000
Total	138,000	134,000	166,000	145,000	215,000	204,000	208,000	213,000	209,000

Source: State Statistical Institute of Turkey; *http://faostat.fao.org/faostat/

11.1.4 Type of Tea Produced and Technologies Used

Turkey is producing fully oxidized tea (black tea) (Anon, 2007). However, more recently, some growers started to produce green tea. On the other hand, Oolong tea is not familiar in Turkey. The modified orthodox method is commonly used in black tea processing in Turkey. The processing steps of the orthodox method are:

Withering: is the process by which the 70% – 80% water content is reduced to 50% – 55% in special baths.

Rolling: is the operation in which the cell extract of the withered tea leaf is spread over the rolled leaf surface, the fresh tea leaves are cut, ground and rolled in various kinds of tea manufacturing machinery and the oxidation process begins.

Fermentation: is the event whereby the black tea acquires the desired color, acridity, brightness, odor and aroma by changing the biological structure of the chemical compounds existing in the cell extract of the rolled fresh tea leaf as a result of the effect of oxidizing enzymes.

Drying: is the process by which the humidity level of tea leaves rolled and fermented in the drying furnaces is reduced to 3% – 4% by stopping the oxidation, so that the tea is rendered storable and packageable.

Sorting: is the process by which the dry teas coming out of the furnace are sorted according to thinness, thickness and quality by screening them through the standard mesh wires.

Turkish people have their own way of making and drinking black tea (çay in Turkish), which has become a way of life in Turkish culture. Wherever you go in Turkey, tea is offered as a sign of friendship and hospitality, anywhere and at any time, before or after any meal. Turkish tea is full-flavored and strong to be served in little tulip-shaped glasses which you have to hold by the rim to save your fingertips from burning, because it's served boiling hot (Fig. 11.4).

Fig. 11.4. Tulip-shaped tea glass in Turkey

You can add sugar to it but no milk, and you can have it either lighter (weaker) or darker (stronger) depending on your taste because Turkish tea is made by pouring some very strong tea into the glass, then adding boiling water to the desired strength. Serious tea-drinking people usually go to a coffee or tea house where they serve it from a samovar (Semaver in Turkish) so they can refill their glasses themselves as much as they want (Fig. 11.5).

Fig. 11.5. Traditional Turkish tea house including samovar

Production of Turkish tea is carried out during a 6-month season between May and October, which offers the best climate.

11.1.5 Climate, Monthly Data on Rainfall and Temperature

The long term monthly climatic data for the main tea growing areas (Rize, Trabzon and Artvin) in Turkey are given in Tables 11.2, 11.3 and 11.4. The Rize region had the highest number of rainy days within a month, followed by Trabzon and Artvin. Among these three regions, Artvin is the coldest one.

Table 11.2 The long term climatic data of the Rize region (from 1975 to 2006)

	Jan.	Feb.	Mar.	Apr.	May	Jun.	Jul.	Aug.	Sep.	Oct.	Nov.	Dec.
Average temperature (°C)	6.5	6.2	7.7	11.8	15.9	20.2	22.8	23.0	19.8	15.8	11.4	8.2
Average highest temperature (°C)	10.5	10.3	11.7	15.7	19.2	23.6	26.1	26.5	24.0	20.1	15.9	12.4
Average lowest temperature (°C)	3.6	3.3	4.7	8.4	12.4	16.4	19.6	19.9	16.7	12.7	8.4	5.2
Average rainy days within month (d)	14.9	14.5	15.9	15.2	15.4	14.5	14.5	15.1	15.1	16.1	14.9	15.3
The lowest temperature (°C)	–5.4	–6.4	–6.1	–2.8	4.2	9.7	12.0	13.8	9.2	3.2	0.4	–3.2

Table 11.3 The long term climatic data of the Trabzon region (from 1975 to 2006)

	Jan.	Feb.	Mar.	Apr.	May	Jun.	Jul.	Aug.	Sep.	Oct.	Nov.	Dec.
Average temperature (°C)	7.4	7.0	8.3	12.0	15.8	20.2	23.1	23.2	20.1	16.2	12.3	9.2
Average highest temperature (°C)	10.9	10.6	12.0	15.9	19.0	23.5	26.3	26.8	23.9	20.0	16.1	12.8
Average lowest temperature (°C)	4.5	4.1	5.4	8.8	12.7	16.7	19.8	20.0	16.9	13.2	9.3	6.3
Average rainy days within month (d)	13.5	13.0	14.0	14.6	13.6	11.5	8.3	10.2	11.6	13.7	12.8	13.5
The lowest temperature (°C)	–4.6	–6.1	–5.0	–2.0	5.4	10.3	13.5	13.8	10.0	3.8	1.0	–3.1

Table 11.4 The long term climatic data of the Artvin region (from 1975 to 2006)

	Jan.	Feb.	Mar.	Apr.	May	Jun.	Jul.	Aug.	Sep.	Oct.	Nov.	Dec.
Average temperature (°C)	2.5	3.2	6.6	11.9	15.4	18.3	20.5	20.5	17.7	13.6	8.6	4.1
Average highest temperature (°C)	6.0	7.5	12.0	17.9	21.2	23.5	25.4	25.6	23.5	19.1	12.8	7.4
Average lowest temperature (°C)	–0.7	–0.4	2.2	6.9	10.5	13.5	16.5	16.6	13.6	9.7	5.0	1.1
Average rainy days within month (d)	12.5	12.8	12.5	12.7	13.8	12.4	8.1	7.6	7.9	11.2	10.9	11.8
The lowest temperature (°C)	–11.9	–11.2	–9.8	–7.1	–0.6	5.2	9.5	9.5	5.5	–1.6	–4.0	–10.8

In Turkey there are 2 to 4 harvesting seasons for tea but, in general, 3 harvests are widely collected. The first harvest occurs in May, the 2nd in June and the 3rd in September (Figs. 11.6 and 11.7).

Fig. 11.6. Tea harvest in Turkey

Fig. 11.7. Tea harvest time in Turkey

11.2 Tea Germplasm Collection, Conservation and Appraisal

The origin of seeds, collection and conservation, appraisal and utilization are given below.

11.2.1 Origin of the Seeds and/or Cuttings

In Turkey, most of the tea plantations were established by using seeds and the origin of the seeds was Georgia. It is accepted that Turkish teas belong to *C. sinensis* (L.) O. Kuntze var. *sinensis* type.

11.2.2 Collection and Conservation

The continuous seed propagation of tea in Turkey over a long period has produced populations with different yield and quality properties reflecting a wide genetic variation in the country. The clonal selection studies have been conducted in the Rize region by Ankara University and Ataturk Tea and Horticultural Research Institute belonging to the Ministry of Agriculture in Turkey from 1965 to 1973 (Sarimehmet, 1987). Among these studies 64 promising tea clones were selected as candidates. A germplasm collection with these promising selections was established and maintained at the Ataturk Tea and Horticultural Research Institute in Rize.

11.2.3 Appraisal and Utilization

These selections have been propagated by cuttings for commercial plantations. Among these selected 64 clones, Muradiye-10, Tuglali-10, Derepazari-7, Fener-3, Gundogdu-3, Komurculer and Pazar-20 are widely used among farmers due to high yield and the capacity to adapt. In the Ataturk Tea and Horticultural Research Institute in Rize, only these clones are propagated with cuttings for commercial production and distribution to farmers.

11.3 Tea Breeding and Selection Techniques

As mentioned before, continuous seed propagation of tea in Turkey over a long period has produced populations with different yields and quality reflecting a wide genetic variation in the country. The clonal selection studies have been conducted in the Eastern Black Sea region and several promising selections have been released. These promising genotypes were selected according to selection criteria: high yield capacity, resistance to pest, diseases and cold in natural growing conditions (Sarimehmet, 1987).

11.3.1 Conventional Breeding Techniques

Only selective breeding studies have been carried out on naturally seed propagated tea plantations in Turkey. In other words, no planned breeding studies such as crossing and mutation breeding have been carried out in Turkey so far. Although tea is of great importance in Turkey's economy, little is known about the pattern of genetic variation among the tea accessions grown in Turkey. Previously, clonal identification of tea clones in Turkey was done by using morphological and biochemical descriptors (Oksuz, 1987). However, as in many out-crossing plants, tea is highly heterozygous, with most of its morphological and biochemical traits showing continuous variation and high plasticity. In addition, most of the morphological and biochemical traits are influenced by environmental factors. More recently, some molecular studies related to clonal identification, genetic relationships and genetic diversity of tea clones have been done by several researchers by using RAPD (Beris *et al.*, 2005) and AFLP (Kafkas *et al.*, 2009) techniques. The AFLP analysis, with 6 primer combinations, generated 835 fragments of which 567 were polymorphic, corresponding to 69.8% polymorphism. Genetic similarity values ranged from 0.68 to 0.92, with an average of 0.76. The dendrogram derived by the unweighted pair group method with the arithmetic mean algorithm (UPGMA) and principal coordinate analysis (PCoA) revealed that all tea genotypes could be clearly divided into 4 distinct clusters. The results of this study will provide valuable information to the tea cultivar breeding program for the purpose of parental selection.

11.3.2 Selection Techniques Used for Yield, Resistance and Quality

In 2008, a new project aiming to select promising new tea clones among tea growing regions in Turkey was initiated. This project was supported by the Ministry of Agriculture of Turkey and will last until 2013. The selection criteria in this project are:

(1) Better vegetative growth and high yield capacity;
(2) Homogenous shoot development;
(3) Shiny shoot color;
(4) Better shoot diameter;
(5) Lower flower initiation;
(6) Short dormancy period between harvesting seasons;
(7) Better vegetative development after pruning;
(8) Lower sucker production;
(9) Resistance to cold, drought and disease;
(10) Longer vegetation period;
(11) Difference in harvest period;
(12) High polyphenol content;

(13) Relevance of different tea products such as: green, black, Oolong.

11.3.3 Local Adaptability Tests and Registration Systems

To date, there are no adaptability tests and registered tea cultivars in Turkey.

11.4 Propagation and Extension System of New Cultivars

Previously, tea gardens were usually developed using seeds. However, new selected promising clones are vegetatively propagated for commercial use.

11.4.1 Propagation Techniques

Tea plants are propagated in Turkey by using cuttings. Twigs are taken from the middle portions of shoots and prepared 7 – 8 cm long, including 1 or 2 nodes with a single leaf. The tea plants are hard pruned in February so that they develop healthy annual shoots. Previously, it was reported that the best cutting collection time for obtaining the highest rooting percentage in tea was July and the best IBA (indole-3-butyric acid) treatments were 2000 mg/kg in the Black Sea region in Turkey (Anon, 2007; Ayfer *et al.*, 1987a, 1987b). Therefore the twigs in general are sampled in July and brought to the greenhouse. The use of plant growth regulators for vegetative propagation studies of tea plants in Turkey is not common (Altindal & Balta, 2002). Previous research has shown that tea cuttings are characterized by genotype dependent variable rooting ability (Ayfer *et al.*, 1987a, 1987b; Sen *et al.*, 1991). Recent studies confirm that the treatment of tea cuttings with non-pathogen PGPR (plant growth promoting rhizobacteria) such as *Agrobacterium*, *Bacillus*, *Comamonas* and *Paenibacillus* induced root formation in tea cuttings (Erturk *et al.*, 2008).

11.4.2 Extension System

Tea is grown in a restricted area in Turkey and therefore the extension system is maintained only by local governmental offices. The governmental organization (Caykur) plays the most important role in extension because most extension activities are carried out by Caykur. In addition, Ataturk Tea and Horticultural Research Institute, located in Rize, are also active in this respect.

11.4.3 Strategy for Promotion of New Cultivars

The only selection technique on tea plantations is applied to promote new clones. In fact the farmers are not well aware of the difference between tea clones.

11.5 Research and Development

As mentioned before, there is only one research and development institute (Ataturk Tea and Horticultural Research Institute) related to tea in Turkey. This institute is only funded by the government.

So far only one breeding study (selection of clones among naturally seed propagated tea plantations) has been conducted. Thus, this research cannot be compared.

11.6 Conclusions

Opening new tea growing areas in Turkey was forbidden by the Turkish government in 1993. Thus the future trend is to increase the yield per plant. Moreover, a new selection project was started in 2008 aimed at obtaining a higher yield and multiple quality clones to distribute to farmers.

References

Altindal E, Balta F (2002) Comparison of rooting capabilities of Turkish tea clones. Turkish Journal of Agriculture and Forestry, 26: 195-200.

Anon (2007) Caykur. http://www.caykur.gov.tr.

Ayfer M, Celik M, Celik H, Erden M, Tutgac T, Mahmutoglu H (1987a). The effect of different medium and propagation techniques on rooting of tea cuttings. Proceedings of International Tea Symposium. 26-28 June, 1987, Rize, Turkey, pp.16-25.

Ayfer M, Celik M, Celik H, Vanli H, Tutgac T, Turna T, Dumanoglu H (1987b) The effect of different shading materials, cutting collection time and cutting types on rooting of tea cuttings. Proceedings of International Tea Symposium, 26-28 June, 1987, Rize, Turkey, pp.26-34.

Beris FS, Sandalli C, Canakci S, Demirbag Z, Belduz AO (2005) Phylogenetic analysis of tea clones (*Camellia sinensis*) using RAPD markers. Biologia, 60: 457-461.

Chen L, Gao QK, Chen DM, Xu CJ (2005) The use of RAPD markers for detecting genetic diversity, relationship and molecular identification of Chinese elite tea genetic resources [*Camellia sinensis* (L.) O. Kuntze] preserved in a tea germplasm repository. Biodiversity and Conservation, 14: 1433-1444.

Erturk Y, Ercisli S, Sekban R, Haznedar A, Donmez MF (2008) The effect of plant growth promoting rhizobacteria (PGPR) on rooting and root growth of tea (*Camellia sinensis* var. *sinensis*) cuttings. Romanian Biotechnological Letters, 13(3): 3747-3756.

Food and Agriculture Organization (FAO) (2005-2010) http://faostat.fao.org.

Kafkas S, Ercisli S, Dogan Y, Erturk Y, Haznedar A, Sekban R (2009) Polymorphism and genetic relationships among tea genotypes from Turkey revealed by AFLP markers. Journal of the American Society for Horticultural Science, 134(3): 428-434.

Mendilcioglu K (2000) Tea growth techniques. Ege University Agricultural Faculty, No. 43, p.28.

Oksuz M (1987) Morphological, yield and quality properties of tea clones in Turkey. Caykur, 8: 87.

Sarimehmet M (1987) The effect of N, P and K fertilization on growth of sapling material of Muradiye-10 and Fener-3 tea clones. Tea Industry Publication, Rize, p.114.

Sen SM, Uzun S, Ozkan Y, Vanli H, Tutgac T, Turna T (1991) The propagation of tea clones by cutting and grafting. Yuzunci Yil University Agricultural Faculty Journal, 1: 67-88.

12

Tea Improvement in Nigeria

Samuel S. Omolaja*, Gerald O. Iremiren

Cocoa Research Institute of Nigeria, P.M.B. 5244, Ibadan, Nigeria
*E-mail: dolaomolaja@yahoo.com

Abstract: The chapter presents an account of tea improvement efforts in Nigeria since its introduction into the country around 1952. Commercial tea planting started in 1982. Nigeria produces black tea with the CTC method, labeled 'Highland tea'. The total land area planted to tea is 1,200 ha. The average annual national production is 1,640 tonnes, which meets only 10% of domestic need. Opportunities thus exist for further local and foreign investments in the Nigerian tea industry. Tea improvement started in 1982 with the acquisition of 33 clones by the Cocoa Research Institute of Nigeria. Since then moderate achievements have been recorded. Five out of the 33 clones, namely 35, 68, 143, 236 and 318, with an average harvest of 2.5 tonnes/(ha· year), were selected and released to farmers. Tea clones that were adaptable to the warm lowland environment (143 and 35) and some that could be profitably intercropped with eucalyptus trees in the Nigerian environment were identified. In the tea/eucalyptus intercrop trial, the average yield from tea planted as sole-crop (2.3 tonnes/(ha· year)) was half that from the tea/eucalyptus intercrop (4.2 tonnes/(ha· year)). The eucalyptus trees stabilized the dry season tea production, enhanced the organic matter in the soil content and increased the returns per unit of farm land. Investigation into the genotypic association of the six yield components, their effects on yield and quality of the plucking, revealed tea clones whose final product after processing their leaves could generate less coarse black tea. Hybrid tea plants, which had the potential to perform better than their higher parents were generated. Currently the priority is to strengthen the Institute's improvement program by molecular characterization of the germplasm, which would reveal accurate genetic diversity within Nigerian tea germplasm, and thus assist in selecting suitable clones as parents in a subsequent hybridization program.

12.1 General Introduction

Tea plant (*Camellia sinensis* (L.) O. Kuntze) belongs to the family Theaceae and was discovered by the Chinese around 2700 B.C. (Banerjee, 1992). It originated in China (Yao, 2009). It is the most popular non-alcoholic beverage in the world (Chen *et al.*, 2007; Yao *et al.*, 2008) and cultivated in the 5 continents of the world. In the genus *Camellia* L., there are over 200 species reported so far (Chang & Bartholomew, 1984), of which the tea plant is the most economically important. It is an evergreen woody plant, which when under cultivation is kept at a low level to facilitate harvesting the young shoots, the part from which tea beverage drink is made (Famaye *et al.*, 2006). Today, compared with most agricultural industries of world importance, the commercial tea industry has achieved a relatively high degree of stability (FAO, 2008).

Tea is a diploid ($2n = 30$ chromosomes) but a number of triploids and tetraploids have been found or been created (Bonheure, 1991). Tea is an outbreeder and self-incompatible (Rogers, 1975). The plants are of 2 major types, the *C. sinensis* (L.) O. Kuntze var. *sinensis* teas which are small-leaved, slow growing, dwarf trees, tolerant to cold weather and other adverse conditions, while the *C. sinensis* var. *assamica* (Masters) Kitamura teas have big leaves, are faster in growth and adapted to warmer conditions (De Costa *et al.*, 2007). Other forms of tea include the Camdodran and Wilson's *Camellia.* A large number of hybrids do exist.

Tea is also classified based on the degree of fermentation and oxidation of polyphenol present in the leaves. The 6 types are green, white, yellow, Oolong, black and dark tea (Bonheure, 1991). Of the current global production of tea, comprises about 78% black tea (FAO, 2008), 20% green tea and 2% Oolong tea. While black tea is popular in Asia and some countries in the West, Oolong and green teas are largely consumed in China, Japan, Korea and a few countries in North Africa and the Middle-east, respectively (Basu, 2003).

In ancient times, tea spread naturally as seed because of its use for drinking and medicinal purposes. It was spread from China along the Silk Road, by traders to other parts of the world such as Arabia, North Africa, Europe and other parts of Asia. In the middle 16th century, China tea was brought to America by the Dutch (Visser, 1969).

12.1.1 Tea Production in Africa

The total area of land under tea cultivation in the world is about 3,015 kilohectares, with 88.9% in Asia, 9.4% in Africa and the remaining distributed between South America, Russia and Oceania (FAO, 2010). Of the current world tea production of about 3,936 kilotonnes in 2009, China produces about 1,359 kilotonnes (ITC, 2010). Indeed, Asia produces approximately 83.9% of world tea, followed by Africa with 15%. Worldwide, the current top 5 leading tea producing countries are

China, India, Kenya, Sri Lanka and Vietnam in descending order.

Widespread commercial cultivation of tea commenced in Africa in the 1900s (Anon, 1962). The major tea producing countries in Africa include Kenya, Malawi, Uganda, Tanzania, Rwanda, Zimbabwe, Burundi and Ethiopia. Kenya is Africa's largest tea producer with an average of 328,000 tonnes annually, as estimated by FAO (FAO, 2009). Nigeria currently produces an average of 1,640 tonnes annually.

12.1.2 Introduction of Tea into Nigeria

The first tea plant was introduced into Nigeria around 1952 (Opeke, 1982). Nigerian Beverages Production Company (NBPC), Mambilla Plateau, however, brought in tea clones from Kenya into Nigeria for commercial planting in 1972 (Hainsworth, 1981). But serious commercial tea cultivation commenced on the Mambilla Plateau, Taraba State, Nigeria in 1982 (Esan *et al.*, 1998). The Mambilla Plateau is a cool, high altitude (1,550 m above mean sea level) local government area in Nigeria (Fig. 12.1) (Omolaja & Esan, 2006). The plateau consists of rolling undulating hills covered by grass, with valleys of medium sized trees. The region extends from between latitude 6.5° N to 8.0° N and from longitude 11.0° E to 12.0° E.

12.1.3 Type of Tea Produced and Its Economic Importance in Nigeria

Nigeria produces black tea using the CTC method. NBPC located on the Mambilla Plateau is the major company currently processing tea in Nigeria, with the label 'Highland tea'. The company is jointly owned by Adamawa State, Taraba State, the Northern Nigerian Development Corporation, Nigeria Agricultural, Cooperative and Rural Development Bank, Adamawa & Taraba indigenes, NBPC Staff Trust Fund and Sardauna Local Government indigenes. The first three are, however, the major shareholders. The company was duly incorporated in 1975 and started processing tea in 1982. The company maintains over 600 ha of tea plantation. Though the company relies on a diesel generator as a source of power, work is currently at an advanced stage to generate hydro-electricity from Tonga dam to supply electricity to the company and nearby communities on the plateau. The small scale farmers hold 600 ha. The company buys tea leaves from the farmers. The total land area planted to tea is about 1,200 ha. The average tea garden per farmer is between 0.1 and 0.6 ha (Fig. 12.2). The average yield on the farmers' plantation is low, at about 1.2 tonnes/(ha· year) fresh leaves, because few farmers' plant improved cultivars. The total average annual income for the farmer from 1 ha of tea is about 60.00 US$, which is too low to sustain a family (Omolaja, unpublished data). In the current situation, the farmers realize a very small return from their harvest. As a result of this, the farmers' economy suffers and their purchasing power is low.

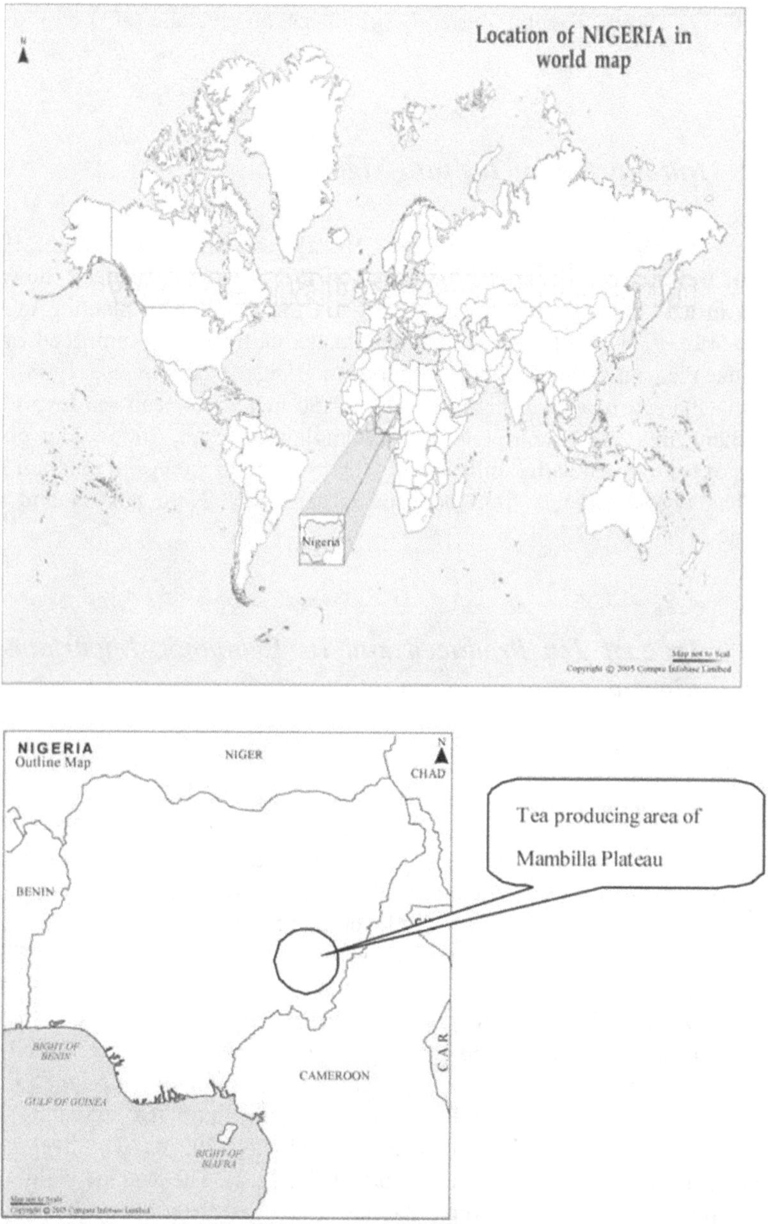

Fig. 12.1. Map of Nigeria showing the tea producing area of the Mambilla Plateau

Fig. 12.2. A small scale tea garden in Nigeria

The average annual national production of tea is currently 1,640 tonnes, which is only able to meet about 10% of domestic need. Hence, Nigeria imports about 12,000 tonnes of tea yearly to meet the needs of the domestic market. With a 140 million population in Nigeria, the opportunity thus exists for further local and foreign investments in the Nigerian tea industry.

12.1.4 Tea Growing Climate in Nigeria

Favorable climatic conditions are critical factors for commercial tea cultivation in any location. Rainfall, temperature and soil pH are the key parameters determining the success of tea cultivation (Filani & Okelana, 1980). Tea requires optimal soil pH of between 4.5 and 5.5 (Ng'etich, 1996). The soil of the Mambilla Plateau is an ultisol (FAO) with a pH of 4.5 to 5.6 (Obatolu, 1984). The soil is red coloured, fertile and well drained. Average annual rainfall is 1,860 mm, which is spread over nine months, from February to October (Esan *et al*., 1998). On Mambilla, the average daily temperature ranges between 17 °C and 26 °C. Sunshine is very important for tea growth. In the dry season, when precipitation is low, overhead sprinklers are installed, which irrigate about 500 ha of a Nigerian tea plantation. An atmospheric relative humidity of between 70% and 90% is prevalent, which is ideal for tea growth.

12.2 Tea Germplasm Collection and Conservation

Tea is a young crop in Nigeria. The introduction of tea into Nigeria and the morphological description of surviving clones are presented.

12.2.1 Collection and Conservation

Tea improvement started in 1982 with the acquisition of 33 tea (*C. sinensis* var. *assamica* (Masters) Kitamura) clones from NBPC by the Cocoa Research Institute of Nigeria (CRIN) (Esan, 1982). The tea clones are conserved in germplasm plots at CRIN Substation, Mambilla, Taraba State. The tea clones were originally introduced from Kenya. Since the inception of tea improvement in Nigeria, moderate achievements have been recorded.

12.2.2 Morphological Description of Surviving Clones under Conservation

Of the 33 tea clones acquired in 1982, only 24 are currently surviving. The understanding of available germplasm is critical to its utilization. Though some of the acquired clones have been characterized for some traits (Esan *et al.*, 1998), a fairly detailed individual clone description of tea grown in Nigeria has not been done.

The morphological descriptions of the surviving clones are presented in Table 12.1. A replicate of surviving clones is currently being re-established in new germplasm.

12.3 Tea Selection and Breeding

In Nigeria, the major objectives of tea improvement were to select for high yield, good quality and lowland adaptability. Planting of eucalyptus trees is popular in the tea growing area of the Mambilla Plateau, Nigeria. This necessitated the need to select for tea clones that can be grown in combination with eucalyptus trees to increase farmers' income per unit area of land and facilitate good soil management. In addition, a major breeding objective was to identify tea clones of good combining ability for yield and quality.

Table 12.1 Morphological descriptions of surviving Nigerian tea clones

Clones	Mature leaf color	Leaf size (cm)	Length of young shoot (cm)	Stem girth (mm)	Flower & fruit productions	Number of surviving plants	Other remarks
Unk	Light green	12.0×5.0	12.0	3.0	Flowers & fruits heavily	50	
370	Light green	10.0×4.0	9.0	2.0	Low flower and fruit productions	10	
19	Yellowish green	13.4×5.2	13.0	3.0	Low flower and fruit productions	16	It flushes poorly
74	Green and flat. The leaf has prominent serration	17.8×8.0	15.0	3.0	Low flower and fruit productions	22	It is slow flushing. Plucking interval may be up to one month
354	Yellowish green	10.0×2.2	9.5	2.0	Low flower and fruit productions	13	
368	Light green	18.0×7.0	12.0	2.0	No flower or fruit	33	Flush regeneration after plucking is slow.
369	Yellowish green	16.0×6.0	12.0	2.0	No flower or fruit	26	It flushes well
33	Yellowish green	14.0×5.0	13.0	3.0	Low flower and fruit productions	56	It flushes well
353	Green	15.0×5.0	13.5	3.0	Flower and fruit productions are high	25	
357	Light green	17.0×8.0	16.5	4.0	Rarely produces flower	34	Flushes regeneration is high. It is a new commercial material
359	Green	15.5×7.0	11.6	3.0	It flowers & fruits	24	Flushes averagely
143	Light green	15.0×6.8	13.0	3.0	It flowers & fruits	53	It flushes well. It is a commercial clone & adaptable to the lowland
14	Yellowish green	14.5×5.0	12.5	3.0	It flowers & fruits	54	
238	Yellowish green	12.0×5.0	13.5	3.0	It flowers & fruits	57	
25	Yellowish green	14.0×6.0	12.5	2.0	It flowers & fruits	57	
108a & b	Green	17.0×7.0	9.5	3.0	It flowers & fruits	70	
363a & b	Light green	12.0×5.0	12.0	3.0	It flowers & fruits heavily	50	

(To be continued)

(Table 12.1)

Clones	Mature leaf color	Leaf size (cm)	Length of young shoot (cm)	Stem girth (mm)	Flower & fruit productions	Number of surviving plants	Other remarks
35a	Light green	18.0×7.3	12.5	3.0	Low flower & fruit productions	58	Flush regeneration rate is high. It is vigorous, hardy, with high pluck weight. It is a commercial clone, but susceptible to drought, leaf spot and insect attack
35b	Light green, bending, with tiny serration	18.0×7.3	12.5	3.0	Low flower & fruit productions	59	Flush regeneration rate is high. It is vigorous, hardy, with high pluck weight. It is a commercial clone, but susceptible to drought, leaf spot and insect attack
68a & b	Light green	18.0×5.0	13.0	3.0	Low flower & fruit productions	32	It is a commercial clone
61a & b	Green (orange-like leaf)	15.0×5.0	13.5	2.0	Heavy flower & fruit productions. Petals were seven in most of the flowers	62	It is most susceptible to bryophyte's attack
228a & b	Yellowish green	11.0×4.5	9.4	2.0	Heavy flower & fruit productions Seven petals were observed in most of the flowers	106	The shoots are low and hard to pluck. The clone is vigorous and may be good for oil production. The clone is susceptible to bryophyte's attack
236a & b	Deep green with purple petiole	11.5×4.0	6.0-11.0	1.0	Heavy flower & fruit productions. The seed is very viable		Flush regeneration rate is high. It is vigorous, hardy, with high pluck weight. It is a commercial clone and tolerant to drought. It has the best flavor. It is the only indigenous variety and a genetic marker
318a & b	Light green	15.0×6.0	12.5	3.0	Low flower & fruit productions	106	Flush regeneration rate is high. It is vigorous, hardy, with high pluck weight. It is a commercial clone

12.3.1 Selection

The main selection objectives of tea cultivars in Nigeria include high yield, lowland adaptability, intercrop with *Eucalyptus* for overcoming the environmental stress, and good tea quality.

12.3.1.1 Selection for Yield Attributes

Preliminary yield evaluation of the 33 tea clones were embarked upon, which finally led to the selection of 5 outstanding ones with an average dry tea leaves harvest of 2.5 tonnes/(ha· year). The 5 clones, namely 35, 68, 143, 236 and 318 are currently planted by some Nigerian tea farmers (Esan, 1989). Further morphological characterization yielded valuable discriminating information of the tea clones (Esan *et al.*, 1998).

12.3.1.2 Clonal Selection for Lowland Adaptability

In Nigeria, tea is planted only in the high altitude region of the Mambilla Plateau, which is limited in land area. The supply of tea from this region alone is inadequate to meet the demand of the local tea processing industries (Obatolu & Ipinmoroti, 2000). In order to allow for the expansion of tea plantations to the lowland areas of the country, a lowland adaptability trial was launched in 1994 with the objective of identifying tea clones that can perform in warm lowland areas. Consequently, between 1994 and 1999, 15 tea clones were evaluated for their adaptability, growth characteristics and yield performance in 6 lowland locations of Nigeria. The 6 lowland locations were: Akwete (Abia State), Ibadan (Oyo State), Ikom (Cross River State), Ikorodu (Lagos State), Iyanomo (Edo State) and Mayo Selbe (Taraba State) (Table 12.2). The tea clones were: 19, 35, 143, 236, 318, 353, 368, 370, UNK-1, 228, 237, 108, 68, 354 and 359. The results obtained showed that clone 143 followed by 35 performed best in yield (Table 12.2) and was most stable across the environment (Table 12.3), hence were adaptable to warm and humid lowland areas of Nigeria (Omolaja & Esan, 2006).

Table 12.2 Soil and climatic characteristics of the 6 locations used for the tea trial

Location	Ecology	Soil type	Soil pH	Average annual rainfall (mm)	Average daily temperature (°C)	Elevation (m amsl)
Ibadan	Rainforest with deciduous trees	Oxic tropudalf (Alfisol)	6.5	1,200 – 2,500	23 – 32	122
Ikorodu	Rainforest	Alfisol	6.1	1,960 – 2,650	21 – 30	50
Iyanomo	Rainforest	Typic paleudult (Ultisol)	5.6	1,800 – 2,400	23 – 34	112
Akwete	Rainforest	Ultisol	4.5 – 5.5	1,800 – 2,800	20 – 32	109
Ikom	Rainforest	Oxic dystropept (Oxisol)	5.1 – 6.5	1,960 – 2,650	20 – 32	119
Mayo Selbe	Guinea Savannah	Ultisol	5.6 – 5.8	931 – 1,400	26 – 38	540

Table 12.3 Stability parameters for the 3 best tea clones

Clone	Survival (%)			Yield (tonne/ha)		
	Mean	Regression coefficient	Deviation sum of squares	Mean	Regression coefficient	Deviation sum of squares
35	27.4	0.68	24.11	1.21	0.52	61.28
143	41.7	1.01	9.82	1.68	0.96	4.45
318	32.8	0.43	17.81	0.99	0.67	22.84
Grand mean	33.97			1.29		

12.3.1.3 Overcoming Environmental Stress by Selection in *Eucalyptus* Intercrop

The general practice by farmers on the plateau is to plant tea as the sole-crop. However, they suffer a reduction in tea yield in the long dry season, which starts in November and lasts till March of the following year. During this period, precipitation is low. In addition, the farmers spend a lot on the procurement of mineral fertilizer in order to obtain a reasonable yield from the tea plantation, even in the rainy season. Eucalyptus (*Eucalyptus falcata* Silver Mallee) is a popular tree planted on the Mambilla as a source of wood (Omolaja *et al*., 2008), but it also provides shade, litters heavily, suppresses weeds and conserves soil moisture. In view of the advantages of this tree in the Mambilla environment, some tea clones were tried in the eucalyptus plantation in 1997 at CRIN Substation, Kusuku, Mambilla Plateau, as a way of identifying tea clones that can be profitably intercropped with eucalyptus trees, for the conservation of soil moisture and increased returns per unit of land area. Rooted tea cuttings were transplanted at 1.0 m between rows and 0.6 m within rows. The planting distance of the eucalyptus was 5.1 m×5.1 m.

Results showed that the average yield from tea planted as the sole-crop (monocrop 2.3 tonnes/(ha· year)) was half that from tea/eucalyptus intercrop (4.2 tonnes/(ha· year)) (Table 12.4). Tea leaves harvested from the intercrop, which was higher in the dry season, ranged between 207 kg/(ha· month) and 342 kg/(ha· month). There was a higher positive correlation of 0.595 (Kendall's tau_b correlation coefficient) between monthly rainfall and yield from the tea/eucalyptus intercrop ($p \leq 0.05$) than that between monthly rainfall and tea yield in the monocrop (0.443) (Fig. 12.3). Planting of the tea/eucalyptus intercrop in tea gardens, besides stabilizing dry season production, increased organic matter content of the soil and reduced costs of irrigation. Tea clones, which could thrive in a eucalyptus-tea combination, were identified (Omolaja *et al*., 2008). A chloroform test was used to analyse the quality of tea leaves from the sole-crop and intercrop. Results showed that there was no difference in the reaction of tea leaves of tea clones in the monocrop and tea/eucalyptus intercrop, indicating that there had not been a reduction in the quality of tea leaves produced with eucalyptus, as compared with those in the monocrop. Since the tea leaves harvest is stabilized in the dry season by the eucalyptus trees, Nigerian tea farmers are encouraged to practice a tea/eucalyptus intercrop for better soil management and to ensure that income from tea leaves does not diminish in the dry season.

Table 12.4 Average monthly rainfall and tea yield for 8 years between January 1999 and January 2007 on the Mambilla Plateau, Nigeria

Month	Rainfall (mm)	Average monthly yield (kg/ha)							
		Monocrop				Tea/*Eucalyptus*			
		236	318	14/3	Mean	236	318	14/3	Mean
Jan.	0.45	65.43	78.53	82.63	75.53	158.80	220.80	240.86	206.84
Feb.	39.85	22.64	24.92	26.69	24.75	200.91	216.92	220.90	212.91
Mar.	107.63	55.60	60.95	56.70	57.75	209.26	206.22	280.21	241.23
Apr.	240.28	194.64	192.92	208.69	237.75	328.89	340.80	390.80	347.83
May	325.28	194.64	192.92	208.69	198.75	351.84	390.81	347.83	384.82
Jun.	457.18	230.62	239.90	251.73	240.75	434.65	398.54	464.64	432.61
Jul.	437.35	215.65	211.84	225.76	217.75	294.82	315.74	348.93	319.83
Aug.	416.13	244.67	220.80	250.78	238.75	409.62	550.58	601.57	520.59
Sep.	417.98	323.03	250.00	360.00	311.01	422.59	510.64	670.63	534.62
Oct.	326.48	226.41	298.64	301.50	275.51	322.60	374.50	384.52	360.54
Nov.	68.05	186.42	280.64	302.50	256.52	305.04	352.02	381.18	346.12
Dec.	0.35	146.44	201.60	220.52	189.52	298.90	319.93	361.90	326.91
Mean	236.41	178.68	191.31	211.10	193.69	311.49	349.79	396.50	352.90
SD	182.25	89.50	88.42	103.48	91.55	88.25	107.83	132.98	106.09
CV (%)	77.15	50.08	46.22	49.02	47.27	28.33	30.83	33.54	30.13
TAY (kg/ha)	/	2,106	2,254	2,496	2,285	3,738	4,198	4,694	4,210

SD: Standard deviation; CV: Coefficient of variation; TAY: Total annual yield

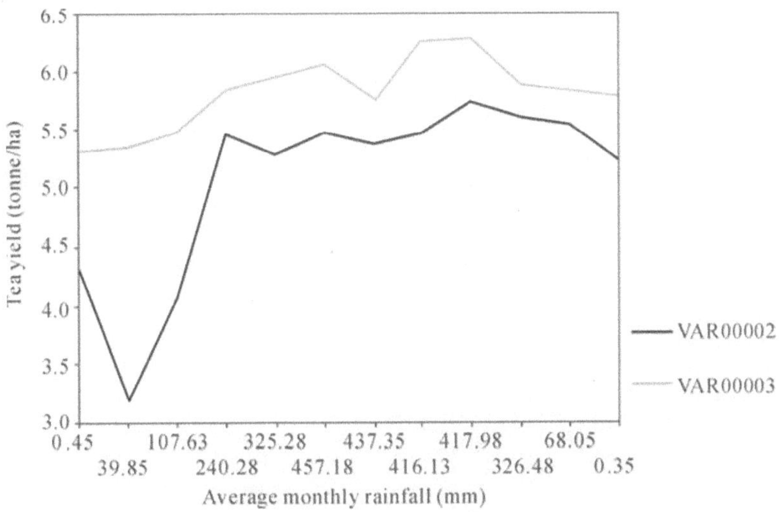

Fig. 12.3. The response of tea yield (tonne/ha) to average monthly rainfall (mm) between 2006 and 2008 in tea monocrop and tea/eucalyptus intercrop fields (VAR00002: tea yield in monocrop; VAR00003: tea yield in intercrop)

12.3.1.4 Selection of Tea Clones for Good Quality

Between 1999 and 2002, 24 tea clones were investigated for the genotypic association of 6 yield components, their effect on yield and plucking quality value. The major objective was to identify tea clones whose final product after processing of the tea leaves would generate less coarse black tea. The experiment was a completely randomized design with 3 replicates. Results showed that 2 yield components, weight of the terminal bud and the first leaf, were positively associated with yield. Path analysis showed that the value of genotypic correlation of terminal bud weight on yield (0.444) was similar to its direct effect (0.446). The weight of the first leaf showed maximum positive effect on yield. Hence, the terminal bud and the first leaf in tea are important components for selection where higher yield and tea quality are major breeding objectives. Differences in plucking quality values were significant ($p \leq 0.05$) among the tea clones studied (Esan & Omolaja, 2002). Clones UNK-1, 68, 228, 237, and 14 had the best plucking quality values in ascending order, while clone 318 had the lowest. This suggested that clones with best quality values will produce black tea with less fibre.

The ratio of polyphenol to amino acid determines tea quality (Ellis & Nyirenda, 1995). In terms of the effect of some chemical components on tea, polyphenol confers bitterness, amino acid makes it fresh while sugar makes it sweet (Owuor, 1992). In an attempt to identify good quality tea clones in the Institute's germplasm, some selection criteria were employed which included possession of leaf hairs, medium to large leaf, which are peculiar botanical characters for good black tea (Esan, 1989), as well as high polyphenol (Omolaja *et al.*, 2008). In 2006, a simple chloroform test was done for some clones to serve as an indication of their polyphenolic content following the procedures of Sanderson (1963) and Ellis & Nyirenda (1995). The tea clones reacted to a chloroform test at different periods. Leaves of clone 143 followed by 236 turned brown faster than other clones after about 20 minutes of exposure to chloroform. At about 45 minutes, when all the clones had been given ample exposure time, clone 143 leaves were evenly brown, while clones 318 and 236 had evolving brown coloration. This indicated that clone 143 was a fast and good fermenter (Fig. 12.4), suggesting that it might produce better quality black tea (Omolaja *et al.*, 2008).

Fig. 12.4. A bush of tea (*Camellia sinensis*) clone 143

12.3.2 Tea Breeding

Tea plant (*C. sinensis* (L.) O. Kuntze) is an allogamous plant. Heterosis or hybrid vigor is the most widely explored phenomenon in both plant and animal improvement programs. It is a practice in which the cross of two stocks produces a hybrid that is superior in growth, size, yield or general vigor (Allard, 1960). It is expressed as the advantage of the F_1, both over the mid-parent as well as the higher parent (Mather & Jinks, 1982). Between 1984 and 1987, observations on the flowering and fruiting patterns of 24 tea clones in the tea germplasm were carried out with the aim of obtaining adequate information on the reproductive pattern of each clone in the gene pool. The result revealed tea clones that synchronized in flowering and these were grouped for potential hybridization in 1985 (Esan, 1989). Records of their fruiting patterns were also taken.

In 2006, a hybridization program was initiated on Nigerian tea with a view to obtain progenies of high heterotic value in yield, quality and tolerance to diseases and insect pest. A half diallel cross was carried out using 6 tea clones, namely 236, 68, 143, 357, 359 and 33, by modifying the pollination procedure of Omolaja and Fawole (2000). The data obtained was transformed and then subjected to an initial test of the significance of differences between clones (Singh & Chaudhary, 1979). Then, comparisons of the performance of parental clones, maternal sibs and the parents' vs F_1 crosses were conducted. Lastly, the heterotic responses of each hybrid cross for leaf number, canopy width and yield were estimated, both on the basis of mid-parents as well as on higher parents, respectively. The heterotic value of half-sibs was also calculated.

The analysis testing differences between parental clones indicated a significant result. The mean performance of the clones is presented in Table 12.5. Clone 143 had the highest yield of 5.03 kg/(plant· year) of fresh leaf. Clones 236, 68 and 357 followed it. Clone 33 had the poorest yield. Yield was from an average of 10 bushes to ensure uniform evaluation. An average yield of 2.5 tonnes/(ha· year) is considered ideal in the Nigerian environment. In terms of canopy width and leaf number, clone 143 still performed best among the clones. The analysis, comparing the performance of the half-sibs, also indicated a significant result. The mean performance of the hybrids of the clones is presented in Table 12.6.

Table 12.5 Mean performance of tea clones in Nigeria

Clones	Leaf number	Canopy width (cm)	Yield (kg/(plant· year))
236	148	60.01	4.41
68	159	64.02	3.21
143	164	65.23	5.03
357	141	40.42	2.61
359	133	28.13	2.42
33	136	29.61	1.13

Table 12.6 Mean performance of tea hybrids

Maternal sibs	Leaf number	Canopy width (cm)	Yield (kg/(plant· year))
236	220	60.40	4.01
68	185	61.06	3.19
143	242	62.05	4.23
357	242	62.05	4.23
359	128	27.76	2.03
33	127	28.22	1.10

Comparison among clones and hybrid combinations showed that those from clone 143, followed by clones 236 and 68, hold some promise both in terms of fresh leaves production as well as canopy width. It was inferred that there are some advantages to be derived by planting some of the hybrids rather than planting the parent clones, since some of the hybrids yielded more than the higher parents. The hybrid tea plants that were produced are currently being evaluated for yield and quality characteristics (Omolaja, unpublished data).

12.3.3 Tea Cultivar Registration Systems

In Nigeria, there is only one national registration body. At the moment, only 5 tea cultivars, namely 35, 68, 143, 236 and 318 (Table 12.7) are registered in 1996 with the National Varietal Release Committee (NVRC) under the National Centre for Genetic Resources and Biotechnology (NACGRAB), Ibadan. The cultivars thrive in the cool high altitude area of the Mambilla Plateau. The average yield is 2.5 tonnes/(ha· year).

12.3.4 Value Added Products and Cultivars Used

Five major value added products were developed at CRIN, some of which received awards at International Conference on O-Cha (tea) Culture and Science in Japan. Some of them are (1) tea wine, characterized by low alcoholic content of 4%, good color, aroma and taste (Aroyeun *et al.*, 2005); (2) tea jam, combined with dietary fibre from tropical fruits, is characterized by low sugar content and high flowability (Aroyeun *et al.*, 2007); (3) tea vegetable oil, characterized by a golden yellow color, is less prone to rancidity with low free fatty acid content; (4) tea yoghurt, has probiotic bacteria, which can help in promoting bowel movement, high shelf stability and good taste; (5) herbal tea was formulated from a combination of tea leaves and mistletoe (Akinwale *et al.*, 2000). It is effective for the control of hypertension. Work is progressing to develop chocolate, animal feeds and soap from tea. All the products used the improved tea cultivars.

Table 12.7 General characteristics of CRIN selected tea clones

Clones	Yield potential (tonne/(ha· year))	Quality potential	Pest tolerance/ resistance	Drought tolerance	Lowland adaptability	Hybrid tea	Other characteristics
35[1]	3.5		Moderately susceptible to leaf spot and insect attack	Tolerant to drought	Adaptable to warm lowland (22 – 27 °C)		
68[1]	2.5	Good pluck quality value					
143[1]	3.5	Fast & good fermenter		Tolerant to drought	Adaptable to warm lowland (22 – 27 °C)	Its sibs have high heterotic value	Can be intercropped with eucalyptus trees
236[1]	3.4						Has purple petiole
318[1]	3.4						Can be inter-cropped with eucalyptus trees
357[2]	3.4						
359[2]	3.4						

[1]Released to farmers; [2]Yet to be released

12.4 Propagation and Extension System of New Cultivars

The propagation method of improved tea cultivars, its extension and strategies to boost tea production in Nigeria are briefly discussed.

12.4.1 Propagation Techniques and Disease Control

The cultivated cultivars are propagated by stem cuttings. In 1994, however, tissue culture was initiated on shoot explants of tea. The objectives were to produce protocols for routine production of disease-free plantlets for ease of germplasm exchange between tea institutions and for the production of somatic embryos. Somatic embryos production is essential to an efficient modern genetic transformation in tea. Somatic embryos were successfully obtained using the Murashige and Skoog medium (Esan *et al.*, 2001).

Tea cuttings stay in the nursery for 18 months before being transplanted to the field. Though some incidences of disease are noticed and some insects are picked on a few clones of tea, there is no serious disease or insect pest attack on commercial tea plantations in Nigeria.

12.4.2 Extension System in Nigeria

Improved tea clones are normally introduced to tea farmers through meetings of the National Coffee and Tea Farmers Association (NACOFTAN) and in workshops at NBPC. NACOFTAN is the umbrella body, to which all tea farmers in Nigeria belong. The national headquarters of NACOFTAN is Gembu, Sardauna Local Government Area of Taraba State.

12.4.3 Strategy for Promotion of New Cultivars

Many tea farmers in Nigeria are poor and are usually located in rural and difficult terrain that limits their access to improved cultivars that are domiciled in the CRIN Substation, Mambilla. The best strategy, therefore, is for the Nigerian government to assist farmers by funding mass production of improved cultivars and distributing it to the farmers so as to increase their yield and make tea planting more profitable.

12.5 Future Strategies and Opportunities

The future strategies for tea improvement in Nigeria are briefly highlighted as below.

12.5.1 Breeding for Pest Resistance

While a few insects are picked on some tea clones, their incidence has never reached an economic threshold. In other words, a serious insurgence of insect pest on tea has not been noticed in Nigeria. Nevertheless, there is a need to continue to closely study the tolerance/resistance of Nigerian tea clones to insect pest attack, to preclude any serious sudden insurgence.

12.5.2 Combining Ability

Work on combining ability among the different clones would continue to guarantee continuous improvement in tea yield and quality.

12.5.3 Genetic Transformation

It is desirable that work on somatic embryogenesis will continue to prepare sufficient grounds for genetic transformation of tea, in the near future.

12.5.4 From Conventional to Molecular Breeding

Though Nigerian tea germplasm has been characterized morphologically, it is desirable that molecular characterization at DNA level, which reveals accurate genetic variance within tea germplasm, be embarked upon. In the light of this, a collaborative work is ongoing between scientists at the Tea Research Institute of the Chinese Academy of Agricultural Sciences and the Cocoa Research Institute of Nigeria, to carry out molecular characterization of Nigerian tea germplasm. It is expected that, at the end of the work, suitable tea clones that are ideal for selection as parents for the hybridization program, would be identified.

12.5.5 Participatory Crop Improvement

Farmers are currently carried along by our improvement program. The scientists regularly attend farmers meeting, while the farmers also are encouraged to visit the Institute to discuss the priority problems facing them in the field. In addition, the Institute organizes stakeholder's workshops annually, during which farmers associations, processors, marketers and other relevant people are invited to brainstorm on the priority problems facing the Nigerian tea industry. The result of such a stakeholder's forum forms the basis for updating the Institute's annual tea improvement strategies.

12.6 Conclusions

All the land potentially suitable for tea farming is not utilized in Nigeria. There is a need to promote the cultivation of tea in other highland locations of Nigeria, such as the Jos and the Obudu Plateau. The Nigerian government needs to provide credit facilities for the tea farmers to enable them to improve their productivity and enhance their earnings potential. There is also the need to establish more tea processing companies in Nigeria in order that the farmers can have greater avenues for marketing their harvest, thereby breaking the current monopoly enjoyed by NBPC, which is the current sole buyer of tea leaves from farmers. The government also needs to assist in the free distribution of improved cultivars of tea to the farmers so as to increase their yield and make tea planting more profitable. Moreover, the government needs to improve farmers' access to fertilizers, which constitute a great production cost for tea farmers.

Acknowledgements

The author is grateful to the Executive Director of the Institute for his support for the review and Prof. Liang Chen for his thorough editing of the manuscript. The contribution of Prof. Darlington, Portugal, is greatly acknowledged. Appreciation is equally extended to other tea scientists, past and present (Prof. Edward Babatunde Esan, Late Dr. Charles Rotimi Obatolu, Dr. Amos Famaye, Dr. Rotimi Ipinmoroti, Mr. Amos Oloyede, etc.).

References

Akinwale TO, Aroyeun SO, Obatolu CR (2000) Physico-chemical, microbiological profiles of blends of tea and mistletoe-a highly medicinal mix. Journal of Food Technology in Africa, 5(4): 123-125.

Allard RW (1960) Principles of Plant Breeding (2nd Edition). New York: John Wiley & Sons. Inc., p.421.

Anon (1962) Historical notes on tea introduction in Africa. In: Tea Estates in Africa (Compiled by Wilson, Smithett & Co) London: Mabey & Fitzclarence Ltd, pp.6-9.

Aroyeun SO, Olubamiwa O, Ogunjobi MAK (2005) Development of wine from infused tea leaves (*Camellia sinensis*). British Food Journal, 107(1): 34-41.

Aroyeun SO, Olubamiwa O, Ogunjobi MAK (2007) Development of jam from tea infusion and tropical fruit dietary fibre. In: Proceedings of the 3rd International Conference on O-Cha (tea) Culture and Science. 2-4 November, 2007, Shizuoka, Japan.

Banerjee B (1992) Selection and breeding of tea. In: Willson KC & Clifford MN (eds.) Tea: Cultivation to Consumption. London: Chapman and Hall, pp.53-81.

Basu B (2003) Drink tea and keep healthy. International Journal of Tea Science, 2(3): 5-7.

Bonheure D (1991) Tea. London & Basingstoke: CTA/Macmillan Education Ltd., p.102.

Chang HT, Bartholomew B (1984) *Camellias*. Portland: Timber Press.

Chen L, Zhou ZX, Yang YJ (2007) Genetic improvement and breeding of tea plant (*Camellia sinensis*) in China: from individual selection to hybridization and molecular breeding. Euphytica, 154: 239-248.

De Costa WAJM, Mohotti AJ, Wijeratne MA (2007) Ecophysiology of tea.

Brazilian Journal of Plant Physiology, 19(4): 299-332.
Ellis R, Nyirenda HE (1995) A successful plant improvement programme on tea (*Camellia sinensis*). Experimental Agriculture, 31: 307-323.
Esan EB (1982) Acquisition of tea germplasm. In: Annual Report of the Cocoa Research Institute of Nigeria (CRIN), p.9.
Esan EB (1989) Progress in tea (*Camellia sinensis* L.) in Nigeria. In: Progress in Tree Crop Research (2nd Edition). Cocoa Research Institute of Nigeria, pp.209-216.
Esan EB, Omolaja SS (2002) Genotypic association, path analysis and pluck quality value in tea (*Camellia sinensis* (L.) O. Kuntze). Tropical Agriculture (Trinidad), 79(2): 1-4.
Esan EB, Aikpokpodion OP, Obatolu CR (1998) Numerical analysis of variations in leaf morphometric characteristics of tea clones (*Camellia sinensis* (L.) O Kuntze) in Nigeria. Nigerian Journal of Tree Crop Research, 2(1): 47-59.
Esan EB, Bright-Agindotan CJ and Omolaja SS (2001) *In vitro* assessment of tea (*Camellia sinensis)* seed ex-plant types for vegetative propagation potential in Nigeria. In: Proceedings of 2001 International Conference on O-Cha (tea) Culture and Science. 5-8 October, 2001, Shizuoka, Japan, 336-338.
Famaye AO, Oloyede AA, Ayegboyin (2006) Handbook on Tea Production in Nigeria. Akure: Pamma Press.
Filani GA, Okelana MAO (1980) The prospects and problems of tea production in Nigeria. In: Proceedings of the Cocoa Board Symposium. 20 August, 1980, University of Ibadan, Nigeria, pp.182-190.
Food and Agriculture Organization (FAO) (2008, 2009, 2011) http://faostat.fao.org/.
Hainsworth E (1981) Tea production on the Mambilla Plateau, Gongola State (now Taraba State), Nigeria. A report on the project by the consultant for the Nigerian Beverages Production Company Limited, p.26.
International Tea Committee (ITC) (2010) Annual Bulletin of Statistics, London.
Mather K, Jinks JL (1982) Biometric Genetics. London: Chapman and Hall.
Ng'etich WK (1996) Pruning in tea: a review focusing on experiments in East Africa. Tea, 17(1): 41-46.
Obatolu CR (1984) Soil of tea (*Camellia sinensis* L.) growing areas of the Mambilla Plateau of Nigeria. CRIN Seminar Series, May, 1984.
Obatolu CR, Ipinmoroti RR (2000) The comparative study of five tea clones under plantain shade in Ibadan, South Western Nigeria. In: Proceedings of Horticultural Society of Nigeria. pp.207-214.
Omolaja SS, Fawole I (2000) Diallel analysis of self- and cross-compatibility in selected clones of *Coffea canephora* Pierre. Nigerian Journal of Science, 34 (4): 417-425.
Omolaja SS, Esan EB (2006) Evaluation of high altitude tea (*Camellia sinensis* (L.) O. Kuntze) for adaptability and yield in lowland ecologies of Nigeria. Bioversity International-FAO, 148: 32-37.
Omolaja SS, Ipinmoroti RR, Adeyemi EA (2008) Effects of tea/eucalyptus intercrop on the yield and quality of some clones of tea [*Camellia sinensis* (L.) O. Kuntze] at Mambilla Plateau, Nigeria. Moor Journal of Agricultural

Research, 9: 45-53.

Opeke LK (1982) Tropical Tree Crops. New York: John Willey and Sons, pp.294-297.

Owuor PO (1992) Comparison of gas chromatographic volatile profiling methods for assessing the flavour quality of Kenya black teas. Journal of the Science of Food and Agriculture, 59: 189-197.

Rogers SS (1975) Preliminary observations on pollen tube incompatibility in some tea clones. Tea Quarterly, 45: 463-470.

Sanderson GW (1963) The chloroform test: A study of its suitability as a means of rapidly evaluating fermenting properties of clones. Tea Quarterly, 34: 193-196.

Singh RK, Chaudhary BD (1979) Biometrical Methods in Quantitative Genetics Analysis (2nd Edition). Ludhiana: Kalyani Publishers, 304p.

Visser T (1969) Tea [(*Camellia sinensis*) (L.) O. Kuntze]. In: Ferwerda FP and Wit F (eds.) Outlines of Perennial Crop Breeding in the Tropics. Miscellaneous Paper 4, Wageningen: The Netherlands, pp.459-493.

Yao MZ, Chen L, Liang YR (2008) Genetic diversity among tea cultivars from China, Japan and Kenya revealed by ISSR markers and its implication for parental selection in tea breeding programmes. Plant Breeding, 127(2): 166-172.

Yao MZ (2009) Studies on genetic diversity and structure of tea germplasm in China based on ISSR and EST-SSR markers. PhD Thesis, Zhejiang University, China, p.90 (in Chinese).

13

Genetics and Chemistry of the Resistance of Tea Plant to Pests

Zong-Mao Chen*, Xiao-Ling Sun, Wen-Xia Dong

Tea Research Institute of the Chinese Academy of Agricultural Sciences, 9 South Meiling Road, Hangzhou 310008, China
*E-mail: chenzm@mail.tricaas.com

Abstract: Insect and mite pests as well as diseases occurring on the tea plant are described in this chapter. The damage caused by these harmful pests and diseases ranges from 5% – 55%, and generally from 10% – 20%. The importance and roles of the resistance of the tea plant to the insect pests and diseases were discussed from the viewpoint of IPM (integrated pest management). The mechanism of resistance of the tea plant including the morphological, physiological and biochemical basis was discussed. By using tea gray blight (*Pestalotiopsis longiseta* (Spegazzini) Dai & Kobayashi), tea anthracnose (*Gloeosporium theae-sinensis* Miyake) and mulberry scale (*Pseudaulacaspis pentagona* Targioni-Tosswtt) as the examples, the genetic basis of resistance of the tea plant to tea diseases and pests was illustrated. The future prospects for the resistance of tea plants were discussed in relation to the development of biotechnology and chemical ecology.

Tea, a favorite beverage of people all over the world, is made from the tender new shoots of the tea plant (*Camellia sinensis* (L.) O. Kuntze). The tea plant is cultivated in more than 52 countries in the world, distributed between latitude 27° S (Corrientes, Argentina) to 43° N (Georgia), in regions including tropical, subtropical and temperate climatic zones. The warm, humid climate typical of most tea-growing countries plus the perennial growth habit of the tea plant provide a relatively steady microclimate and food supply for insect pests and diseases. More than one thousand insect pests (including mites) and more than 500 fungal, bacterial and nematode diseases have been described on the tea plant (Chen & Chen, 1990). Damage to yield and quality caused by these pests and diseases are especially serious in the subtropical and tropical tea areas. The damage in yield was reported to be from 5% – 55% (Hazarika *et al.*, 2009), and generally 10% – 20% (Banerjee, 1983; Cranham, 1966; Chen & Chen, 1990; Muraleedharan, 1992; Muraleedharan & Chen, 1997; Hazarika *et al.*, 2009).

13.1 Situation of Insect Pests and Diseases in Tea Production

Insect pests cause serious damage to the tea shoots and a direct reduction in tea production. Some insect pests cause a weakening of the tea plant and an indirect reduction in tea production. Tea diseases are serious in tropical climate zones with warm weather and a humid environment.

13.1.1 Insect Pests

The insect pests found on the tea plant can be divided into the following four kinds: (1) Sucking pests are the most serious pests in the tea ecosystem, such as the tea leafhopper (*Empoasca vitis* Gothe) and tea black spiny whitefly (*Aleurocanthus spiniferus* Quaintance) in China, the tea mosquito bug (*Helopeltis theivora* Waterhouse) in Africa, the mulberry scale (*Pseudaulacaspis pentagona* Targioni-Tosswtt) in Japan, the tea mirid (*Helopeltis schoutedeni* Reuter) and tea aphid (*Toxoptera aurantii* Boyer de Fonscolombe) in India and Sri Lanka. (2) The defoliating chewing pests are the most destructive pests in tea production and may cause serious damage to the shoots and leaves of the tea plant, such as the tea looper (*Ectropis oblique* Warren) and tea tussock (*Euproctis pseudoconspersa* Strand) in China, the tea small leafroller (*Adoxophyes honmai* Yasuda) and tea leafroller (*Homona magnanima* Diakonoff) in Japan, and tea tortrix (*Homona coffearia* Nietner) in India and Sri Lanka. (3) Tea mites are serious kinds of pests in various tea producing countries in the world. They include the tea pink mite (*Acaphylla theae* Watt) and scarlet mite (*Brevipalpus obovatus* Donnadieu) in China, the tea kanzawai mite (*Tetranychus kanzawai* Kishida) in Japan, the tea red spider mite (*Oligonychus coffeae* Nietner) and scarlet mite (*Brevipalpus phoenicis* Geijskes) in India and Sri Lanka. (4) Some stem-boring pests on the tea branch also cause serious damage to the tea plant, such as tea shot-hole borer (*Xyleborus fornicatus Eichhoff*) in Sri Lanka (Muraleedharan & Chen, 1997; Hazarika *et al.*, 2009).

13.1.2 Diseases

Tea disease also causes serious damage. In particular, shoot and leaf diseases are of more concern to tea growers than to the growers of other crops, for the obvious reason that the tea plant is cultivated for its leaves. Blister blight (*Exobasidium vexans* Massee) is the most serious tea disease in China, India, Sri Lanka and Indonesia. In Japan, they have their specific blister blight called Japanese *Exobasidium* blight (*Exobasidium reticulatum* Ito & Sawada). It is caused by a different *Exobasidium* fungus to tea blister blight. In addition, tea anthracnose disease (*Gloeosporium theae-sinensis* Miyake) and brown blight (*Guignardia camelliae* (Cooke) Butler) are the common leaf diseases in China and Japan. *Phomopsis* stem blight (*Phomopsis theae* Petch) is a common but not serious disease in Japan, India and Sri Lanka. Root disease mainly prevails in the tropical tea growing areas including India, Sri Lanka and Indonesia. The nematode disease includes the root knot (*Meloidogyne incognita* (Kofoid et White) Chitwood, *Meloidogyne javanica* (Treub) Chitwood) and is serious in those tropical tea producing countries. In addition, red root rot (*Poria hypolaterita* (Berk) Cooke, *Ganoderma* sp.) and brown root rot (*Phellinus lamaensis* (Murrill) Heim) are important in India, Sri Lanka, Indonesia and the southern part of China (Chen & Chen, 1990; Ezuka & Ando, 1994).

13.2 Resistance as an Efficient Part of IPM in the Tea Ecosystem

Plant resistance as a part of agricultural control in IPM (integrated pest management) has attracted more attention in agricultural practice. Tea being a healthy beverage crop, chemical control was limited because of the pesticide residue, taint of chemicals and resistance of pests to chemical control. Thus, the utilization of plant resistance in IPM was emphasized in tea production.

13.2.1 The Type of Resistance

Tea is a beverage crop. People drink tea by using made tea infused by hot water. Water is a polar solvent, so it may extract the pesticides on/in the tea leaves in the infusion. From the viewpoint of safety, it is recommended to minimize the use of pesticides as much as possible so to produce non-pollution tea. The tea ecosystem is a mixture of many species. There are obvious differences in the resistance to pests and diseases. According to the degree of resistance of the tea plant, the resistance can be divided into the following 5 types (Chen & Chen, 1990).

Immunity: The immune cultivar is expressed in full immunity to the damage of pests and infestation of pathogens. However, it is rare to discover an immune tea cultivar in tea production.

Resistance: The resistant cultivar is characterized by less infestation when compared to a susceptible cultivar under the same conditions. For example, the cultivars Longjing 43 and Fuding Dabaicha are resistant to brown blight disease (Chen *et al*., 1996). The incubation period of brown blight on these 2 resistant cultivars is longer (7 – 14 days and 9 – 11 days) than that of susceptible cultivars (4 – 7 days) (Chen & Chen, 1990). According to the evaluation of the resistance of 276 tea cultivars in Japan on the mulberry scale (*Pseudaulacaspis pentagona* Targioni-Tosswtt), there are great varietal differences in infestation to mulberry scale. There were more resistant cultivars in large-leaf tea varieties (*Camellia sinensis* var. *assamica* (Masters) Kitamura) than in small-leaf varieties (*Camellia sinensis* (L.) O. Kuntze var. *sinensis*) (Furuno *et al*., 2001).

Susceptibility: The susceptible cultivar is characterized by the greater damage caused by pests and diseases. As an example, the cultivar Yunnan Daye is highly susceptible to the *Guignardia camelliae* (Cooke) Butler fungus. The cultivar Yabukita is highly susceptible to the *Gloeosporium theae-sinensis* Miyake fungus. So anthracnose disease is epidemic in a large area of Japanese tea production due to the fact that the cultivation area of this susceptible cultivar occupies approximately 77% of the total tea acreage in Japan (Mondal, 2008).

Escaping: The mechanism of escape, when the plant is susceptible, is not consistent with the period when pests and diseases occur. As an example, the occurrence of tea bud blight (*Phyllosticta gemmiphliae* Chen et Hu) on Georgia cultivars is less than on other cultivars due to their late sprouting time (Chen & Chen, 1990).

Tolerance: Some cultivars have the same degree of infestation when compared to other cultivars. However, the decrease in yield is not significant due to the former tolerant cultivar compensating for the effects of pests or diseases. For example, the yield of tolerant cultivars is much higher than that of susceptible cultivars when infested by tea shot borer in Sri Lanka due to the root system recovering faster.

13.2.2 Resistant Tea Cultivars are Necessary to IPM

In IPM of tea production, all tea producing countries pay more attention to the resistance of tea plant to pests and diseases. During the development of a tea garden, it is important to know the most important pests and diseases and then the resistant plant cultivars. In the research work on tea science, the resistance to major tea pests and diseases is the important target of tea breeding projects and many resistant cultivars have been successfully fostered. Table 13.1 lists some resistant tea cultivars developed in major tea producing countries (Chen & Chen, 1990; Mizuta & Nagatomo, 2004; Mondal, 2008).

Table 13.1 Some resistant tea cultivars bred in major tea producing countries

Character	Cultivar	Country	References
Mite tolerance	TRFK 7/9	Kenya	TRFK, 1999
Scale resistance	EPK TN 14-3	Kenya	TRFK, 1999
Mulberry scale resistance	Hatsumomiji Yumekaori	Japan	Mizuta *et al.*, 2004 Nagatomo *et al.*, 2007
Anthracnose resistance	Yamatomidori, Fujimidori	Japan	Ikeda & Amma, 2004
Gray blight resistance	Benihomare, Cha Chuukanbohon Nou 6	Japan	Takeda, 2003
Brown blight resistance	Longjing 43, Fuding Dahao	China	Chen & Chen, 1990
Blister blight tolerance	TRI 2043, DT 1	Sri Lanka	Hajra, 2001

13.3 Mechanism of Resistance of Tea Plant to Pests and Diseases

Different tea cultivars are variable in morphological and genetic characteristics, differentially impacting on the damage levels of pests and diseases that attacked (Banerjee, 1987; Hajra, 2001). Understanding the mechanism of tea plant resistance to pests and diseases is the basis of breeding the resistant cultivar. It can be discussed from the following aspects: morphological, physiological and biochemical mechanisms.

13.3.1 Morphological Mechanism

Some morphological features including cuticularization on the undersurface, the pubescence density on tea leaves, amounts of stomata, thickness of the cuticle, color and shape of leaves, may influence the penetration of pathogens and deposition of insects.

Anthracnose is a major disease of mature and old leaves of the tea plant in China and Japan. In an investigation into the resistance of the tea plant to anthracnose, it was discovered that the pathogen of anthracnose (*Gloeosporium theae-sinensis* Miyake) penetrates into the leaf tissue via the pubescence on the undersurface of tea leaves and spreads along the cavity of pubescence into the leaf tissue. The cell wall of pubescence gradually thickens with the ageing of tea leaves until the cavity is fully blocked up. Hence, the pathogen penetrates and infects the tender tea leaves on only one bud and one leaf shoot or one bud and two leaf shoots. Investigation indicated that the resistance of tea cultivars is closely related to the amount of hair on the undersurface of tea leaves. The thickening rate of the lateral wall of tea hair in resistant tea cultivars is faster than that in the susceptible cultivars. These secondary cells with a thicker wall could be transformed into a lignified structure, and fill the cavity with hair, thus making it impossible for the pathogen to penetrate the tea leaves (Ezuka & Ando, 1994). Those cultivars that lack hair or have less pubescence are generally highly resistant.

Those plants with more pubescence are generally unfavorable for pests that feed with sucking and piercing mouthparts. For example, for the leafhopper it is difficult to reach the parenchyma tissue of the leaf surface with a layer of dense pubescence. And the thickness and rigidity of leaves were suggested to be feeding barriers for the tea leafhopper, which is difficult to overcome (Zeng *et al.*, 2001). For lepidopterous pests, it is also difficult to feed and oviposit on a leaf surface with dense pubescence. Besides, if the pubescence contained high amounts of lignin, the insect pests would be prematurely dead if excessive amounts of these tea leaves were ingested. And pubescence on the surface of tea leaves could also make it hard for the insect to cling to them.

Tea pink mite (*Acaphylla theae* Watt) is a serious pest in China. Investigations into the mechanism of resistance of tea cultivars showed that the resistant cultivars possessed a higher density of pubescence, a thicker cuticular layer on the undersurface and lower stomatal density than susceptible tea cultivars (Chen *et al.*, 1996; Xu *et al.*, 1996) (Table 13.2). The Chinese tea cultivars found to be resistant to tea pink mite included Fuding Dahao, Maoxie and Yunqi among 6 tested cultivars.

Table 13.2 Morphological characteristics on the tea leaf surface between resistant and susceptible tea cultivars

Items	Average of resistant cultivars ($n = 4$)	Average of susceptible cultivars ($n = 5$)	T value	Degree of difference
Stomata amounts (stomata/mm^2)	92.75	376.50	11.88	Extreme significance
Pubescence density (hair/mm^2)	6.32	2.14	7.15	Extreme significance
Hair length (μm)	739.86	660.61	1.02	Non-significance

$t(0.05) = 2.306$, $t(0.01) = 3.355$

Those anatomical characteristics, including the thickness of the cuticle and the amounts of stomata on tea leaves, are related to the disease resistance of tea plants. Generally, the leaf structure of small-leaf tea cultivars is characterized in the thicker cuticle layer, with 2 – 3 layers of palisade tissue, thus unfavorable to the penetration of pathogens. According to the investigation, these cultivars are generally more resistant to tea brown blight (*Guignardia camelliae* (Cooke) Butler). The leaf structure of susceptible large-leaf cultivars is characterized by the thin cuticle, with one layer of palisade tissue, so favorable to the penetration and spread of the pathogen (Chen & Chen, 1990).

Investigations revealed that the occurrence of the pathogen and insect were related to the spatial position of leaves and leaf structure. Tea red leaf spot (*Phyllosticta theicola* Petch) is a popular tea pathogen in China, always infecting tender leaves of the tea plant. Tender leaves such as the first and the second leaves below the buds, especially those which spread horizontally, were easily infected by the disease, but it was difficult for the pathogen to form appressoria on the surface of the fourth and the fifth leaves below the buds. Cultivars with a thick leaf cuticle, lax spongy cell arrangement and more palisade layers possessed a relatively strong resistance to the pathogen (Gao, 1997). *Polyphagotarsonemus latus* Banks is another instance, which addicts the tea plant with leaves spread horizontally and less density of pubescence (Liu & Zhou, 1994). The *Polyphagotarsonemus latus* Banks resistant cultivars possessed a higher density of pubescence, lower stoma density and a thicker cuticle on the lower surface (Liu *et al.*, 1999).

According to the investigation conducted in Sri Lanka, the structure of the tea plant is related to the feeding habits of insect pests. For example, the wood hardness of a tea branch is related to damage by termites. The harder the tea stems and branches, the less the damage appears on the cultivar.

The other physical factors, such as the leaf color, leaf size and shape also influence the resistance to pests and diseases of the tea plant. For example, leaves of kelly green are easier to find by tea leafhoppers than any other leaf colors, especially fuscous leaves. So, tea plants with leaves of kelly green are favored by tea leafhoppers, whereas tea plants with early sprouting, curly leaves, a long period of tenderness or rich fuzz on the leaf are susceptible to leafhoppers (Zhu, 1992; Zhang *et al.*, 1994).

13.3.2 Physiological Mechanism

After the pests and pathogens have penetrated and been deposited, the tea plant will continue to display resistance via the physiological mechanism.

(1) Hypersensitive necrosis reaction. The pathogen of tea blister blight (*Exobasidium vexans* Massee) invades the tea leaves, the phenomenon of 'Green Island' will appear in the penetrated area of infested leaves. The pathogen of tea blister blight is an obligate fungus, so further infection will be limited. It is a hypersensitive guard reaction (Chen & Chen, 1990). According to investigation of the tea root lesion nematode disease, it was shown that the high resistance clones possessed a rapid necrosis reaction and a corky structure formed surrounding the infested area.

(2) Slow developing reaction. After infection by the pathogen, the incubation period in the resistant cultivars is rather longer than that in susceptible cultivars. It is a kind of relative resistance, a quantitative resistance. Investigations conducted in China showed that the incubation period of tea brown blight in various cultivars is varied. Those resistant cultivars such as Longjing 43 and Fuding Dahao possessed a longer incubation period than that in susceptible cultivars such as Yunnan Daye. In addition, the amount of spots on resistant cultivars was more than that on susceptible cultivars (Chen *et al.*, 1996; Hazarika *et al.*, 2009).

Mulberry scale is a serious insect pest in Japan. It has spread over the tea area of Japan since 1985, and showed an epidemic tendency since 1993. According to survey information, this pest occurred in 73% of the total tea garden in Miyazaki,

Japan (Sudoi, 1997). It is a phloem feeding pest. A breeding program for the selection of mulberry scale resistant clone was conducted in Japan in the 21st century. Research indicated that there is an obvious difference in the sap pressure in phloem between the resistant and susceptible cultivars (Mizuta & Nagatomo, 2004; Sudoi, 1997). The nutritional sap sucked from the phloem of resistant tea cultivars and susceptible tea cultivars showed a significant difference (Table 13.3) (Nagatomo *et al.*, 2007). The nymph of mulberry scale sucked much less nutritional sap from the resistant tea plant than that from the susceptible tea plant. Related to this phenomenon, the adults of scale feed on the resistant tea plants of small body size and oviposited far less eggs than on the susceptible tea plants due to insufficient nutrition. According to analysis based on the electronic monitoring system (EMS) in Japan, the B waveform was shorter on the resistant cultivars and the C and E waveforms were shorter on the susceptible cultivars. Wax secretion was not observed on any female scale on the resistant cultivars, while it was observed in almost all females on the susceptible cultivars (Mizuta & Nagatomo, 2004). By means of this mechanism of feeding activity, a new cultivar, Yumekaori, was successfully fostered in Japan recently (Mizuta & Nagatomo, 2004).

Table 13.3 Mechanism and evaluation of the resistance to mulberry scale of different tea cultivars

Cultivar	Number of eggs per female scale (average ±SE)*	Infestation degree under natural conditions #	Evaluation on resistance
Yumekaori	4.6±0.8 a	1.0	Strongly resistant
Sayamakaori	32.4 ±4.7 b	1.2	Strongly resistant
Yabukita	80.8±6.6 c	3.3	Susceptible
Yudakamedori	101.7±7.3 cd	2.8	Susceptible
Kanayamedori	120.4 ±9.9 d	3.7	Susceptible

* Ratio of amounts of eggs and the total of oviposited eggs and eggs in ovary.

According to the amounts of male cocoons in the 4th generation in 2001. The value means 1(zero)-5 (many cocoons)

13.3.3 Biochemical Mechanism

The biochemical mechanism of resistance is a complicated and comprehensive consequence. The relationship between the tolerance of tea cultivars to *Xyleborus fornicatus* Eichhoff, a stem borer insect, and the availability of a kind of sterol, the α-spinasterol, is an interesting biochemical process. It is essential for the normal growth and metamorphosis of insects. *Xyleborus fornicatus* Eichhoff is unable to synthesize the sterol nucleus itself. However, the α-spinasterol can be converted by *Xyleborus fornicatus* Eichhoff insects to moulting hormones that are required for pupation of the beetle larvae. Investigation showed that the resistance of the tea plant to tea stem borer is related to the contents of α-spinasterol (Wickremasinghe *et al.*, 1976; Wickremasinghe & Thirugnanasuntharan, 1980). Analysis showed that the susceptible tea cultivars contained more α-spinasterol than that in the tolerant cultivars. The other factor is the availability of α-spinasterol. It depends on two factors: one is the concentration of the sterol *per se*, the other is the level of saponin. Saponin would bind the sterol and so reduce the α-spinasterol availability. It is also a factor in determining the degree of susceptibility to insect infestation. In addition, saponin could have an adverse effect on the development of stem borer. Analysis showed that the tolerant cultivars contained a higher saponin concentration than that in susceptible cultivars. Investigation indicated that the degree of tolerance of the tea plant to stem borer pest is closely related to the combined effects of α-spinasterol and saponin (Wickremasinghe *et al.*, 1976; Wickremasinghe & Thirugnanasuntharan, 1980). Generally, the tolerant cultivar contained lower α-spinasterol and higher saponin.

Another investigation, related to the resistance of the tea cultivar to red spider mite (*Oligonychus coffeae* Nietner) and the rhodoxanthin content was conducted in Sri Lanka (Muraleedharan, 1992). Results showed that the contents of rhodoxanthin in tea leaves are closely related to the degree of infestation of the tea plant by red spider mite. A further investigation showed that rhodoxanthin is a phagostimulant and promoter of propagation of mites (Muraleedharan & Chen, 1997).

The amino acid content in tea leaves is closely related to the infestation degree of tea diseases and insect pests. Research showed that the cultivars with higher arginine content are usually a target for infestation by tea Kanzawai red spider (*Tetranychus kanzawai* Kishida) (Takeshi, 1976; Takeda, 2003). Tea cultivars possessing higher contents of acidic amino acids such as theanine, glutamic acid, aspartic acid etc., are resistant to *Tetranychus kanzawai* Kishida (Kaneko, 1981). Investigation showed that this mite prefers arginine as the oviposition stimulator. On the contrary, those cultivars with higher contents of theanine are resistant to Kanzawai spider mite. So the arginine/theanine ratio was used in the evaluation of the resistance of tea cultivars to Kanzawai spider mite in Japan (Takeda, 2003). Analytical results showed that *Acaphylla theae* Watt resistant cultivars possessed higher contents of total amino acids and theanine, glutamic acid, aspartic acid, caffeine, and a lower content of reduced sugar and water soluble sugar (Chen *et al.*, 1996; Xu *et al.*, 1996).

The resistance of the tea plant to pink mite is related to the contents of caffeine, soluble sugar and total amino acids in tea shoots (Chen *et al.*, 1996). Investigation showed that the contents of caffeine and amino acids in tea shoots are closely related to the resistance of tea cultivars to pink mite, and the contents of soluble sugars in tea shoots are negatively related to the resistance of the tea plant (Chen *et al.*, 1996). The average contents of caffeine, total amino acids and soluble sugar of resistant tea cultivars and susceptible cultivars were 3.07% and 2.67% of dry matter, 1,488.29 mg/100 g and 759.72 mg/100 g tea shoot as well as 3.51% and 4.34% of dry matter, respectively. A further investigation showed that caffeine and amino acids possess an escaping effect on pink mite. The higher the concentration of caffeine and amino acids, the stronger the escaping effect revealed (Chen *et al.*, 1996).

Polyphenols are the characteristic compounds of the tea plant. The contents of polyphenolic compounds are as high as 35% – 40% in the leaves of the tea plant. Virus disease is generally less common in the tea plant compared with other crops, and the reason is estimated to be the high content of polyphenols in tea leaves. Polyphenol-rich clones are believed to be resistant to mites (Lang'at *et al.*, 1998) and were also used as a potential indicator for drought tolerance in the tea plant in Kenya recently (Cheruiyot *et al.*, 2005), while polyphenol-rich but nitrogen-poor clones are resistant not only to *Brevipalpus phoenicis* Geijskes scarlet mite but also to *Ologonychus coffeae* Nietner spider mite (Sudoi, 1997). It was reported that catechins, especially the ether type catechins (epicatechin gallate (ECG), epigallocatechin gallate (EGCG)), showed escaping effects on tea Kanzawai spider mites. An evaluation of the resistance of tea clones to tea gray blight was conducted in Kenya. Results showed that the leaf content of total polyphenols inversely correlated to the size of lesions developed following inoculation with *Pestalotiopsis theae* (Sawada) Steyaert pathogen. The cultivar with higher polyphenol content developed a smaller size of lesion on tea leaves (Lang'at *et al.*, 1998). An increase in the phenol as well as orthodihydroxy phenol contents were always noticed in resistant cultivars following inoculation with *Corticium theae* Barnard, *Pestalotiopsis theae* (Sawada) Steyaert and *Glomerella cingulata* (Stoneman) Spaulding and von Schrenk in India (Chakraborty *et al.*, 2005a). Investigation indicated a greater accumulation of orthodihydroxy phenol in the resistant interaction of tea cultivars against *Corticium theae* Barnard, *Pestalotiopsis theae* (Sawada) Steyaert and *Glomerella cingulata* (Stoneman) Spaulding and von Schrenk pathogens. This indicated that it plays a role in the disease resistance mechanism. It is because orthodihydroxy phenol is easily oxidized to highly reactive quinones which are effective inhibitors of sulphydryl enzymes, thereby preventing the metabolic activities of host and parasite cells.

Some antifungal compounds that showed toxic activity to pathogens were detected in tea leaves after challenge by foliar fungal pathogens. An investigation conducted in India showed that, two days after inoculation, the maximum accumulation of pyrocatechol was detected in *Pestalotiopsis theae* (Sawada) Steyaert, *Corticium theae* Barnard and *Glomerella cingulata* (Stoneman) Spaulding and von Schrenk inoculated leaves of resistant cultivars (Table 13.4) (Chakraborty *et al.*, 2005a). Healthy leaf tissue contained a very low amount of this compound.

Table 13.4 Accumulation of pyrocatechol in tea cultivars after challenge and inoculation with foliar pathogens

Cultivar	Pyrocatechol (mg/g fresh tissue)	
	Healthy	Inoculated
CP 1	96	612 [a]
TV 20	90	610 [b]
TV 9	80	484 [b]
TV 27	78	477 [a]
TV 22	71	346 [b]
TV 17	62	325 [a]
TV 23	75	375 [b]

Inoculated with: [a] Pestalotiopsis theae (Sawada) Steyaert and [b] Corticium theae Barnard

Epicatechin gallate, a component of tea, and its analogues were reported to be against methicillin resistant *Staphylococcus aureus* Rosenbach (Hamilton-Miller & Shah, 2000). The higher level of total sugars in tea cultivars can be correlated with the susceptibility of the tea plants to *Phomopsis* disease (Ponmurugan & Baby, 2007). Tridecanone and undecanone are two compounds related to tea plant resistance to tea weevils (*Myllocerinus aurolineatus* Voss and *Basilepta melanopus* Lefevre), and the resistant cultivars have a high content of these methyl ketones in the leaves (Zheng *et al.*, 2008), while chlorlgenic acid is poisonous to the tea aphid (*Toxoptera aurantii* Boyer de Fonscolombe) and some Lepidopterou' insects, so the content in the tea plant is one of the characteristics of susceptible tea plant cultivars (Zheng *et al.*, 2008).

The activity of some enzymes was reported to be in close relation to the resistance of cultivars to tea pests, especially to tea diseases. An investigation of the determination of enzyme activity on healthy and inoculated tea plants showed that peroxidase activity increased only in resistant cultivars after 48 h of inoculation. This confirmed the role of this enzyme in the resistance mechanism in fungal infection. Peroxidase isozyme pattern analysis revealed that the new isozymes are induced after inoculation and 2 isozymes were found to be specifically associated with resistance (Chakraborty *et al.*, 2005a). An analysis of the phenylalanine ammonia lyase (PAL) enzyme in tea leaves showed the activity of this enzyme increased after the invasion of a fungal pathogen, but only in the resistant cultivars. The highest activity was found in TV 18, a resistant cultivar. The susceptible cultivar, UP26, exhibited a very low increase in activity after inoculation. The authors considered that the induction of higher peroxidase and PAL enzymes activity, along with some novel peroxizymes, play an important role in the case of tea brown blight infection (Chakraborty *et al.*, 2005a). In an investigation of the resistance mechanism of the tea plant to blister blight, the accumulation of chitinas, β-1,3-glucanase and peroxidase following the inoculation of resistant tea cultivar (TV 20) was reported (Chakraborty *et al.*, 2005b).

13.4 Genetics of Resistance of Tea Plant to Diseases and Insect Pests

Significant morphometric and genetic variability existed among tea cultivars to which pests react differentially. Tea is a perennial and a cross pollination crop. Although, over the past 40 years, vegetative clones occupied a large area of world tea production, a relatively large part of the world tea production is still grown from seed. Crossing creates diversity and variation in the progeny. The resistance or susceptibility of the tea plant to diseases and insect pests largely depends on the heredity of the plant, but is also influenced by the environment and variation induced by crossing between the various cultivars. Thus firstly, various tea cultivars showed a significant variance in the degree of resistance and susceptibility. Secondly, the resistance of the same cultivar to various pests may be not constant forever; it will vary due to the environment, the change in biotype or physiological race of pathogen and insect pests.

13.4.1 Phenotype and Genotype Analysis in Resistance Breeding

With the advance in science and technology, the resistance of the tea plant to diseases and insect pests is not investigated just for the evaluation of the degree of resistance, but also to conduct the analysis of the genetic basis, thus providing the rules and information for further selection. In an investigation of the genetic basis of resistance to gray blight, it was shown that the resistance of the tea plant to gray blight (*Pestalotiopsis longiseta* (Spegazzini) Dai and Kobayashi) was found to be controlled by 2 independent dominant resistance genes $\boldsymbol{Pl_1}$ *and* $\boldsymbol{Pl_2}$ (Takeda, 2002; Tanaka, 2006). The $\boldsymbol{Pl_1}$ gene is genetically epistatic in relation to $\boldsymbol{Pl_2}$, therein the $\boldsymbol{Pl_1}$ gene was assumed to confer a high level of resistance to the tea plant, whereas the $\boldsymbol{Pl_2}$ gene only conferred a moderate level of resistance. There were 9 genotypes tested for resistance to *Pestalotiopsis longiseta* (Spegazzini) Dai and Kobayashi in tea plants. Six genotypes $\boldsymbol{Pl_1Pl_1Pl_2Pl_2}$, $\boldsymbol{Pl_1Pl_1Pl_2}pl_2$, $\boldsymbol{Pl_1Pl_1}pl_2pl_2$, $\boldsymbol{Pl_1}pl_1\boldsymbol{Pl_2Pl_2}$, $\boldsymbol{Pl_1}pl_1\boldsymbol{Pl_2}pl_2$, $\boldsymbol{Pl_1}pl_1pl_2pl_2$ were resistant, 2 genotypes $pl_1pl_1\boldsymbol{Pl_2Pl_2}$, $pl_1pl_1\boldsymbol{Pl_2}pl_2$ were moderately resistant and 1 genotype $pl_1pl_1pl_2pl_2$ was susceptible (Takeda, 2002; Tanaka, 2006). Table 13.5 shows the 5 tea cultivars with various resistances to gray blight disease and their genotypes. According to the genetic analysis of various cultivars with different level of resistance, it is possible to predict the genotype of the tea plants with resistance to tea gray blight in the F_1 generation after crossing between the parents. Analysis showed that the cultivars with 2 $\boldsymbol{Pl_1}$ genes are very important materials for the breeding of resistant cultivars to gray blight disease (Tanaka, 2006).

Table 13.5 Resistance degree of 5 cultivars to tea gray blight disease and the proposed genotype

Cultivar	Resistance degree	Proposed genotype
Yabukita	Susceptible	$pl_1pl_1pl_2pl_2$
Fujimidori	Moderate resistance	$pl_1pl_1\boldsymbol{Pl_2}pl_2$
Sayamakaori	Resistant	$\boldsymbol{Pl_1}pl_1pl_2pl_2$
Yamatomidori	Resistant	$\boldsymbol{Pl_1}pl_1pl_2pl_2$
Z-1	Resistant	$\boldsymbol{Pl_1}pl_1\boldsymbol{Pl_2Pl_2}$

According to genetic analysis, the phenotypes and genotypes related to the resistance to tea gray blight of 89 main tea cultivars in Japan were identified (Table 13.6) (Takeda, 2003). Among these, 65 cultivars were derived from Japanese native plants, 9 cultivars originated from China and 15 cultivars were hybrids between the var. *assamica* and var. *sinensis*.

Table 13.6 Phenotypes and genotypes of 89 main tea cultivars in Japan

Phenotype	Genotype*	Cultivars#
Resistant	**Pl_1 Pl_1**??	Benihomare, Benihikari, Benifuuki, Indo, Benifuji, Inzatsu 131, Tadanishiki, Fushun, Kuritawase, Minamisayaka, Karabeni, Chin-Shin-Oolong, San-Cha-Tsi-Lan, Chin-Shin-Da-Pan, Huang-Gan
	Pl_1 pl_1**Pl_2Pl_2**	Benitachiwase, Akane, Houryoku, Satsumabeni, Benikaori, Yaeho, Miyoshi, Izumi, Takachiho, Himemidori, Tamamidori, Unkai, Hoshinomidori, Yamanami, Asagiri, Komakage, Z-1, Okuhikari, Kanaya-15, Makurazaki-4, Makurazaki-5
	Pl_1 pl_1**Pl_2**pl_2	Hatsumomiji, Rukuro, Koyanishi, Shunmei, ME-52, Nka-03, Makurazaki-7, Makurazaki-8, S-6
	Pl_1 $pl_1pl_2pl_2$	Sayamakaori, Yamatomidori, Surugawase, Okumidori, Kanayamidori, Minamikaori, Satouwase, Asanoka, Ooiwase, NN-27, Shizu-zai-16, Saitama-9, Kanaya-7, Miyakei-2, Makurazaki-11, Makurazaki-13, Makurazaki-16
Moderate Resistance	pl_1pl_1**Pl_2Pl_2**	Makizono-dai-chaju, Makurazaki-1, Nagasaki-2
	pl_1pl_1**Pl_2**pl_2	Yutakamidori, Meiryoku, Fujimidori, Minekaori, Kurasawa, Yamakai, Nka-O-278, Unryuu-cha, Makurazaki-18, Makurazaki-23
Susceptible	$pl_1pl_1pl_2pl_2$	Yabukita, Asatsuyu, Saemidori, Toyaka, Fukumidori, Natsumidori, Sayamamidori, Okumusashi, Okuyutaka, Hokumei, Harumidori, NN-12, Mie-260, Kanaya-5

* **Pl_1** gene confers a high level of resistance; **Pl_2** gene confers a moderate level of resistance

\# Cultivars written in bold letters are Assam hybrids between var. *assamica* and var. *sinensis*. Underlined cultivars are of Chinese origin. Cultivars without marks are suitable for Sencha, a kind of Japanese green tea

In the investigation of the genetic basis of the resistance of the tea plant to anthracnose disease, it was indicated that the resistance degree was related to the amounts of pubescence on the leaf of the tea plant (Ezuka & Ando, 1994; Ikeda & Amma, 2004). The large-leaf tea cultivars (var. *assamica*) are generally highly resistant to anthracnose disease. By using the large-leaf varieties as the parent in the crossing system, 50% of the F_1 progeny showed high resistance, the disease severity index will lower than 1.5 (the lower the index shows the higher the resistance). Results showed that the different cultivars harbored different alleles. For example, the susceptible cultivar—Sayamakaori (disease severity index = 3.78 ± 0.51) harbored mostly dominant alleles, while the other resistant cultivar—Yamatomidori (disease severity index = 1.30 ± 0.25) harbored mostly recessive alleles (Ikeda & Amma 2004).

In an investigation of the *Curvularia eragrostidis* (P. Hem.) J. A. Meyer pathogen on the tea leaf, it was shown that the pathogenicity of *Curvularia eragrostidis* (P. Hem.) J. A. Meyer to different cultivars of tea was found to be related to the level of common antigens present between the tea plant and pathogen (Dasgupta *et al.*, 2005).

13.4.2 Utilization of DNA Marker in Resistance Breeding

Based on the genetic analysis and the investigation of the mechanism of resistance, physiological, biochemical and DNA markers were used in tea breeding, so as to speed up the breeding procedure. Nagatomo *et al.* (2007) recommended that the amounts of eggs of mulberry scale on tea plants could be used as the physiological marker to evaluate the resistance degree of tea cultivars (Takeda, 2003). The less the eggs due to insufficient nutrition obtained from the tea plant, the higher the resistance of the tea plant. Tanaka (2006) in Japan developed a simple, rapid, low-cost DNA extraction method and the original 'e-RAPD' (emphasized random amplified polymorphic DNA) system. Based on these, marker-assisted seletion (MAS) method was put forward in tea breeding. In the mulberry scale resistant breeding system, the *MSR1-S1* and *MSR1-S8* resistance gene and some allele-specific markers for the genes were established (Tanaka, 2006).

13.5 Future Prospects

Pest resistant cultivars are the least expensive, safest and most practical way to combat pest damage. Pest resistant cultivars that have the combined attributes of high yield and good quality help to curtail the build-up of noxious pests, at no additional cost. In particular, the strict maximum residue limit (MRL) of pesticide residue was established in the world, the pesticide application is minimized, so as to provide a no-residue or low residue tea. The use of resistant/tolerant cultivars in tea production, therefore, serves as the foundation for evolving ecologically a compatible pest management program.

Resistant cultivars could be obtained from the breeding system including field selection, hybridization and purposeful genetic manipulation. The investigation of the chemo-ecological behavior of pests and the morphological, anatomical and biochemical characteristics of the tea plant, as well as the interaction between pests and the tea plant, provide the selection target in resistance breeding. Fig. 13.1 is an example of this relationship. The development and introduction of biotechnology speed up the process of tea breeding significantly. Management of the tea garden also influences the resistance of the tea plant. Excessive application of nitrogen fertilizer may favor the growth and development of mites and scale and decrease the resistance of the tea plant to the pathogen of diseases. Potassium and potash nutrition may increase the resistance of the tea plant generally.

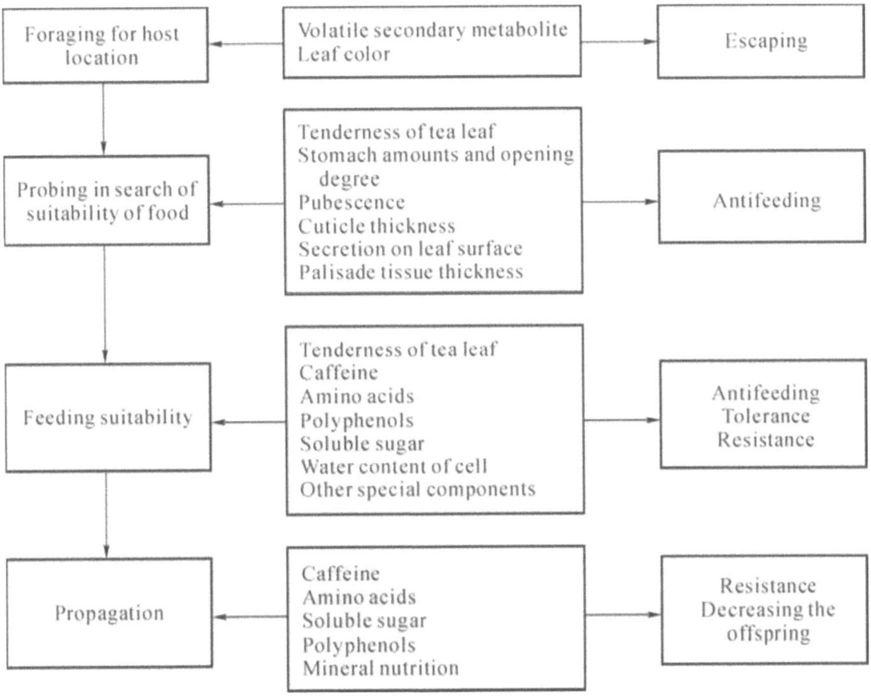

Fig. 13.1. Relationship between the behavior of the pest, the character of the tea plant and the resistance in breeding

The development of chemical ecology provides a new pathway for increasing the resistance of the plant to pests. In the past 20 years, a lot of investigations have contributed to the relationship between the host plant-herbivore-natural enemies. Results showed that there is a chemical communication between the tea plant-tea pests-natural enemies. The pests are located on the tea plant via the volatiles emitted from the tea plant. Every plant has its characteristic fingerprint volatile that differs from the volatiles of other plants. When pests damage the tea plant, the metabolic pathway changes and the damaged plant will emit specific volatiles which are different from the volatiles emitted before damage. These new specific volatiles play a role of "calling for help" (Baldwin *et al*., 2006). It is attractive to the natural enemies—the foe of the pest, and generally escapes to the tea pests. According to the investigation, the volatile is the carrier of chemical communication. It regulates the density and the kind of species in the ecosystem. On the basis of mastering the knowledge of the attracting spectrum of volatiles, scientists started to apply the formulation of various volatiles to attract the natural enemies of the pest and to search for plants that emit more volatiles which attract natural enemies, as well as the gene(s) responsible for volatile emission. Then they will make the genetic transformation to produce the transgenic plant that contains the target gene(s). These plants will emit more specific volatiles which could attract the natural enemies to protect the crop and reduce the density of pests. It is a special method to increase the "resistance" of the plant to pests. Some achievements have been obtained with some crops. The investigation of the chemical communication between tea plant-pest-natural enemies was conducted in China and Japan, and the breeding of highly specific volatile tea cultivars will be conducted in the near future.

References

Baldwin I, Halitschke R, Paschold A, von Dahl CC, Preston C (2006) Volatile signaling in plant-plant interactions: "talking trees" in the genomics era. Science, 311: 812-815.

Banerjee B (1983) Arthropod accumulation on tea in young and old habitats. Ecological Entomology, 8: 117-123.

Banerjee B (1987) Can leaf aspect affect herbivory? A case study with tea. Ecology, 68: 839-843.

Chakraborty BN, Rana S, Das S, Das G, Som R, Datta S, Chakraborty U (2005a) Defense strategies of tea towards foliar pathogens. In: Chakraborty U, Chakraborty B (eds.) Stress Biology. New Delhi: Narosa Publishing House, pp.208-215.

Chakraborty BN, Sharma M, Das Biswas R (2005b) Defense responses in tea plants triggered by *Exobasidium vevans.* In: Chakraborty U, Chakraborty B (eds.) Stress Biology. New Delhi: Narosa Publishing House, pp.226-232.

Chen HC, Xu Ning, Chen XF, Chen ZM, Yu FL (1996) On the resistance mechanisms of tea clones to pink tea rust mite. Acta Phytophylacica Sinica, 23: 137-142 (in Chinese).

Chen ZM, Chen XF (1990) The Diagnosis of Tea Diseases and Its Control. Shanghai: Shanghai Science & Technology Publishers, pp.32-41 (in Chinese).

Cheruiyot EK, Mumera LM, Ng'etich WK, Hassanali A, Wachira F (2007) Polyphenols as potential indicators for drought tolerance in tea (*Camellia sinensis* (L.)). Bioscience, Biotechnology and Biochemistry, 71: 2190-2197.

Cranham JE (1966) Tea pests and their control. Annual Review of Entomology, 11: 491-514.

Dasgupta S, Saha D, Saha A (2005) Levels of common antigens in determining pathogenicity of *Curvularia eragrostidis* in different tea varieties. Journal of Applied Microbiology, 98: 1084-1092.

Ezuka A, Ando Y (1994) Tea diseases in Japan. Japan Plant Protection Association, pp.182-204.

Furuno T, Nagatomo S, Nonaka T, Shige M, Tanka T (2001) Varietal difference of tea plant to the infestation of *Pseodaulaspis pentagoma.* Tea Research Journal, 91: 5-12 (in Japanese).

Gao XH (1997) Relationship between *Phylosticta theicola* Petch and leaf structure and different points in space. Journal of Tea Science, 17(1): 21-26 (in

Chinese).
Hajra NG (2001) Advances in selection and breeding of tea—a review. Journal of Plantation Crops, 29: 1-17.
Hamilton-Miller JMT, Shah S (2000) Activity of the tea component epicatechin gallate and analogues against methicillin resistant *Staphylococcus aureus*. Journal of Antimicrobial Chemotherapy, 46: 847-863.
Hazarika LK, Bhuyan M, Hazarika BN (2009) Insect pests of tea and their management. Annual Review of Entomology, 54: 267-284.
Ikeda N, Amma S (2004) Diallel analysis of resistance to anthracnose in tea. Breeding Research, 6: 135-141 (in Japanese).
Kaneko T (1981) Relationship between the content of free argininemia and ovipisition position of *Tetranychus kanzawai* Kishid. Tea Technology Research, 60: 12-16.
Lang'at Jk, Othieno W, Musau JM (1998) Evaluation of some Kenyan tea clones for resistance/susceptibility to *Pestalotiopsis theae* as influenced by some chemical attributes of mature green leaf. Tea, 19: 6-10.
Liu YQ, Xu Z, Zhou ZK, Xie DX, Yang XH (1994) Morphological and biochemical parameters of tea varieties resistance to *Polyphagotarsonemus latus* Banks. Journal of Sichuan Agricultural University, 17(2): 187-191 (in Chinese).
Liu ZS, Zhou JG (1994) Progress in the field of tea breeding researchers in the past 30 years in China. Journal of Tea Science, 14(2): 89-94 (in Chinese).
Mizuta T, Nagatomo H (2004) Analysis of feeding activity of *Pseodaulaspis pentagoma* (Hemiptera: Diaspididae) on the resistant cultivars of tea plants. Tea Research Report, 98: 21-32.
Mondal TK (2008) Tea breeding. In: Jain SM, Priyadarshan PM (eds.) Breeding Plantation Tree Crops: Tropical Species. Berlin: Springer Science + Business Media, pp.547-589.
Muraleedharan N (1992) Pest control in Asia, In: Wilson KC, Clifford NM (eds.), Tea: Cultivation to Consumption, London: Chapman and Hall, pp.735-412.
Muraleedharan N, Chen ZM (1997) Pests and diseases of tea plant and their management. Journal of Plantation Crops, 25: 15-43.
Nagatomo H, Sato K, Furuno T, Hirakawa I, Nagatomo H, SatoK (2007) The cultivation of 'Yumekaoro' of green tea which is resistant to *Pseodaulaspis pentagoma*. Tea Research Report, 104: 1-14.
Ponmurugan P, Baby UI (2007) Evaluation of fungicides and biocontrol agents against *Phomopsis* canker of tea under field conditions. Australasian Plant Pathology, 36: 68-72.
Sudoi V (1997) Tea pests with special reference to mites: Research achievements and future thrusts. Tea, 18: 156-165.
Takeda Y (2002) Genetic analysis of tea gray blight resistance in tea plants. Japan Agricultural Research Quarterly, 36: 143-150.
Takeda Y (2003) Phenotypes and genotypes related to tea gray blight disease resistance in the genetic resources of tea in Japan. Japan Agricultural Research Quarterly, 37: 31-35.

Takeshi K (1976) Resurgence of the sucking pests in tea production and its affecting factor. Tea, 29: 49-53.

Tanaka J (2006) Study on the utilization of DNA markers in tea breeding. Bulletin of National Institute of Vegetable and Tea Science, 5: 113-155 (in Japanese).

TRFK (1999) Annual Technical Report. Tea Research Foundation of Kenya (TRFK), Tea Board of Kenya.

Wickremasinghe RL, Thirugnanasuntharan K (1980) Biochemical approach to the control of *Xyleborus fornicatus*. Plant & Soil, 55: 9-15.

Wickremasinghe RL, Perera BPM, Perera PWC (1976) α-Spinasterol, temperature and moisture content as determining factors in the infestation of *Camellia sinensis* by *Xyleborus fornicatus*. Biochemical Systematics and Ecology, 4: 103-110.

Xu N, Chen XF, Chen HC, and Chen ZM (1996) Morphological and biochemical parameters of tea varieties resistance to pink mite (*Acaphylla theae* Watt). Journal of Tea Science, 16: 125-130 (in Chinese).

Zhang YL, Zhang JW, Yang Y, Huang YH, Wang YJ (1994) Investigation of tea resistant cultivar resources and resistant mechanisms. Tea communication, 4-6 (in Chinese).

Zheng GY, Liang L Y, YangYQ, Gao X H (2008) Material base of tea resistance to pest insects. Journal of Tea Business, 30: 16-18 (in Chinese).

Zhu JQ (1992) Primary investigation of the resistance of different tea cultivars to *Empoasca vitis* Gothe. Acta Phytophylacica Sinica, 19(1): 29-32 (in Chinese).

14

Germplasm and Breeding Research of Tea Plant Based on DNA Marker Approaches*

Sui Ni[1§], Ming-Zhe Yao[2§], Liang Chen[2※], Li-Ping Zhao[2], Xin-Chao Wang[2]

[1] Faculty of Life Science and Biotechnology, Ningbo University, Ningbo 315211, China

[2] Research Center for Tea Germplasm and Improvement, Tea Research Institute of the Chinese Academy of Agricultural Sciences; National Center for Tea Improvement, Hangzhou 310008, China

※ E-mail: liangchen@mail.tricaas.com

§ These authors contributed equally to this work

Abstract: Tea is the most popular non-alcoholic and healthy beverage worldwide. Tea production contributes greatly to the economy and provides job opportunities for many countries in Asia and Africa. Meanwhile, the germplasm of tea, with a huge potential for the future of the whole tea industry, is presently one of the most valuable and fundamental materials for tea breeding and tea biotechnology. DNA molecular markers have been proven to be robust and valuable approaches in the studies of genetic diversity and variation, molecular identification, molecular phylogenetics, genetic stability and integrity of tea germplasm and the genetic linkage map for breeding of tea. In this paper, a brief description of the molecular marker studies of tea has been summarized. The purpose is to provide an effective way of undertaking a massive tea germplasm appraisal and evaluation, to develop new applicable and cheap DNA markers, to establish a high density genetic linkage map and analyze the agronomically important QTLs and, finally, to facilitate the markers assisted early selection and shorten breeding procedures in tea.

* Reprinted with the kind permission from the Higher Education Press of China, first published in *Frontiers of Agriculture in China*, 2008, 2(2): 200-207

14.1 Introduction

Tea, *Camellia sinensis* (L.) O. Kuntze, is the most popular non-alcoholic and healthy beverage around the world. It contributes to massive wealth and job opportunities in many Asian, African and South American countries, including China, India, Kenya, Sri Lanka, etc. For example, in Kenya, tea contributes about 26% of the export earnings, accounting for 4% of the gross domestic product (GDP) annually (Wachira & Ronno, 2004). In Sri Lanka, about 95% of the total tea produced in 2003 was exported, contributing to about 14% of the total foreign exchange earnings, accounting for about 2.3% of its GDP (Wijeratne, 2004). Currently, the annual value of the tea industry is estimated to have reached about 5 billion US dollars in China. Furthermore, tea is a highly efficient, key agricultural product in the tea-suitable mountainous regions. In 2004, 861,000 tonnes of tea were produced in China, surpassing the amount (850,500 tonnes) produced in India, making China the No.1 tea producing country in the world (FAO, 2005). With 977,600 tonnes produced in 2006, the production of tea in China reached a higher level (Feng & Wang, 2007).

Tea includes 5 species and 2 varieties (Chen *et al*., 2000), belonging to Sect. *Thea* (L.) Dyer, genus *Camellia* L. and family Theaceae. The origin of tea was Yunnan Province in southwestern China (Yu, 1986; Hasimoto & Takasi, 1978; Hasimoto, 2001; Yamaguchi *et al.,* 1999). Usually, only *C. sinensis* and the varieties derived from it are widely cultivated and utilized nowadays. Self-incompatibility and long-term allogamy make the tea plant highly heterogeneous and consequently with broad genetic variation (Chen *et al*., 2005).

Currently, tea germplasm is becoming one of the most valuable fundamental materials for tea breeding and tea biotechnology with huge potential for the whole tea industry in the future. The achievements in tea germplasm collection, preservation, exploitation, utilization and enactment of present and long-term breeding programs of tea all depend largely on the knowledge and understanding of the genetic background, diversity, relationship and identification. Molecular markers, such as RFLP (restriction fragment of length polymorphism), RAPD (random amplified polymorphic DNA), AFLP (amplified fragment length polymorphism), CAPS (cleaved amplified polymorphic sequence), ISSR (inter simple sequence repeat), SSR (simple sequence repeat), EST-SSR (expressed sequence tag based SSR) and ALPs (amplicon length polymorphisms) etc. were initiated by using RAPD markers with the aim of assessing the genetic diversity of the Kenyan and Korean tea cultivars (Wachira *et al*., 1995; Lee *et al*., 1995) and the elite tea cultivars in China (Chen *et al*., 1998). Presently, these markers have been proven to be robust and valuable in the research for genetic diversity and variation, introduction and spread, molecular identification and DNA fingerprinting, molecular phylogenetics, genetic stability and integrity and the establishment of the genetic linkage map for tea breeding.

14.2 Assessment of Genetic Diversity and Variation

The success of any breeding or genetic conservation program is dependent on the amount and distribution of the genetic variation present in the gene pool. But our knowledge of the genetic diversity in tea and its related species is still limited. Previously, the germplasm of tea has routinely been characterized by using morphological and biochemical descriptors. Although these descriptors are valuable for tea group identification at the varietal level, they are limited at the levels of inter- and intra-varietal polymorphism and may not account for all the diversity in the species. Therefore, recent attention has been focused on the use of different molecular markers for their potential value in the estimation of the genetic relationship and determination of the genetic diversity of the tea germplasms. For evaluation of genetic diversity in 38 clones belonging to 3 tea varieties, *C. assamica*, *C. sinensis* and *C. assamica* ssp. *lasiocalyx*, RAPD analysis was adopted by which to show that the polymorphism in a single population ranged from 50.3% in the *lasiocalyx* population to 88.5% in the var. *sinensis* types. The accessions derived from the Kenyan collection, which contains representatives of all 3 types, was shown to be 93.6% of the variability. On average, 70% and 30% of genetic variability were distributed within and between populations, respectively (Wachira *et al.*, 1995). Other similar studies also supported these results (Paul *et al.*, 1997; Kaundun *et al.*, 2000; Wachira *et al.*, 2001; Mishra & Sen-Mandi, 2004). These findings are in agreement with the observation that out-breeding woody plants retain considerable variability and the greatest variation is exhibited within populations (Hamrick, 1990).

Using RAPD and AFLP markers, the diversity of tea plants from different geographical origins was analyzed. It was found that the population diversity (H) decreased as follows: wild *Camellia* > Indian tea > Chinese tea > Kenyan tea > Sri Lankan tea > Vietnamese tea > Japanese tea > Chinese Taiwan tea (Wachira *et al.*, 2001). Among the populations, the tea plants from China and India are most likely to have captured a greater proportion of diversity. The DNA markers of RAPD, AFLP, RFLP and ISSR show that the genetic resources of Chinese tea possess more genetic diversity than any other region in the world (Chen *et al.*, 1998, 2005; Chen & Yamaguchi, 2002; Paul *et al.*, 1997; Luo *et al.*, 2002; Shao *et al.*, 2003; Kaundun & Matsumoto, 2003a; Duan *et al.*, 2004; Huang *et al.*, 2004; Huang JA *et al.*, 2006; Yao *et al.*, 2007; Hou *et al.*, 2007). This is not surprising in view of the fact that China is thought to be the center of origin of tea and a complex genetic background may have been resulted in long-term out-crossing and cultivation in China (Chen *et al.*, 2005). In contrast, a low diversity of Japanese tea cultivars was identified in many studies (Lee *et al.*, 1995; Kaundun *et al.*, 2000; Kaundun & Park, 2002; Park *et al.*, 2002; Matsumoto *et al.*, 2004), which could be caused by the long and intensive selection program of tea in this country.

Establishment of vegetative propagation techniques has resulted in widespread replacement of the diverse traditional seedling tea populations with improved and uniform clonal cultivars in many countries. This has led to the loss of some valuable genetic resources. In Japan, for example, improved clonal cultivars now account for more than 90% of all tea gardens with one dominant clone, Yabukita, accounting for 85% of them (Kaundun & Matsumoto, 2004). Planting areas of clonal tea cultivars are also increasing in other tea producing countries, such as China (from less than 10% in 1980s to 33.1% in 2006), India (currently about 50%), Sri Lanka (about 55%) and Kenya (about 59%). AFLP analysis of 49 tea cultivars from South India show that the var. *sinensis* type shows maximum diversity, with an index of 0.612, while the var. *assamica* type shows a minimum diversity, with an index of 0.285, among the populations characterized (Balasaravanan *et al*., 2003). This study revealed that it has a narrow genetic diversity among tea germplasms commonly grown in South India at present.

Another study related to the diversity of 29 Darjeeling-grown tea clones in India was carried out via AFLP analysis. The results show that the similarities between the plants ranged from 68% to 92% (Mishra & Sen-Mandi, 2004), indicating that tea clones growing in Darjeeling belong to a narrow gene pool that originated from intra- as well as inter-specific hybridization. In China, ISSR analysis covering 36 clonal tea cultivars revealed that the diversity within tea regions was lower than that in the whole nation. The level of diversity in the tea areas of Jiangnan and South China is higher than those of Southwest China. Analysis of molecular variance (AMOVA) shows that the variance component within regions (94.4%) was larger than that among regions (5.6%). Meanwhile, the genetic similarity among all clonal cultivars ranged from 0.58 – 0.84 (average 0.69) indicating that the genetic basis was comparatively narrow (Yao *et al*., 2008). Therefore, it is necessary to enlarge the diversity of cultivated tea clones in tea breeding by introducing broad genetic resources. The breeding of tea cultivars needs a sustainable effort from the preservation of tea germplasms to the development of superior varietal material through wide genetic crosses.

The analysis of CAPS markers in tea was focused on 3 key genes in the phenylpropanoid pathway, i.e., *PAL* (phenylalanine ammonia lyase), *CHS* (chalcone synthase) and *DFR* (dihydroflavonol 4-reductase). The differentiation between *assamica* and *sinensis* varieties was analyzed using PCR-RFLP analysis (Kaundun & Matsumoto, 2003a). The polymorphism among 24 tea genotypes was detected based on the primers for heterologous chloroplast and nuclear microsatellite (SSR) (Kaundun & Matsumoto, 2002). Using the same genotypes, the authors conducted a factorial correspondence analysis with STS (sequence tagged site)-RFLP markers. It indicated that the nuclear SSR alleles could classify the *assamica* and *sinensis* genotypes into 2 groups, demonstrating these markers to be useful in establishing the genetic relationship, future mapping, population genetics and fingerprinting studies for tea (Kaundun & Matsumoto, 2002; 2003a).

Genetic diversity analysis also revealed that the nuclear SSR was much more effective than other types of molecular markers, offering prospects for its use in fingerprinting, mapping and population genetics studies, whereas polymorphisms detected at a cpSSR locus will allow the determination of plastid inheritance in the species. Until now, 15 SSR loci in total were developed for the tea plant (Freeman *et al*., 2004). These have been evaluated for polymorphism in a set of tea clones to determine their usefulness for authentication purposes. The majority of the SSRs developed were proven to be highly polymorphic both between and within different geographical origins and offered potential value for investigating the population genetics and genetic origins of the tea plant.

14.3 Evidence of Introduction and Spread

The distribution of tea from its origin to the main tea producing countries had previously been studied using morphological traits (Yamaguchi *et al.*, 1999), pollen characteristics (Lee *et al.*, 2001) and the terpene index (Takeo *et al.*, 1992). Recently, many studies for elucidating the differentiation of tea germplasms in different countries have been developed by applying molecular marker approaches. These studies not only provided the information about genetic relationships between tea germplasms, but also offered molecular evidence to support the assumption and suggestion of the introduction and spread of tea in certain countries. RAPD marker analysis showed that Korean tea has undergone little genetic diversification after being introduced from China (Yamaguchi *et al.*, 1999). A comparison of PAL-RFLP DNA diversity analysis between Japanese and Chinese-Korean tea germplasms clarified that limited populations of tea from China have been introduced into Japan. The distribution of tea in Japan first resulted from the limited populations and then evolved into the local tea varieties. Finally, more and more Japanese cultivars were produced by selections from local tea plants and crossing (Matsumoto, 2001; Matsumoto *et al.*, 2002; Matsumoto, 2006). Similarly, the tea plants in Korea were largely from China and partially from Japan 50 to 100 years ago (Matsumoto *et al.*, 2004). RAPD analysis provided the molecular evidence that both the tea cultivars in Korea and the dominant Japanese cultivar, Yabukita, had a close relationship with 'Jiukengzhong', an endogenous cultivar in Zhejiang Province, China, from the co-segregation markers among them. It is concluded that the ancestor of both the Korean and Japanese cultivar 'Yabukita' is most possibly from Zhejiang, China (Kim *et al.*, 2001). AFLP principal coordinate analysis supported the suggestion that the Kenyan clones have been derived from collections made in the Assam region, India (Paul *et al.*, 1997).

14.4 Molecular Identification and DNA Fingerprinting

During the past century, a number of descriptors, such as morphological characteristics, phytochemicals and the terpenoid index, have been used to characterize tea species. However, it has been argued that these descriptors may not reflect the correctness of genetic differentiation because most of them are largely affected by environmental and developmental factors. As a heterogenous plant with many overlapping morphological, biochemical and physiological attributes, the tea plant shows continuous variation and a high degree of plasticity. It has been proven difficult to identify the discrete taxonomic groups in tea.

As an efficient, auxiliary means to distinguish intra- or inter-specific variations of tea germplasms, the molecular markers have been proven to be useful and sufficient to characterize and discriminate the tea varieties and cultivars, even those which cannot be distinguished on the basis of morphological and phenotypic traits, as well as commercial teas from the market (Wachira *et al.*, 1995, 1997, 2001; Kaundun *et al.*, 2000; Liang *et al.*, 2000; Kaundun & Matsumoto, 2003a, 2003b; Lai *et al.*, 2001; Mondal, 2002; Chen & Yamaguchi, 2005; Chen *et al.*, 2005; Yao *et al.*, 2005). A RAPD marker, OPN-03-1400, known to identify species of section *Thea* (Wachira *et al.*, 1997), could be used as a species-specific marker and be useful in elucidating the gene flow via inter-specific introgression. Similar studies also show that RAPD was a very effective tool in discriminating Korean "wild tea" (Lee *et al.*, 1995). With 2 specific markers of 1,200 bp band WKA-24a and 610 bp band WKA-24b from RAPD analysis, the *C. sinensis* var. *assamica* and *C. sinensis* could be correctly identified (Liang *et al.*, 2000). With the development of DNA markers, in total 24 wild tea germplasms, which are composed of different species and varieties originating from China, could be easily discriminated by: (1) unique RAPD markers; (2) specific band patterns; (3) a combination of the band patterns; (4) DNA fingerprinting provided by different primers (Chen & Yamaguchi, 2005).

DNA markers also have been used to identify the parentage of tea cultivars. The cytoplasmic DNA, which is inherited maternally, could be used for RAPDs to determine the seed parent (Tanaka, 2006). The registered parents of 2 Japanese cultivars, 'Yutakamidori' and 'Meiryoku', were found to be incorrect and genuine parents were identified by RAPD analysis (Tanaka & Yamaguchi, 1996). After a RAPD analysis of 78 clones, Tanaka *et al.* (2001) did not think that the pollen parents of tea cultivar 'Sayamakaori' exist at present. The RAPD and SSR analysis also verified the variety 'Robiraki' to be the progeny of a hybrid in which *C. sinensis* and *C. japonica* are the pollen parent and the seed parent, respectively (Tanaka *et al.*, 2003). In China, parentages of a few F_1 plants were identified by RAPD (Li *et al.*, 2001) and ISSR (Hou *et al.*, 2006). These studies show that molecular markers are effective tools to authenticate the released cultivars and their parents, and are helpful to protect the cultivars for intellectual property rights.

Japanese green tea cultivars and 463 local tea plants, including the mountainous tea '*yama-cha*', were analyzed using PAL as a DNA marker to determine the process of differentiation in Japanese tea plants (Matsumoto *et al.*, 2002). The main DNA fragments detected by RFLP analysis, named as A, B and D, were inherited as multiple allelic genes at one locus. Japanese tea cultivars were divided into 5 groups according to RFLPs: AA, AB, AD, BD and DD. The AA group included many cultivars selected from local tea plants. The BD group consisted of cv. 'Yabukita' and its descendants by artificial crossing. There was no

BB group of cultivars. Allelic frequencies of A, B and D were 0.66, 0.08 and 0.22, respectively, and these values were the same in local tea plants as those collected from all regions of Japan. Since the frequency in '*yama-cha*' was the same as that of local tea plants, it is thought that these tea plants have the same origin. These results indicate that a process of differentiation from the ancestral material presumably introduced from China to the local tea plants occurred and resulted, finally, in the cultivars by selection from local tea plants and crossings.

The variability in the repeats of 5S rDNA with specific restriction endonucleases (*Sau*3AI, *Bam*H I, and *Apo*I) in 28 different tea clones representing 3 types of tea in India was analyzed. The results clearly showed that the 5S rDNA gene in tea could be used as a molecular marker to distinguish *C. sinensis* tea from the other important types of tea, such as *assamica* and *lasiocalyx* (Singh & Ahuja, 2006).

DNA markers have been applied in not only tea plants, but also in dry or processed tea tissues. A modified CTAB (cetyl trimethylammonium bromide) procedure for DNA isolation from processed dried commercial tea samples was described by Singh *et al.* (1999), which provided a basic and prerequisite protocol to obtain high quality DNA from dry market obtained teas. In Japan, STS-RFLP and ALPs were developed based on coding and non-coding DNA regions of the 3 genes, PAL, CHS and DFR. These genes were successfully used to detect the polymorphism of 46 main tea cultivars. It was found that 'Yabukita', a highly productive cultivar in Japan, representing 85% of clonal tea acreage, showed a unique profile when PAL intron was digested with *Dde*I, thus allowing its rapid authentication at low cost. DNA fingerprinting may be useful to identify and label the dry commercial tea products with their respective cultivars, so as to avoid blended tea being sold and discourage fraudulent commercialization (Kaundun & Matsumoto, 2003b; 2004).

14.5 Studies on Molecular Phylogenetics

A good understanding of the genetic relationships within and between species of tea plants is not only critical for the effective management of germplasm collections, but also useful for selection of parents for hybridization in tea breeding programs by which to maintain a broad range of genetic variability. However, the classification of tea and its related species has been controversial for many years. Up to now, conventional classification systems were generally established based on a combination of morphological and agronomic traits (Chang & Bartholomew, 1984). Forty-six species and varieties in the section *Thea* were released and further revised into 12 species and 6 varieties (Ming, 1992), or 5 species and 2 varieties (Chen *et al.*, 2000), respectively. Isozymes have also been used extensively to characterize tea genetic resources (Lu *et al.*, 1992) but they are limited by the relatively low levels of polymorphism detectable.

Therefore, molecular markers at the genomic DNA level may overcome the above problems and provide an efficient way to estimate genetic relationships and molecular phylogenetics in tea. It was found that RAPD and STS markers could be used to assess the relationships among cultivated tea and allied wild species of the genus *Camellia* (Wachira *et al*., 1997), providing useful information on genetic relationships at the subgenus and sectional level in *Camellia*. Results indicated that the subgenus *Thea* has a strong affinity to the subgenus *Metacamellia* and the section *Archecamellia* was the most distant section from *Thea* and probably the most ancestral. The genetic relatedness at the sectional level was discovered in decreasing order as follows: *Thea* > *Theopsis* > *Camelliopsis* > *Camellia* > *Paracamellia* > *Oleifera* > *Furfuracea* > *Archecamellia*, to be in general agreement with relationships derived from the use of compatibility studies (Takeda, 1990). At the same time, within the section *Thea*, *C. irrawadiensis*, a wild tea and traditionally non-cultivated, was clustered with *C. sinensis* var. *sinensis* accession 'K/Purple', a commercial tea clone in Kenya that contained the purple/brick red anthocyanin pigments characteristic. It was deduced that cultivar 'K/Purple' was most probably a natural hybrid between *C. sinensis* var. *sinensis* and *C. irrawadiensis*.

In China, 24 species and varieties classified by Chang's taxonomic system (Chang & Bartholomew, 1984) in section *Thea*, genus *Camellia* were basically clustered into 2 groups using RAPD analysis, the 5-locule ovary group and the 3-locule ovary group (Chen & Yamaguchi, 2002), which was largely consistent with the classification results by phytochemistry (Du *et al*., 1990), karyotype (Liang *et al*., 1994) and morphology (Chen *et al*., 2000).

In addition, a comparison of cpDNA sequences with maternal inheritance is an efficient means for investigation of phylogenetic relationships among taxa. Borthakur *et al*. (1998) developed a novel method for the isolation of cpDNA from tea leaves and the characteristics of cpDNA restriction fragments in Korean wild tea plants were studied (Lee & Nou, 1999). Nucleotide sequences of ribosomal RNA maturase (*mat*K) regions in cpDNA were determined to assess the genetic difference within cultivated teas from India, Bangladesh, Myanmar, Thailand, Laos, Vietnam, China and Japan (Katoh *et al*., 2003). By nucleotide alignment analysis, 10 different types were defined and clustered into 3 groups. The results of the *mat*K nucleotide sequence analysis indicated that the cultivated teas of India, Bangladesh, Eastern China and Japan belonged to the group of *C. sinensis* and so did the cultivated teas in the estates of South East Asia. However, the native cultivars in Myanmar and Southern China had a genetic similarity to *C. taliensis* and *C. irrawadiensis*, whereas the native cultivars of Thailand and Vietnam were associated morphologically with taxa. This study demonstrated that members of *C. irrawadiensis* and *C. taliensis* are popular cultivars found widely in South East Asia.

14.6 Detection of Genetic Stability and Integrity

The tea plant is propagated either through seeds or cuttings. To avoid extensive genetic non-uniformity resulting from seed propagation and, consequently, dilution of a quality product from a particular clone, tea clones are traditionally multiplied by cuttings. Genetic stability of clones is vital for tea germplasm preservation, breeding and production. RAPD analysis of some Chinese elite cultivars and their cutting offspring showed no variation occurring at a DNA level (Chen *et al.*, 1999).

However, vegetative propagation is limited by several factors, such as (1) slower rates of propagation, (2) unavailability of suitable planting material due to winter dormancy and drought, (3) poor survival rate in nurseries due to poor root formation of some clones and, (4) seasonally dependent rooting ability of the cuttings (Mondal & Chand, 2002; Mondal *et al.*, 2004). Therefore, micropropagation appears to be an alternative ideal choice for the circumvention of the problems related to conventional propagation. Additionally, micropropagation is also essential for tea transgenic research. But the genetic stability derived from organized meristem cultures in plants is retained or does not remain questionable. The occurrence of cryptic genetic defects arising via somaclonal variation in regeneration can seriously limit the broader utility of micropropagation systems.

Therefore, it is important to establish the suitability of a particular micropropagation protocol for a particular clone, with respect to the production of genetically identical and stable plants, before it is released for commercial purposes. Studies indicate subtle genetic variation at the DNA sequence level in organized meristem derived from micropropagated plants of tea (Devarumath *et al.*, 2002; Mondal & Chand, 2002; Singh *et al.*, 2004; Thomas *et al.*, 2006). Clearly, it is demonstrated that organized meristem cultures are not always genetically true-to-type. Molecular markers have the potential to be profitably utilized for the fast and unambiguous assessment of genetic integrity in micropropagated plants.

14.7 Establishment of Genetic Linkage Map

Genetic linkage maps provide a great potential for increasing the speed of cultivar improvement programs in perennial crops, which often have a long juvenile period, large plant size and are predominantly allogamous and highly heterozygous. Genetic linkage maps will effectively overcome all of these factors which hinder conventional breeding, reducing large investments in time and land for progeny trials in cultivar breeding. However, few such maps exist for tree crops until now, owing to the development of specific genetic stocks and tester lines containing important traits in the perennial crops being time-consuming and self-incompatible if compared to annual crops. The availability of the multigenerational pedigrees required for genome mapping is therefore limited. This resulted in the pedigrees in tea and many outbred tree species only being available in those involving two parents and their full-sib progeny, or maternal half-sib families.

The breeding strategies for tea usually take into consideration the following factors of a highly outbred nature, long generation time from seed to adult plants and the potential for vegetative propagation. The harvestable yield of tea is confined to the terminal two top leaves and a bud. Leaf characteristics therefore form the basis of selection for the two most important agronomic traits: yield and quality. These traits require several years to develop and are not amenable to early selection. Thus, selection based on one or more molecular markers linked to a quantitative trait locus (QTL) could potentially shorten the tea-breeding cycle.

A first linkage map for the tea plant was constructed with RAPD markers by Tanaka (1996) and the markers related to theanine content, date of bud sprouting, resistance to anthracnose and tolerance to cold were detected (Tanaka, 1996, 2000). Another linkage map from the female parent, SFS150, was established with RAPD and AFLP markers (Hackett *et al.*, 2000). There were 126 markers, covering 1,349.7 cM, with an average distance of 11.7 cM between loci on the map. Recently, an AFLP linkage map for the tea plant was also constructed in China. The map of a female parent included 17 linkage groups and contained 208 markers, covering a total length of 2,457.7 cM. The average distance between markers was 11.9 cM. A map from a male parent included 16 linkage groups and located 200 markers, covering a total length of 2,545.3 cM and the average distance between markers was 12.8 cM (Huang *et al.*, 2005). Most recently, a partial genetic map of backcross F_1 generation between 'Zhenong 12' (selected from the open pollination of 'Fuding Dabaicha'×'Yunnan Dayecha') and 'Fuding Dabaicha' was also generated using RAPD and ISSR markers (Huang FP *et al.*, 2006). However, in previous studies the number of individuals for mapping was limited (only, 46 – 94) and the density was not high enough to meet the demand of precise mapping. The genome size of the tea plant is estimated to be 4.0 Giga bp (Tanaka *et al.*, 2006). These constructed maps were still limited to locate QTLs linked with some important traits due to their low distribution of molecular markers. Thus, no practical genetic map has been available for tea until now.

14.8 Prospects

Great advances have been achieved with molecular markers for tea recently, though this is still far behind cereal crops and some other industrial wood crops. Although the molecular marker can be directed in several ways, currently, and in the near future, more attention and priority should be placed on the following recommendations:

(a) Undertake a massive tea germplasm appraisal and evaluation project around the world, through a common 'tea germplasm characterization consortium', which already exists for several similar crops (Mondal *et al.*, 2004).

(b) Develop new applicable and cheap DNA markers for tea plants, especially functional markers associated with traits, such as EST-SSR (Jin *et al.*, 2006; Zhao *et al.*, 2008) and SNP (single nucleotide polymorphism).

(c) Establish a high density genetic linkage map and analyze the agronomically important QTLs to facilitate marker-assisted early selection and shorten the tea breeding procedures.

(d) Generate early selective markers at the nursery stage for various biotic and abiotic stresses which will revolutionize tea breeding where we suffer from the lack of selection criteria and long gestation periods.

Acknowledgements

This work was supported in part by the Hi-Tech Research and Development Program of China (863 Plan) (No. 2006AA10Z171), Qianjiang Personnel Program of Zhejiang Province, China (No. 2006R10042), the R&D Infrastructure and Facility Development Program of China (No. 2005DKA21002-08), the National Key Technologies R&D Program of China (No. 2006BAD13B06) and the NSF of Zhejiang Province, China (Nos. Y305105, Y305124).

References

Balasaravanan T, Pius PK, Raj Kumar R, Muraleedharan N, Shasany AK (2003) Genetic diversity among South Indian tea germplasm (*Camellia sinensis, C. assamica* and *C. assamica* ssp. *lasiocalyx*) using AFLP markers. Plant Science, 165: 365-372.

Borthakur S, Mondal TK, Parveen SS, Guha A, Sen P, Borthakur A, Deka PC (1998) Isolation of chloroplast DNA from tea, *Camellia* sp. India Journal of Experimental Biology, 36: 1165-1167.

Chang HT, Bartholomew B (1984) *Camellias*. Portland: Timber Press.

Chen L, Yamaguchi S (2002) Genetic diversity and phylogeny of tea plant (*Camellia sinensis*) and its related species and varieties in the section *Thea* genus *Camellia* determined by randomly amplified polymorphic DNA analysis. Journal of Horticultural Science & Biotechnology, 77(6): 729-732.

Chen L, Yamaguchi S (2005) RAPD markers for discriminating tea germplasms at the inter-specific level in China. Plant Breeding, 124: 404-409.

Chen L, Yang YJ, Yu FL, Gao QK, Chen DM (1998) Genetic diversity of 15 tea (*Camellia sinensis* (L.) O. Kuntze) cultivars using RAPD markers. Journal of Tea Science, 18(1): 21-27 (in Chinese).

Chen L, Yu FL, Yang YJ, Chen DM, Xu CJ, Gao QK (1999) A study on genetic stability of excellent tea germplasms (*Camellia sinensis* (L.) O. Kuntze) using RAPD markers. Journal of Tea Science, 19(1): 13-16(in Chinese).

Chen L, Yu FL, Tong QQ (2000) Discussions on phylogenetic classification and evolution of section *Thea.* Journal of Tea Science, 20(2): 89-94 (in Chinese).

Chen L, Gao QK, Chen DM, Xu CJ (2005) The use of RAPD markers for detecting genetic diversity, relationship and molecular identification of Chinese elite tea genetic resources [*Camellia sinensis* (L.) O. Kuntze] preserved in a tea germplasm repository. Biodiversity and Conservation, 14(6): 1433-1444.

Devarumath RM, Nandy S, Rani V, Marimuthu S, Muraleedharan N, Raina SN (2002) RAPD, ISSR and RFLP fingerprints as useful markers to evaluate genetic integrity of micropropagated plants of three diploid and triploid elite tea clones representing *Camellia sinensis* (China type) and *C. assamica* ssp. *assamica* (Assam-India type). Plant Cell Reporters, 21: 166-173.

Du QZ, Li MJ, Liu WH, Wang HS (1990) Chemical and numerical taxonomies of plants *Thea* section plants. Journal of Tea Science, 10(2): 1-12 (in Chinese).

Duan HX, Shao WF, Wang PS, Xu M, Pang RH, Zhang YP, Cui WR (2004) Study on the genetic diversity of peculiar tea germplasm resources in Yunnan by RAPD. Journal of Yunnan Agricultural University, 19(3): 246-254 (in Chinese).

Feng WS, Wang GQ (2007) Chinese tea output and selling quantities reached high peaks in 2006. China Tea, 29(2): 4-5 (in Chinese).

Food and Agriculture Organization (FAO) (2005) http://faostat.fao.org/.

Freeman S, West J, James C, Lea V, Mayes S (2004) Isolation and characterization of highly polymorphic microsatellites in tea (*Camellia sinensis*). Molecular Ecology Notes, 4(3): 324-326.

Hackett CA, Wachira FN, Paul S, Powell W, Waugh R (2000) Construction of a genetic linkage map for *Camellia sinensis* (tea). Heredity, 85(4): 346-355.

Hamrick JL (1990) Isozymes and the analysis of genetic structure in plant populations. In: Soltis ED, Soltis PS (eds.) Isozymes in Plant Biology. London: Chapman and Hall, pp.87-105.

Hasimoto M (2001) The origin of the tea plant. In: Proceedings of 2001 International Conference on O-Cha (tea) Culture and Science (Session II). 5-8 October, 2001, Shizuoka, Japan, pp.J5-J7.

Hasimoto M, Takasi S (1978) Morphological studies on the origin of the tea plant. V. A proposal of one place of origin by cluster analysis. Japanese Journal of Tropical Agriculture, 21: 93-101.

Hou YJ, He Q, Liang GL, Li PW, Peng P, Deng M (2006) ISSR analysis of the hybrid of the descendants of tea camellias. Journal of Southwest Agricultural University (Natural Science), 28(2): 267-270 (in Chinese).

Hou YJ, He Q, Li PW, Liang GL, Peng P, Deng M (2007) Genetic diversity of tea camellias germplasm by ISSR molecular marker. Southwest China Journal of Agricultural Sciences, 20(3): 462-465 (in Chinese).

Huang FP, Liang YR, Lu JL, Chen RB, Mamati G (2004) Evaluation of genetic diversity in Oolong tea germplasms by AFLP fingerprinting. Journal of Tea Science, 24(3): 183-189 (in Chinese).

Huang FP, Liang YR, Lu JL, Chen RB (2006) Genetic mapping of first generation of backcross in tea by RAPD and ISSR markers. Journal of Tea Science, 26(3): 171-176 (in Chinese).

Huang JA, Li JX, Huang YH, Luo JW, Gong ZH, Liu ZH (2005) Construction of AFLP molecular markers linkage map in tea plant. Journal of Tea Science, 25(1): 7-15 (in Chinese).

Huang JA, Li JX, Huang YH, Luo JW, Gong ZH, Liu ZH (2006) Genetic diversity of tea (*Camellia sinensis* (L.) O. Kuntze) cultivars revealed by AFLP analysis. Acta Horticulturae Sinica, 33(2): 317-322 (in Chinese).

Jin JQ, Cui HR, Chen WY, Lu MZ, Yao YL, Xin Y, Gong XC (2006) Data mining for SSRs in ESTs and development of EST-SSR marker in tea plant (*Camellia sinensis*). Journal of Tea Science, 26(1): 17-23 (in Chinese).

Katoh Y, Katoh M, Takeda Y, Omori M (2003) Genetic diversity within cultivated teas based on nucleotide sequence comparison of ribosomal RNA maturase in chloroplast DNA. Euphytica, 134: 287-295.

Kaundun SS, Matsumoto S (2002) Heterologous nuclear and chloroplast microsatellite amplification and variation in tea, *Camellia sinensis*. Genome, 45: 1041-1048.

Kaundun S S, Park Y G (2002) Genetic structure of six Korean tea populations as revealed by RAPD-PCR markers. Crop Science, 42: 594-601.

Kaundun SS, Matsumoto S (2003a) Development of CAPS markers based on three key genes of the phenylpropanoid pathway in tea, *Camellia sinensis* (L.) O. Kuntze, and differentiation between *assamica* and *sinensis* varieties. Theoretical and Applied Genetics, 106: 375-383.

Kaundun SS, Matsumoto S (2003b) Identification of processed Japanese green tea based on polymorphisms generated by STS-RFLP analysis. Journal of Agriculture and Food Chemistry, 51: 1765-1770.

Kaundun SS, Matsumoto S (2004) PCR-based amplicon length polymorphisms (ALPs) at microsatellite loci and indels from non-coding DNA regions of cloned genes as a means of authenticating commercial Japanese green teas. Journal of the Science of Food and Agriculture, 84: 895-902.

Kaundun SS, Zhyvoloup A, Park YG (2000) Evaluation of the genetic diversity among elite tea (*Camellia sinensis* var. *sinensis*) accessions using RAPD markers. Euphytica, 115: 7-16.

Kim H, Liang YR, Lu JL (2001) Comparative study on genomic DNA diversity between Korean and Chinese tea cultivars by RAPD technique. Journal of Tea Science, 21(2), 103-107 (in Chinese).

Lai JA, Yang WC, Hsiao JY (2001) An assessment of genetic relationships in cultivated tea clones and native wild tea in Taiwan using RAPD and ISSR markers. Botanical Bulletin of Academia Sinica, 42: 93-100.

Lee EK, Luo YP, Tong QQ (2001) Study on the comparative morphology of pollen between Chinese Mt, Tiantai and Korean Mt, Jiri tea plants. In: Proceedings of 2001 International Conference on O-Cha (tea) Culture and Science (Session II). 5-8 October, 2001, Shizuoka, Japan, pp.33-36.

Lee SH, Nou IS (1999) Characteristics of chloroplast DNA restriction fragments in *Camellia sinensis*. Journal of the Korean Tea Society, 5(1): 33-44.

Lee SH, Choi HS, Kim RH, Lee HY, Nou IS (1995) Identification of Korean wild tea plants and Japanese green tea germplasms using RAPD markers. Journal of the Korean Tea Society, 1(1): 129-148.

Li XH, Shi ZP, Liu CL, Luo JW, Shen CW, Gong ZH (2001) Parentage identification of filial generation tea plants from "Yunnan Daye" and "Rucheng Baimao" with RAPD method. Journal of Tea Science, 21(2): 99-102 (in Chinese).

Liang GL, Zhou CQ, Lin MJ, Chen JY, Liu JS (1994) Karyotype variation and evolution of sect *Thea* in Guizhou. Acta Phytotaxonomica Sinica, 32: 308-315 (in Chinese).

Liang YR, Tanaka J, Takeda Y (2000) Study on diversity of tea germplasm by RAPD marker. Journal of Zhejiang Forestry College, 17(2): 215-218 (in Chinese).

Lu CY, Liu WH, Li MJ (1992) Relationship between the evolutionary relatives and the variation of esterase isozymes in tea plant. Journal of Tea Science, 12(1): 15-20 (in Chinese).

Luo JW, Shi ZP, Shen CW, Liu CL, Gong ZH, Huang YH (2002) Studies on genetic relationships of tea cultivars (*Camellia sinensis* (L.) O. Kuntze) by RAPD analysis. Journal of Tea Science, 22(2): 140-146 (in Chinese).

Matsumoto S (2001) Analysis of the differentiation of Japanese green tea cultivars using DNA markers. In: Proceedings of 2001 International Conference on O-Cha (tea) Culture and Science (Session Ⅱ). 5-8 October, 2001, Shizuoka, Japan, pp. J13-J16.

Matsumoto S (2006) Studies on the differentiation of Japanese tea cultivars (*Camellia sinensis* var. *sinensis*) according to the genetic diversity of phenylalanine ammonia-lyase. Bulletin of National Institute of Vegetable and Tea Science, 5: 63-111.

Matsumoto S, Kiriiwa Y, Takeda Y (2002) Differentiation of Japanese green tea cultivars as revealed by RFLP analysis of phenylalanine ammonia-lyase DNA. Theoretical and Applied Genetics, 104: 998-1002.

Matsumoto S, Kiriiwa Y, Yamaguchi S (2004) The Korean tea plant (*Camellia sinensis*): RFLP analysis of genetic diversity and relationship of Japanese tea. Breeding Science, 54: 231-237.

Ming TL (1992) A revision of *Camellia* Sect. *Thea*. Acta Botanica Yunnanica, 14(2): 115-132 (in Chinese).

Mishra RK, Sen-Mandi S (2004) Genetic diversity estimates for Darjeeling tea clones based on amplified fragment length polymorphism markers. Journal of Tea Science, 24(2): 86-92.

Mondal TK (2002) Assessment of genetic diversity of tea (*Camellia sinensis* (L.) O. Kuntze) by inter-simple sequence repeat polymerase chain reaction. Euphytica, 128: 307-315.

Mondal TK, Chand PK (2002) Detection of genetic variation among micropropagated tea (*Camellia sinensis* (L) O. Kuntze) by RAPD analysis. In Vitro Cellular & Developmental Biology-Plant, 38: 296-299.

Mondal TK, Bhattacharya A, Laxmikumaran M, Ahuja PS (2004) Recent advances of tea (*Camellia sinensis*) biotechnology. Plant Cell, Tissue and Organ Culture, 76: 195-254.

Park YG, Kaundun SS, Zhyvoloup A (2002) Use of the bulked genomic DNA-based RAPD methodology to assess the genetic diversity among abandoned Korean tea plantations. Genetic Resource and Crop Evolution, 49(2): 159-165.

Paul S, Wachira FN, Powell W, Waugh R (1997) Diversity and genetic differentiation among populations of Indian and Kenyan tea [*Camellia sinensis* (L.) O. Kuntze] revealed by AFLP markers. Theoretical and Applied Genetics, 94: 255-263.

Shao WF, Pang RH, Wang PS, Xu M, Duan HX, Zhang YP, Li JH (2003) RAPD analysis of tea trees in Yunnan. Scientia Agricultura Sinica, 36(12): 1582-1587 (in Chinese).

Singh D, Ahuja PS (2006) 5S rDNA gene diversity in tea (*Camellia sinensis* (L.) O. Kuntze) and its use for variety identification. Genome, 49: 91-96

Singh M, Bandana, Ahuja PS (1999) Isolation and PCR amplification of genomic DNA from market samples of dry tea. Plant Molecular Biology Report, 17: 171-178.

Singh M, Saroop J, Dhiman B (2004) Detection of intra-clonal genetic variability in vegetatively propagated tea using RAPD markers. Biologia Plantarum, 48: 113-115.

Takeda Y (1990) Cross compatibility of tea (*Camellia sinensis*) and its allied species in the genus *Camellia.* Japan Agricultural Research Quarterly, 24: 111-116.

Takeo T, You XQ, Wang HF, Kinukasa H, Li MJ, Cheng QK, Wang HS (1992) One speculation on the origin and dispersion of tea plant in China—One speculation based on the chemotaxonomy by using the content-ration of terpen-alcohols found in the tea aroma composition. Journal of Tea Science, 12(2): 81-86.

Tanaka J (1996) RAPD linkage map of tea plant and the possibility of application in tea genetics and breeding. Tea Research Journal, 84(S): 44-45 (in Japanese).

Tanaka J (2000) Construction of linkage and QTL analysis of tea plant. In: Liang YR, Liu ZS, Park YG, Takeda Y, Tanaka J, Lu JL, Zhao D (eds.) Proceedings of the International Symposium on Molecular Biology and Tea Breeding. 20-28 November, 2000, Hangzhou, China, pp.23-26.

Tanaka J (2006) Study on the utilization of DNA markers in tea breeding. Bulletin of National Institute of Vegetable and Tea Science, 5: 113-155 (in Japanese).

Tanaka J, Yamaguchi S (1996) Use of RAPD markers for the identification of parentage of tea cultivars. Bulletin of National Research Institute of Vegetable, Ornamental Plant and Tea (B), 9: 31-36 (in Japanese).

Tanaka J, Yamaguchi N, Nakamura Y(2001) Pollen parent of tea cultivar Sayamakaori with insect and cold resistance may not exist. Breeding Research, 3: 43-48 (in Japanese).

Tanaka J, Metoku S, Takeda Y (2003) Garden-variety camellia 'Robiraki' derived from crossing between *Camellia japonica* as seed parent and *C. sinensis* as pollen parent. Application of RAPD and SSR marker analysis to tea breeding by interspecific hybridization. Breeding Research, 5: 149-154 (in Japanese).

Tanaka J, Taniguchi F, Hirai N, Yamaguchi S (2006) Estimation of the genome size of tea (*Camellia sinensis*), camellia (*C. japonica*), and their inter-specific hybrids by flow cytometry. Tea Research Journal, 101: 1-7.

Thomas J, Vijayan D, Joshi SD, Lopez SJ, Kumar RR (2006) Genetic integrity of somaclonal variants in tea (*Camellia sinensis* (L.) O. Kuntze) as revealed by inter simple sequence repeats. Journal of Biotechnology, 123: 149-154.

Wachira FN, Ronno W (2004) Current research on tea in Kenya. In: Proceedings of 2004 International Conference on O-Cha (tea) Culture and Science. 4-6 November, 2004, Shizuoka, Japan, pp.59-65.

Wachira FN, Waugh R, Hackett CA, Powell W (1995) Detection of genetic diversity in tea (*Camellia sinensis*) using RAPD markers. Genome, 38: 201-210.

Wachira FN, Powell W, Waugh R (1997) An assessment of genetic diversity among *Camellia sinensis* L. (cultivated tea) and its wild relatives based on randomly amplified polymorphic DNA and organelle-specific STS. Heredity, 78: 603-611.

Wachira FN, Tanaka J, Takeda Y (2001) Genetic variation and differentiation in tea (*Camellia sinensis*) germplasm revealed by RAPD and AFLP variation. Journal of Horticultural Science & Biotechnology, 76(5): 557-563.

Wijeratne MA (2004) Tea industry in Sri Lanka. In: Proceedings of 2004 International Conference on O-Cha (tea) Culture and Science. 4-6 November, 2004, Shizuoka, Japan, pp.51-54.

Yamaguchi S, Matsumoto S, Tanaka J (1999) Genetic dispersal of tea plant. In: Jain NK (eds.) Global Advances in Tea Science. New Delhi: Aravali Books International (P) Ltd., pp.413-426.

Yao MZ, Huang HT, Yu JZ, Chen L (2005) Analysis on applicability of ISSR in molecular identification and relationship investigation of tea cultivars. Journal of Tea Science, 25(2): 153-157 (in Chinese).

Yao MZ, Chen L, Wang XC, Zhao LP, Yang YJ (2007) Genetic diversity and relationship of clonal tea cultivars in China revealed by ISSR markers. Acta Agronomica Sinica, 33(4): 598-604 (in Chinese).

Yao MZ, Chen L, Liang YR (2008) Genetic diversity among tea cultivars from China, Japan and Kenya revealed by ISSR markers and its implication for parental selection in tea breeding programs. Plant Breeding, 127: 166-172.

Yu FL (1986) Discussion on the originating place and the originating center of tea plants. Journal of Tea Science, 6(1): 1-8 (in Chinese).

Zhao LP, Liu Z, Chen L, Yao MZ, Wang XC (2008) Generation and characterization of 24 novel EST derived microsatellites from tea plant (*Camellia sinensis*) and cross-species amplification in its closely related species and varieties. Conservation Genetics, 9(5): 1327-1331.

Appendix

Table A1 The world tea acreage, production and exports in 2010

Country*	Acreage (ha)	Production (tonnes)	Exports (tonnes)	Value (1,000 US$)
China	1,970,200	1,475,060	302,525	784,169
India	579,400	966,403	183,700	535,260
Kenya	171,916	399,006	441,021	1,233,576
Sri Lanka	188,007	331,427	298,587	1,314,257
Vietnam	132,000	157,000	98,000	184,000
Turkey	77,500	148,000	4,000	5,900
Indonesia	124,400	129,200	87,101	178,549
Japan	47,500	93,000	2,287	50,161
Argentina	40,000	90,000	102,323	110,146
Bangladesh	54,900	59,272	913	2,100
Uganda	26,000	56,468	50,834	95,500
Malawi	18,600	51,591	48,579	76,612
Tanzania	22,721	31,646	25,388	47,993
Rwanda	12,300	22,249	25,300	60,000
Myanmar	78,400	19,000	NA	NA
Iran	17,500	16,800	4,700	NA
Nepal	17,127	16,608	8,600	NA
Zimbabwe	5,800	14,293	8,498	NA
Burundi	8,400	6,800	6,000	NA
Papua New Guinea	3,830	6,800	5,800	7,000
Mozambique	3,470	6,500	2,200	NA
Ethiopia	2,630	5,600	2800	NA
Brazil	5,340	5,400	2,544	8,563
Cameroon	1,600	4,400	4,100	NA
Korea	3,730	4,300	270	NA
Georgia	37,200	4,300	1,000	NA

(To be continued)

(Table A1)

Country*	Acreage (ha)	Production (tonnes)	Exports (tonnes)	Value (1,000 US$)
Peru	2,910	3,950	130	NA
Congo	NA	3,400	2,700	NA
South Africa	4,100	3,400	1,800	NA
Russia	1,780	3,300	NA	NA
Malaysia	3,350	2,400	270	NA
Azerbhaijan	5,470	2,100	NA	NA
Ecuador	1,020	2,050	1,180	NA
Australia	900	1,670	NA	NA
Mauritius	698	1,467	35	305
Zaire	6,500	NA**	NA	NA
Grand total	3,691,938	4,162,327	1,728,812	/

Data sources: International Tea Committee (ITC) (2011) Annual Bulletin of Statistics, London.
* Country is sorted by the production. **NA: Not available

Index

FSC
www.fsc.org
MIX
Papier | Fördert
gute Waldnutzung
FSC® C083411

Kolibri 360 GmbH
Ferdinand-Jühlke-Straße 7
99095 Erfurt, Deutschland
produktsicherheit@kolibri360.de